AF412876

Femtosecond Biophotonics

Core Technology and Applications

The introduction of femtosecond pulse lasers has provided numerous new methods for non-destructive diagnostic analysis of biological samples. This book is the first to provide a focused and systematic treatment of femtosecond biophotonic methods. Each chapter combines theory, practice and applications, walking the reader through imaging, manipulation and fabrication techniques. Beginning with an explanation of nonlinear and multi-photon microscopy, subsequent chapters address the techniques for optical trapping and the development of laser tweezers. In a conclusion that brings together the various topics of the book, the authors discuss the growing field of femtosecond micro-engineering. The wide range of applications for femtosecond biophotonics means this book will appeal to researchers and practitioners in the fields of biomedical engineering, biophysics, life sciences and medicine.

Min Gu is Director of the Centre for Micro-Photonics at Swinburne University of Technology in Australia. He is an established researcher in the fields of nanophotonics and biophotonics, and has internationally renowned expertise in three-dimensional optical imaging theory. Professor Gu is a Fellow of the Australian Academy of Science and the Australian Academy of Technological Sciences and Engineering. He has published over 550 articles, along with two reference books in the field of optics.

Damian Bird is Senior Development Scientist at Universal Biosensors Pty. Ltd. In 2007 he was awarded the Kaye Merlin Brutton Prize from the Faculty of Science at the University of Melbourne. Dr Bird has worked in academia and industry in America and Australia, and has co-authored several chapters in books on fluorescence applications.

Daniel Day is a Senior Research Fellow and is currently Deputy Director of the Centre for Micro-Photonics at Swinburne University of Technology. He has co-authored chapters in numerous books in the field of optical data storage. Dr Day's research interests have led to the creation of two companies, 3DCD Technology Pty. Ltd. and Biosurfaces Pty. Ltd., which aim to develop technology in optical data storage and cell culture ware.

Ling Fu is an Associate Professor at Huazhong University of Science and Technology in China, and is Principal Investigator of Medical Optoelectronics in Wuhan National Laboratory for Optoelectronics. Dr Fu has published in many scientific journals and serves as an assistant editor for the Journal of Innovative Optical Health Sciences.

Dru Morrish is a researcher in the field of microscopy at Swinburne University of Technology. He also has three years' industrial research experience in biosensing based on resonance detection methods. Dr Morrish has published articles in internationally referred journals and has been involved in the organisation of various international conferences.

Femtosecond Biophotonics

Core Technology and Applications

MIN GU

Swinburne University of Technology, Victoria

DAMIAN BIRD

Universal Biosensors Pty Ltd, Victoria

DANIEL DAY

Swinburne University of Technology, Victoria

LING FU

Huazhong University of Science and Technology, P.R. China

DRU MORRISH

Swinburne University of Technology, Victoria

CAMBRIDGE UNIVERSITY PRESS
Cambridge, New York, Melbourne, Madrid, Cape Town, Singapore,
São Paulo, Delhi, Dubai, Tokyo

Cambridge University Press
The Edinburgh Building, Cambridge CB2 8RU, UK

Published in the United States of America by Cambridge University Press, New York

www.cambridge.org
Information on this title: www.cambridge.org/9780521882408

First published 2010

Printed in the United Kingdom at the University Press, Cambridge

A catalogue record for this publication is available from the British Library

ISBN 978-0-521-88240-8 Hardback

To our families

Contents

Preface

In 1995, the first author of this book joined Victoria University. Immediately after that, he established a new research group called the Optoelectronic Imaging Group (OIG), with a focus on the introduction of femtosecond lasers into optical microscopy. While the first two-photon fluorescence microscope was reported in 1990, it was not until 1996 that the first two-photon fluorescence microscope in Australia was constructed by a group of OIG Ph.D. students with a femtosecond laser supported by the major equipment fund of Victoria University. It was this new instrument that gave the OIG research students and staff a powerful tool to conduct biophotonic research. At the beginning of 2000, most of the OIG members moved to Swinburne University of Technology to form a new research centre called the Centre for Micro-Photonics (CMP). Since 1995, research students of the OIG and the CMP, including four of the authors of the book, Damian Bird, Daniel Day, Ling Fu and Dru Morrish, have made many significant contributions to femtosecond biophotonic methods. The aim of this book is to provide a systematic introduction into these methods. Chapters 1–3, 6 and 8 were completed by Min Gu and Chapters 4, 5, 7 and 9 were written by Damian Bird, Ling Fu, Dru Morrish and Daniel Day, respectively. All the authors participated in the final editing of the book.

It is apparent that many people have made important contributions to the contents described in this book. We would like first to acknowledge the significant contributions by the previous and current members of the OIG and the CMP, including Dr Ze'ev Bomzon, Dr James W. M. Chon, Dr Charles Cranfield, Dr Xiaoyuan Deng, Associate Professor Xiaosong Gan, Dr Djenan Ganic, Mr Jean-Baptiste Haumonte, Mr Aernout Kisteman, Dr Smitha Kuriakose, Mr Yoan Micheau, Professor Sarah Russell, Dr Steven Schilders and Dr Mingun Xu. We would like to thank our collaborators Professor Erik W. Thompson and Dr Elizabeth D. Williams from St Vincent's Institute of Medical Research and Department of Surgery, University of Melbourne in the area of two-photon autofluorescence microscopy, Dr Blessing Crimeen-Irwin and Dr Mandy Ludford-Menting from the Peter MacCallum Cancer Institute (Centre) in the area of two-photon fluorescence resonance energy transfer microscopy, Dr Sarah Cartmell from Institute for Science and Technology in Medicine, Keele University (UK) in the area of femtosecond laser targeting.

We are grateful to Swinburne University of Technology and Victoria University for their important support to our research. We would like to thank the Australian Research

Council for its support to the biophotonic research projects at the OIG and CMP. It is important to acknowledge the significant support of Swinburne University of Technology through its strategic initiatives program since 2000. Without its generous support, it would not have been possible for us to complete this book. Last but not least, we are most grateful to our families for their understanding, support and encouragement during the course of completing this book.

1   Introduction

This chapter serves as an introduction to this book. Section 1.1 gives a brief review on the development of biophotonics and summarises the main achievements in biophotonics due to the introduction of femtosecond pulse lasers, while Section 1.2 defines the scope of the book.

1.1   Femtosecond biophotonics

Biophotonics involves the utilisation of photons, quanta of light, to image, sense and manipulate biological matter. It provides the understanding of the fundamental interaction of photons with biological media and the application of this understanding in life sciences including biological sciences and biomedicine [1]. In that sense, biophotonics research dates back to times when biologists started to use optical microscopy and spectroscopy with a conventional light source such as a lamp. These two forms of classic biophotonic instrument revolutionised biological research and are the classic bridge between photonics and life sciences because they provide a non-destructive way to view the two-dimensional (2D) microscopic world that human eyes cannot, as well as the function of microscopic samples through colour or spectroscopic information.

Biophotonics became a recognised new discipline after the laser was invented in 1960. Laser light is fundamentally different from conventional light in the sense that it possesses high brightness in a narrow spectral window, is highly directional, and exhibits a high degree of coherence. Since 1960, these unique features have facilitated many important applications of laser technology in biological and biomedical studies [2, 3]. One of the important milestones in this area is the combination of laser light with an optical microscope, which led to laser scanning confocal microscopy [4, 5]. The main advantage of confocal microscopy is its three-dimensional (3D) imaging capability [5, 6] with sub-micrometer resolution and selective spectroscopic information, which is highly desired in biological and biomedical observation of live objects. Another revolutionary milestone of this combination is the invention of the laser tweezers technique that provides a tool to manipulate and modify microscopic objects [7–9]. These two types of laser-based instrument form a modern bridge between photonics and life sciences as they provide a tool not only to observe a 3D live sample but also to manipulate a microobject under investigation.

Recently, these two biophotonics techniques have advanced to an exciting stage with inclusion of a femtosecond pulse laser [10]. This type of laser operates in a pulsed mode with a pulse duration of approximately 100 fs (1 s $= 10^{15}$ fs). As a result, the peak power of a femtosecond laser is so high that the nonlinear optical response of biological samples can be induced for imaging. This opened a revolutionary era of optical microscopy – nonlinear optical microscopy for biological and biomedical research. A multi-photon fluorescence process is one of the nonlinear processes caused by the simultaneous absorption of two or more incident photons under the illumination of an ultrashort pulsed beam. The energy of the resulting fluorescence photons is slightly less than the sum of the energy of the absorbed incident photons.

A milestone of nonlinear optical microscopy is two-photon fluorescence microscopy [11,12]. The original idea of two-photon fluorescence scanning microscopy was proposed by Sheppard *et al.* in the late 1970s along with other nonlinear scanning microscope modes [13,14]. Two-photon fluorescence microscopy was first demonstrated by Denk *et al.* in 1990 [11]. It has been reported that strong three-photon fluorescence can also be generated in some organic solutions [15] where three incident photons are absorbed simultaneously and the radiating photon has energy approximately three times as large as the incident one. It was Hell and his colleagues who recorded the first three-photon fluorescence microscopic image using a nonlinear crystal [16], which demonstrated the 3D imaging ability of this microscopic method [17]. There are a number of advantages of multi-photon fluorescence microscopy over single-photon fluorescence microscopy. With multi-photon absorption, one can access ultraviolet (UV) photon excitation without using a UV source. Due to the cooperative multi-photon excitation, the photobleaching process associated with single-photon fluorescence is confined only to the vicinity of the focal region, and 3D imaging becomes possible without the necessity for a confocal pinhole mask [12, 17]. However, axial resolution in multi-photon fluorescence microscopy can be improved if a confocal mask is used [18–20]. As a result of using a long wavelength for excitation, unwanted scattering including Mie and Rayleigh scattering can be effectively reduced, which in turn leads to a deep penetration depth [21–23]. The introduction of femtosecond lasers into microscopy has also revolutionised the way that biologists conduct research in fluorescence resonance energy transfer (FRET) and fluorescence lifetime imaging (FLIM).

Another innovative nonlinear optical microscopy technique is based on second-harmonic generation from biological samples because it provides the morphology information of fine structures at a cellular level [24–26].

While nonlinear optical microscopy has been applied to many biological applications for cellular imaging [24–28], there has been little investigation into medical applications of this new optical microscopy technique. In fact, a main advantage of nonlinear optical microscopy is its capability of localised excitation and penetration of thick tissue in a non-invasive way [11, 24, 25, 27]. This unique feature is particularly important for developing a non-invasive diagnosis tool for early cancer detection. The recent advances in developing compact nonlinear optical microscopy [29, 30] make it possible to achieve a fundamental breakthrough for early cancer detection.

Although femtosecond lasers have revolutionised laser scanning microscopy since their first experimental demonstration in 1990 [11], the combination of femtosecond lasers with laser tweezers was not explored until the recent independent publications by the group at Swinburne University of Technology [31] and the group at St. Andrews University [32]. Recently, the concept of femtosecond tweezing has been extended to near-field optics [33, 34]. It should be emphasised that femtosecond laser tweezers offer some distinct advantages: we can significantly enhance multi-photon processes in conjunction with the tweezing action with the potential of targeted DNA transfection of cells [35]. To achieve this, one should develop a new instrument on which both multi-photon microscopy and femtosecond tweezing can be conducted. As such, it can be used to address questions of fundamental cell biology and cell mechanics to facilitate the assessment of protein–protein interactions and modifications in intact, living cells and organisms [36].

The impact of using a femtosecond laser in optical microscopy on biotechnology is far beyond its revolution in optical microscopy and laser tweezers. In fact, the combination of a femtosecond laser and optical microscopy has led to a cutting-edge photonic technology, 3D micro-/nano-fabrication which can generate 3D microstructures that are impossible to generate using conventional laser lithography. There are two femtosecond laser 3D micro-/nano-fabrication methods called 3D lithography [37, 38] and 3D microexplosion [39–41], respectively. Such 3D micro-/nano-fabrication methods will revolutionise cell biology research by providing 3D cellular imaging, manipulation and engineering chips [42]. The combination of femtosecond lasers with other photonics technology such as microfluidics [43–45] and photonic crystal [46] techniques is further leading to a wide range of new technologies that can be used for cellular imaging, manipulation and engineering. These devices will play fundamental roles in 3D controllable bioreaction and tissue/cell growth [47–51] in stem cell research.

It is therefore clear that the introduction of femtosecond lasers not only revolutionised modern laser scanning microscopy and laser tweezers technology but also, more importantly, opens a new avenue to biological and biomedicine studies, as shown in Fig. 1.1. In that sense, femtosecond biophotonics creates a horizon and is a new bridge between photonics and life sciences. The new development of compact femtosecond pulse lasers will continue to have a significant impact on biological research in the future, potentially as a result of some of the biophotonics technology described in this book.

1.2   Scope of the book

This book looks specifically at some of the major technologies that have been developed in recent years as a result of the incorporation of femtosecond pulse lasers into optical microscopy. In particular the book addresses the evolution of optical microscopy due to the introduction of the femtosecond laser and its effect on imaging of biological samples, to a range of novel and emerging technologies that all have applications.

The book is the culmination of our work to date with femtosecond lasers and their applications in photonics, specifically biophotonics [52–60]. Although there are a

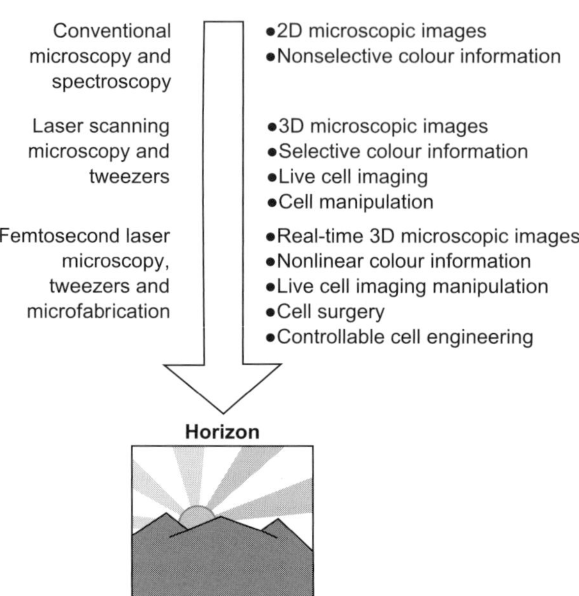

Fig. 1.1 A new horizon created by femtosecond biophotonics devices.

number of books that are focused on individual topics on optical microscopy and laser tweezers, there has been no book that systematically covers the topics in femtosecond laser-based biophotonics. The book not only covers the basic knowledge of femtosecond biophotonics but also describes the latest research which aims to develop more advanced tools.

The book is designed to act as a reference point for scientists and engineers working in the fields of photonics, biophotonics, biomedical optics, bioengineering, life sciences and applied physics. It is not the intention of the authors for the book to be used as a text book. The book is aimed at readers with at least a postgraduate level of knowledge in the fields of applied physics, photonics, biomedical optics, biophotonics, bioengineering and life sciences. The book will also be useful for engineers and scientists who are or will be involved in biophotonics research. The authors assume that the readers have at least an understanding of modern optics [61], optical imaging theory [62] and optical microscopy [4–6].

The chapters in this book follow what is essentially a timeline of advances in the field of biophotonics as a result of the new tools available through the use of femtosecond pulse lasers. The chapters are categorised to illustrate how the field of biophotonics has evolved from 'imaging' to 'manipulation' to 'fabrication' or 'engineering'. The content of the chapters strikes a balance between the theoretical and experimental developments.

Each of the chapters deals with a specific technology that has recently emerged or made significant advances as a result of the introduction of femtosecond pulse lasers. Each chapter incorporates a mini review of the broader scope of the technology being discussed as well as the detailed investigation of a particular aspect of that technology which is relevant to biophotonics applications.

Apart from this introductory chapter, the rest of this book can be divided into four parts. Chapters 2–3 are the first part of the book and are dedicated to nonlinear optical microscopy with a focus on two-photon fluorescence microscopy through tissue. After a general review of nonlinear optical microscopy, Chapter 2 presents two-photon autofluorescence microscopy through tissue and two-photon autofluorescence microspectroscopy [52]. The principle of two-photon FRET and FLIM is also introduced in Chapter 2. One of the major issues for two-photon fluorescence microscopy through tissue is the effect of multiple scattering due to the turbid property of tissue. This problem is addressed both experimentally and theoretically in Chapter 3 [52].

As a logical extension to imaging through tissue, it is necessary to develop compact multi-photon microscopy for biomedical studies. The second part of the book, comprising Chapters 4 and 5, gives rise to the recent development of miniaturised multi-photon microscopy, which is of particular importance for the medical applications of this technology including early cancer detection [55, 59]. Chapter 4 covers the investigation into the utilisation of single-mode fibre optics for multi-photon microendoscopy [55]. The recent development of multi-photon microendoscopy based on double-clad photonic crystal fibre optics and microelectric-mechanical system (MEMS) devices is described in Chapter 5 [59].

Chapters 6–8 constitute the third part of the book, which is centred on laser trapping and laser tweezers. The introduction to laser tweezers is first presented in Chapter 6, where a particular application of this technology in near-field optical microscopy is given [53]. In this new approach, a trapped microparticle is adopted to convert the evanescent wave to propagating waves that can be detected in the far-field region for imaging. This effect is manifested by the so-called morphology-dependent resonance phenomenon and can be significantly enhanced by the use of a femtosecond laser beam for simultaneous laser trapping and two-photon fluorescence excitation. The detail of femtosecond laser trapping and relevant two-photon induced morphology-dependent resonance is the topic of Chapter 7 [57]. Chapter 8 extends the idea of laser trapping and femtosecond laser tweezers into the near-field region where an evanescent wave is dominant [56, 58, 60].

Finally, the last part of the book, which is Chapter 9, is focused on the recent field called femtosecond microengineering [54]. The use of femtosecond-based microfabrication methods has facilitated the development of many new technologies for life sciences. Chapter 9 exemplifies some of the new devices including microfluidic devices, photonic-crystal-based sensing, bioreactors for stem cell growth and cellular engineering and microgrids for controlling cells. This chapter also presents a novel combination of laser trapping and multi-photon microscopy.

References

[1] P. Prasad. *Introduction to Biophotonics*. New Jersey, John Wiley & Sons, 2003.

[2] V. V. Tuchin. *Handbook of Optical Biomedical Diagnostics*. Bellingham, SPIE Press, 2002.

[3] T. Vo-Dinh. *Biomedical Photonics Handbook*. London, CRC Press, 2003.

[4] J. B. Pawley. *Handbook of Biological Confocal Microscopy*. New York, Plenum, 1994.

[5] T. Wilson and C. J. R. Sheppard. *Theory and Practice of Scanning Optical Microscopes*. London, Academic Press, 1984.

[6] M. Gu. *Principles of Three-Dimensional Imaging in Confocal Microscopes*. Singapore, World Scientific, 1996.

[7] A. Ashkin. Applications of laser radiation pressure. *Science*, **210**:1081–1088, 1980.

[8] D. Grier. A revolution in optical manipulation. *Nature*, **424**:810–816, 2003.

[9] K. Dholakia, P. Reece and M. Gu. Light takes hold: optical micromanipulation. *Chem. Soc. Rev.*, **37**:42–45, 2008.

[10] J. Diel and W. Rudolph. *Ultrashort Laser Phenomena: Fundamentals, Techniques and Applications on a Femtosecond Time Scale*. London, Academic Press, 1996.

[11] W. Denk, J. H. Strickler and W. W. Webb. Two-photon laser scanning fluorescence microscopy. *Science*, **248**:73–75, 1990.

[12] C. J. R. Sheppard and M. Gu. Image formation in two-photon fluorescence microscopy. *Optik.*, **86**:104–106, 1990.

[13] C. J. R. Sheppard, R. Kompfner, J. Gannaway and D. Walsh. The scanning harmonic optical microscope. *IEEE J. Quant Elect.*, **13**:912, 1977.

[14] C. J. R. Sheppard and R. Kompfner. Resonant scanning optical microscope. *Appl. Opt.*, **17**:2879–2882, 1978.

[15] G. S. He, J. B. Bhawalker, P. Prasad and B. A. Reinhardt. Three-photon-absorption-induced fluorescence and optical limiting effects in an organic compound. *Opt. Lett.*, **20**:1524–1526, 1995.

[16] S. Hell, S. K. Bahlmann, M. Schrader, A. Soini, H. Malak, I. Gryczynski and J. R. Lakowicz. Three-photon excitation in fluorescence microscopy. *J. Biomed. Opt.*, **1**:71–74, 1996.

[17] M. Gu. Resolution in three-photon fluorescence scanning microscopy. *Opt. Lett.*, **21**:988–990, 1996.

[18] M. Gu and C. J. R. Sheppard. Effects of a finite-sized pinhole on 3d image formation in confocal two-photon fluorescence microscopy. *J. Mod. Opt.*, **40**:2009–2024, 1993.

[19] M. Gu and C. J. R. Sheppard. Comparison of three-dimensional imaging properties between two-photon and single-photon fluorescence microscopy. *J. Microscopy*, **177**:128–137, 1995.

[20] M. Gu and X. S. Gan. Effect of the detector size and the fluorescence wavelength on the resolution of three- and two-photon confocal microscopy. *Bioimaging*, **4**:129–137, 1996.

[21] M. Gu, X. Gan, A. Kisteman and M. Xu. Comparison of penetration depth between single-photon excitation and two-photon excitation in imaging through turbid tissue media. *App. Phys. Lett.*, **77**:1551–1553, 2000.

[22] X. S. Gan and M. Gu. Fluorescence microscopic imaging through tissue-like turbid media. *J. App. Phy.*, **87**:3214–3221, 2000.

[23] X. Deng, X. Gan and M. Gu. Multi-photon fluorescence microscopic imaging through double-layered turbid tissue media. *J. App. Phy.*, **91**:4659–4665, 2002.

[24] Y. R. Shen. Surface properties probed by second-harmonic and sum-frequency generation. *Nature*, **337**:519–525, 1989.

[25] P. J. Campagnola and L. M. Loew. Second harmonic imaging microscopy for visualizing biomolecular arrays in cells, tissues and organisms. *Nat. Biotechnol.*, **21**:1356–1360, 2003.

[26] X. Deng, E. D. Williams, E. W. Thompson, X. Gan and M. Gu. Second-harmonic generation from biological tissues: effect of excitation wavelength. *Scanning*, **24**:175–178, 2002.

[27] W. Zipfel, R. Williams and W. W. Webb. Nonlinear magic: multiphoton microscopy in the biosciences. *Nat. Biotechnol.*, **21**:1369–1377, 2003.

[28] P. So and B. Masters. *Handbook of Biological Nonlinear Optical Microscopy*. London, Oxford University Press, 2008.

[29] D. Bird and M. Gu. Fibre-optic two-photon scanning fluorescence microscopy. *J. Microscopy*, **208**:35–48, 2002.

[30] L. Fu and M. Gu. Fibre-optical nonlinear optical microscopy and endoscopy. *J. Microscopy*, **226**:195–206, 2007.

[31] D. Morrish, X. Gan and M. Gu. Morphology-dependent resonance induced by two-photon excitation in a micro-sphere trapped by a femtosecond pulsed laser. *Opt. Express*, **12**:4198–4202, 2004.

[32] B. Agate, C. T. A. Brown, W. Sibbett and K. Dholakia. Femtosecond optical tweezers for in-situ control of two-photon fluorescence. *Opt. Express*, **12**:3011–3017, 2004.

[33] M. Gu, J.-B. Haumonte, Y. Micheau, J. Chon and X. Gan. Laser trapping and manipulation under focused evanescent wave illumination. *App. Phys. Lett.*, **84**:4236–4238, 2004.

[34] S. Kuriakose, D. Morrish, X. Gan, J. Chon, K. Dholakia and M. Gu. Near-field optical trapping with an ultrashort pulsed laser beam. *App. Phys. Lett.*, **92**:081108, 2008.

[35] U. Tirlapur and K. Konig. Cell biology: targeted transfection by femtosecond laser. *Nature*, **418**:290–291, 2002.

[36] D. Day, C. Cranfield and M. Gu. High speed fluorescence imaging and intensity profiling of femtosecond induced calcium transients. *Int. J. Biomedical Imaging*, **2006**:93438, 2006.

[37] S. Wu, J. Serbin and M. Gu. Two-photon polymerisation for three-dimensional microfabrication. *J. Photochem. Photobio. A: Chemistry*, **181**:1–11, 2006.

[38] B. Jia, J. Li and M. Gu. Two-photon polymerisation for three-dimensional photonic devices in polymer and nanocomposites. *Australian J. Chem.*, **60**:484–495, 2007.

[39] D. Day and M. Gu. Formation of voids in a doped polymethylmethacrylate polymer. *Appl. Opt.*, **41**:1852–1956, 2002.

[40] M. Ventura, M. Straub and M. Gu. Void-channel microstructures in resin solids as an efficient way to photonic crystals. *App. Phys. Lett.*, **82**:1649–1651, 2003.

[41] M. Straub, M. Ventura and M. Gu. Multiple higher-order stop gaps in infrared polymer photonic crystals. *Phys. Rev. Lett.*, **91**:0434901, 2003.

[42] D. Day, K. Pham, M. Ludford-Menting, J. Oliaro, S. Russell and M. Gu. A method for prolonged imaging of motile lymphocytes. *Immunol. Cell Bio.*, **87**:154–158, 2009.

[43] D. Day and M. Gu. Microchannel fabrication in pmma based on localized heating by nanojoule high repetition rate femtosecond pulses. *Opt. Express*, **13**:5939–5946, 2005.

[44] D. Day and M. Gu. Rapid prototyping of microfluidic devices with femtosecond laser fabricated photomasks. *Opt. Express*, **14**:10753–10758, 2006.

[45] D. Therriault, S. R. White and J. Lewis. Chaotic mixing in three-dimensional microvascular networks fabricated by direct-write assembly. *Nat. Mat.*, **2**:265–271, 2003.

[46] J. Wu, D. Day and M. Gu. A microfluidic refractive-index sensor based on three-dimensional photonic crystals. *App. Phys. Lett.*, **92**:071108, 2008.

[47] C. Nelson, R. Jean, J. Tan, W. Liu, N. Sniadecki, A. Spector and C. Chen. Emergent patterns of growth controlled by multicellular form and mechanics. *Proc. Natl. Acad. Sci. USA*, **102**:11594–11599, 2005.

[48] S. Hollister. Porous scaffold design for tissue engineering. *Nat. Mat.*, **4**:518–524, 2005.

[49] L. Griffith and M. Swartz. Capturing 3d tissue physiology *in vitro*. *Nat. Rev. Mol. Cell Bio.*, **7**:211–224, 2006.

[50] D. Albrecht, G. Underhill, T. Wassermann, R. Sah and S. Bhatia. Probing the role of multicellular organization in three-dimensional microenvironments. *Nat. Methods*, **3**:369–375, 2006.

[51] A. Engler, S. Sen, H. Sweeney and D. Discher. Matrix elasticity directs stem cell lineage specification. *Cell*, **126**:677–689, 2006.

[52] S. Schilders. *Microscopic Imaging in Turbid Media*. Ph.D. thesis, Department of Physics, Victoria University of Technology, Australia, 1999.

[53] P. C. Ke. *Near-Field Scanning Optical Microscopy with Laser Trapping*. Ph.D. thesis, Department of Physics, Victoria University of Technology, Australia, 2000.

[54] D. J. Day. *Three-dimensional Bit Optical Data Storage in a Photorefractive Polymer*. Ph.D. thesis, School of Biophysical Sciences and Electrical Engineering, Swinburne University of Technology, Australia, 2001.

[55] D. Bird. *Fiber-Optic Two-Photon Fluorescence Microscopy*. Ph.D. thesis, Faculty of Engineering and Industrial Research, Swinburne University of Technology, Australia, 2002.

[56] D. Ganic. *Far-Field and Near-Field Optical Trapping*. Ph.D. thesis, Faculty of Engineering and Industrial Research, Swinburne University of Technology, Australia, 2005.

[57] D. Morrish. *Morphology Dependent Resonance of a Microsphere and Its Application in Near-Field Scanning Optical Microscopy*. Ph.D. thesis, Faculty of Engineering and Industrial Research, Swinburne University of Technology, Australia, 2005.

[58] B. Jia. *A Study on the Complex Evanescent Focal Region of a High Numerical Aperture Objective and Its Applications*. Ph.D. thesis, Faculty of Engineering and Industrial Research, Swinburne University of Technology, Australia, 2006.

[59] L. Fu. *Fibre-Optical Nonlinear Endoscopy*. Ph.D. thesis, Faculty of Engineering and Industrial Research, Swinburne University of Technology, Australia, 2006.

[60] S. Kuriakose (Varghese). *Characterisation of Near-Field Optical Trapping and Biological Applications*. Ph.D. thesis, Faculty of Engineering and Industrial Research, Swinburne University of Technology, Australia, 2007.

[61] B. E. A. Saleh and M. C. Teich. *Fundamentals of Photonics*. New York, John Wiley & Sons, 1991.

[62] M. Gu. *Advanced Optical Imaging Theory*. Heidelberg, Springer Verlag, 2000.

2 Nonlinear optical microscopy

This chapter first gives an introduction to the principle of nonlinear optical microscopy including two-photon fluorescence and second harmonic microscopy (Section 2.1). A comparison of single-photon, two-photon and reflection second harmonic generation (SHG) microscopy through tissue is given in Section 2.2. The focus of Sections 2.3 and 2.4 is two-photon autofluorescence spectroscopic microscopy through skeletal muscle tissue. Sections 2.5 and 2.6 are mainly devoted to the demonstration of three-dimensional (3D) fluorescence resonance energy transfer (FRET) and fluorescence lifetime imaging (FLIM) techniques of living cells under two-photon excitation, respectively.

2.1 Nonlinear optical microscopy

Nonlinear optics has been growing at a prodigious rate after the first demonstration of SHG with a pulsed ruby laser in 1961 [1]. Over the last two decades, the development of imaging techniques based on nonlinear optical effects has been propelled by rapid technological advances in mode-locked femtosecond lasers, laser scanning microscopy and fluorescence probe synthesis. Taking advantage of the high spatial resolution due to the higher order dependence on the excitation intensity, nonlinear optical microscopy has been a powerful tool for bioscience [2].

Nonlinear optical effects used as the imaging contrast occur when a biological tissue interacts with an intense laser beam and exhibits a nonlinear response to the applied field strength. In the light–matter interaction, the induced polarisation vector P of the material subject to the vectorial electric field E can be expressed as [1]

$$P = \chi^{(1)} \otimes E + \chi^{(2)} \otimes E \otimes E + \chi^{(3)} \otimes E \otimes E \otimes E + \cdots, \qquad (2.1)$$

where $\chi^{(i)}$ is the ith order nonlinear susceptibility tensor and $\otimes$ represents a combined tensor product and integral over frequencies. The bulk nonlinear optical susceptibilities $\chi^{(2)}$ and $\chi^{(3)}$ are obtained from the corresponding high-order molecular nonlinear optical coefficients (hyperpolarisability) β and γ by using a sum of the molecular coefficients over all molecule sites. Typically materials with conjugated π-electron structures give large optical nonlinearities.

The usual linear susceptibility $\chi^{(1)}$ contributes to the single-photon absorption and reflection of the light in tissues. The $\chi^{(3)}$ corresponds to third-order processes such as two-photon absorption, third harmonic generation (THG), and coherent anti-Stokes Raman scattering (CARS), while SHG results from $\chi^{(2)}$. The unique features of each nonlinear optical imaging contrast mechanism are summarised in Table 2.1. Not surprisingly, the combination of these nonlinear optical imaging modes extends the useful range of nonlinear optical microscopy.

2.1.1   Multi-photon fluorescence microscopy

Multi-photon excitation was predicted by Göppert-Mayer in 1931 [3], theoretically showing that photons of less energy together can cause an excitation usually produced by the absorption of a single photon of higher energy (Table 2.1). The first experimental demonstration of two-photon absorption was to excite fluorescence emission in CaF_2:Eu^{3+} after the invention of pulsed ruby lasers [4]. Three-photon excitation was first achieved in naphthalene crystals in 1964 [5]. Subsequently, multi-photon excitation has been applied to molecular spectroscopy. The unprecedented biological applications of multi-photon excitation began with the invention of two-photon laser scanning microscopy by Denk *et al.* in 1990 [6], originally developed for localised photochemical activation of caged biomolecules. Five years after that, three-photon fluorescence microscopy was also reported and has shown its imaging capability for cellular processes by harnessing the ultraviolet (UV) fluorescence of molecules [7].

The efficiency of multi-photon absorption depends on the multi-photon absorption cross section of the molecule, and on two or more photons interacting with the molecule simultaneously ($\sim 10^{-16}$ s). Therefore, the maximum fluorescence output available for image formation is obtained by using ultrashort pulse lasers. In an n-photon process excited with a pulsed laser beam of pulse width τ and repetition rate f, the fluorescence strength η is enhanced by a factor given by

$$\eta = \frac{1}{(\tau f)^{n-1}}, \qquad (2.2)$$

compared with continuous-wave illumination [6]. Particularly in microscopic applications, the fluorescence (and potential photobleaching and photodamage related to fluorescence excitation) drops off rapidly away from the focal plane, so that inherent 3D resolution is achieved without confocal detection optics. This spatial localisation enables 3D resolved activation of caged bioactive molecules by photochemical release from an inactivating chromophore within femtolitre volumes [6].

In multi-photon microscopy, mainly ballistic (non-scattered) photons contribute efficiently to the fluorescence generation in the focal volume. The strength of the two-photon excited fluorescence signal at imaging depth z is proportional to $e^{\frac{-2z}{l_s}}$, where l_s is the mean free path describing the scattering strength of tissue. In general, near-infrared light used by multi-photon microscopy experiences less scattering in most biological tissues, and therefore gives relatively deep penetration compared with confocal microscopy with a UV or visible wavelength. Furthermore, multi-photon microscopy uses wide-field detection to collect both non-scattered fluorescence photons from the focal region and multiply scattered fluorescence photons contributing to the image formation. Thus, the

Table 2.1 Features of nonlinear optical microscopy using various nonlinear optical phenomena. 2PEF and 3PEF represent two- and three-photon excited fluorescence.

Features	2PEF	3PEF	SHG	THG	CARS
Energy level description					
Corresponding susceptibility	$\chi^{(3)}$	$\chi^{(5)}$	$\chi^{(2)}$	$\chi^{(3)}$	$\chi^{(3)}$
Discovery in optics	1961	1964	1961	1962	1965
Introduction into bioscience	1990	1995	1986	1997	1982
Advantages	• Deeper imaging with less phototoxicity; • Spatial localisation for fluorescence excitation; • Intrinsic fluorescence from NADH, flavins, and green fluorescence protein.	A new window to excite deeper ultraviolet fluorophores such as amino acids, neuro transmitters with less background and photodamage to live cells.	• Coherent process, symmetry selection; • Probing well-ordered structures, functions of membranes, nonfluorescence tissues; • No absorption of light.	• Coherent process, no symmetry requirement; • No absorption of light; • Imaging both in bulk and at surfaces for extended conjugation of pi electrons.	• Coherent process; • Inherent vibrational contrast for the cellular species, requires no endogenous or exogenous fluorophores; • Vibrational and chemical sensitivities.
Main applications	Depth imaging in brain slices, combined with disease models and fluorescence indicators.	General cells and tissues morphology, redox state by use of intrinsic ultraviolet emissions.	Structural protein arrays, collagen-related diseases, membrane potential with styryl dyes.	General cells and tissues, developmental biology neuroscience.	C-H stretching band, amide I band, phosphate stretching band in cells and tissues.
Laser sources	Tunable Ti:sapphire laser over 700–1000 nm with pulse width of 100 femtoseconds			Optical parametric oscillator (OPO), Cr:forsterite at 1230 nm	Picosecond Ti:sapphire laser, OPO, Nd:vanadate at 1064 nm

efficient fluorescence excitation and collection in multi-photon microscopy make it an ideal tool to image deep into turbid biological tissue. With a femtosecond laser of pulse width 100 fs and average power 1 W, an imaging depth of approximately 600–800 μm in the neocortex is achievable. A deeper penetration of up to 1 mm in the neocortex is possible if a regenerative amplifier is used [8]. The effect of multiple scattering in tissue-like turbid media on multi-photon fluorescence microscopy will be studied in detail in Chapter 3 [9].

It should be mentioned that multi-photon absorption spectra exhibit significant deviations from single-photon counterparts due to their different selection rules, though emission spectra are generally identical. Multicolour imaging is allowed to excite different fluorophores simultaneously through different order processes with a single wavelength, in which emissions are spectrally shifted by hundreds of nanometres and uninterrupted for collection. Multi-photon microscopy therefore offers a great flexibility to study physiology and pathology of tissue by exciting intrinsic indicators such as autofluorescence.

Advantages of multi-photon microscopy, combined with animal cancer models and gene-expression, have provided unprecedented morphological and functional insights into tumour studies and have revealed new approaches to develop novel therapeutics that target not only the tumour surface but also internal organs. In particular, the recent advances in semiconductor nano-particles and metal nano-particles, such as quantum dots [10] and gold nano-rods [11, 12], have opened up another exciting possibility for multi-photon microscopy to study tumour pathophysiology.

2.1.2 Harmonic generation microscopy

Nonlinear optical microscopy can be further extended to the use of higher harmonic light (SHG and THG), in which the energy of incident photons, instead of being absorbed by a molecule, is scattered via a process of harmonic up-conversion. The second-order susceptibility $\chi^{(2)}$ responsible for SHG only exists in non-centrosymmetric materials [1]. In contrast, THG can be applied to image general cellular structures due to the non-vanishing $\chi^{(3)}$ for all materials [1].

Second harmonic generation was the first demonstration of nonlinear optical phenomena after the invention of the laser. Shortly thereafter, SHG was applied to spectroscopy for interface characterisation and laser physics for frequency doubling [1]. The first SHG microscope and the first scanning SHG microscope were demonstrated in 1974 [13] and 1978 [14], respectively. In 1965, SHG in biological tissue was first reported independently by Deer *et al.* [15] and Zaret [16], while the first biological SHG microscope was demonstrated in 1986 [17], wherein SHG was used to study the orientation of collagen fibres in rat tail tendon.

Due to the invention of multi-photon microscopy and advances in commercial femtosecond lasers, only in the last few years has harmonic generation microscopy been established as a powerful imaging modality for visualisation of structural proteins in biological tissue [18]. The application of THG microscopy has also been emerging recently [19]. Harmonic generations, especially SHG, have been used to obtain high-resolution 3D images of endogenous arrays of collagen, microtubules and muscle myosin in a wide variety of cells and tissues.

Since harmonic generation microscopy is based on nonlinear optical processes, it retains the benefits of multi-photon excitation microscopy, such as the intrinsic 3D sectioning ability and relative depth penetration. As different imaging mechanisms from multi-photon absorption, harmonic generations do not involve an excited state (see Table 2.1), and therefore leave no energy deposition in tissue, permitting non-invasive imaging which is desirable for clinical applications. Furthermore, harmonic generations are coherent processes, where the phase of the radiated harmonic light is tightly matched to the phase of the applied fundamental light, so that harmonic light exhibits a significant dependence on the spatial distributions of both the molecules and the field of the incident fundamental light.

This coherence-conserving feature enables the polarisation dependence of harmonic light that provides information about molecular organisation and nonlinear susceptibility not available from fluorescence light with random phase [20]. Different from the conventional polarisation microscopy examining the linear birefringence of samples, harmonic generation microscopy can obtain the absolute orientation of molecules by the use of arbitrary combinations of fundamental and harmonic polarisations. Another physical implication during the coherent processes is that the harmonic emission is propagated in the forward direction. Therefore, a transmission–collection geometry has been used in most of the harmonic generation microscopy. However, in turbid tissue imaging with a focused beam, if the scatterer size is approximately equal to the illumination wavelength or much less, the orientation and distribution of scatterers and the off-axis scattering have an impact on both the forward- and backward-directed emission profile. Consequently, backward SHG becomes pronounced, making the epi-collection for *in vivo* tissue imaging practical.

2.1.3 Coherent anti-Stokes Raman scattering microscopy

In addition to harmonic generation microscopy, CARS microscopy is another 3D high resolution imaging technique that circumvents exogenous probes. As shown in Table 2.1, CARS is a four-wave mixing process in which a pump beam at frequency ω_p, a Stokes beam at frequency ω_s and a probe beam at frequency ω_p' are interacted with a sample to result in an anti-Stokes signal at $\omega_{as} = \omega_p - \omega_s + \omega_p'$. In most experiments, the pump and the probe beams are derived from the same laser. The vibrational contrast in CARS microscopy arises from the resonant oscillation when the beat frequency $(\omega_p - \omega_s)$ matches the frequency of a particular Raman active molecular vibration mode. Furthermore, due to its coherence nature, CARS signal generation only occurs when the field–sample interaction length is less than the coherence length. The generated CARS signal is proportional to $(\chi^{(3)})^2 I_p^2 I_s$, having a quadratic dependence on the pump field intensity and a linear dependence on the Stokes field intensity. It therefore provides a 3D sectioning capability in CARS microscopy.

The first systematic study of CARS was carried out in 1965 [21]. Since typical CARS signals are orders of magnitude stronger than the corresponding spontaneous Raman

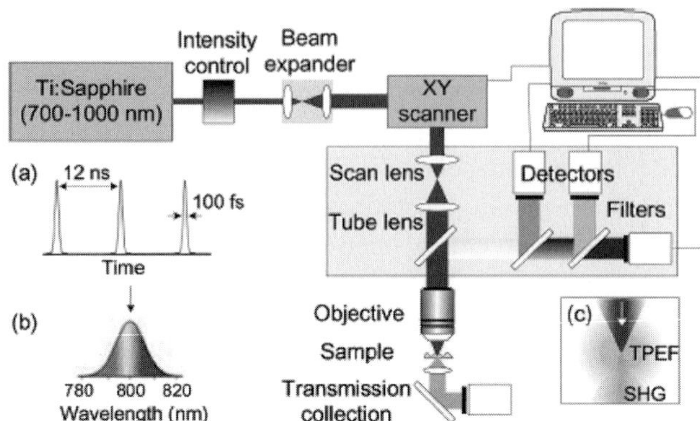

Fig. 2.1 Schematic diagram of a nonlinear optical microscope. (a) Ultrashort pulses from a Ti:sapphire at repetition 80 MHz. (b) Pulses have a pulsewidth of approximately 100 fs and a bandwidth of approximately 10 nm. (c) Two-photon excited fluorescence (TPEF) is isotropically emitted, whereas SHG is mainly forward-directed.

response, CARS spectroscopy has been the most extensively used nonlinear Raman technique. Although the first report on CARS microscopy [22] was more than two decades ago, rapid development has been made only in the past few years [23], contributing to the advances in pulsed laser technology. Unlike multi-photon and harmonic generation microscopy, typical CARS microscopy makes use of two synchronised mode-locked picosecond lasers. The picosecond pulses enable high spectral resolution with a narrow bandwidth for a specific vibrational mode and an improved signal-to-background ratio.

2.2 Two-photon fluorescence and second harmonic generation microscopy

Since this book is mainly focused on two-photon and SHG microscopy, we depict a typical nonlinear optical microscope in Fig. 2.1. A femtosecond pulse laser source is usually a Ti:sapphire oscillator with a tunable wavelength range over 700–1000 nm. The beam is scanned by the x–y scanner (usually a pair of galvanometric mirrors) before being focused by the microscope objective. In this system, the SHG signal is collected in the forward direction where the SHG is efficiently propagated, while the fluorescence signal is imaged in the backward direction. Using an appropriate filter, one can also detect the SHG signal in the backward direction.

2.2.1 Comparison of single-photon and two-photon fluorescence imaging

The difference between single-photon and two-photon fluorescence excitation is further illustrated in Fig. 2.2. For single-photon excitation a fluorophore absorbs a single photon of the illumination light (typically in the UV or visible spectrum), which results in

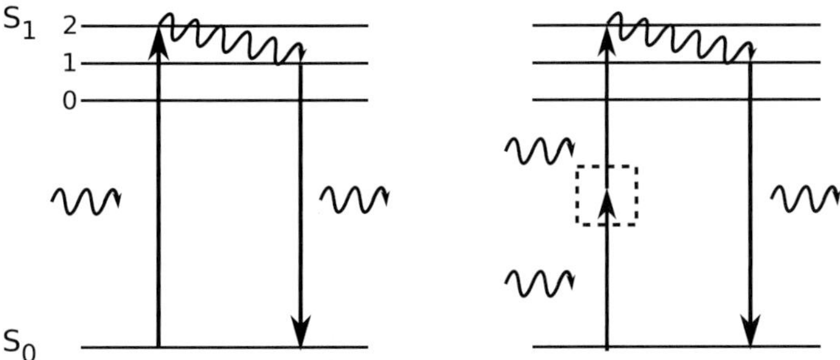

Fig. 2.2 Energy level diagram for (a) single-photon and (b) two-photon excitation.

transition of the electrons occupying the ground state (S_0) to a higher electronic state (S_1). The energy of each photon absorbed is given by

$$E = h\nu, \tag{2.3}$$

where h is Planck's constant and ν is the frequency of an illumination photon.

Excited states are metastable and have typical lifetimes of the order of 10^{-9} seconds [1]. Therefore, after absorption, electrons in excited states (S_1) return to the ground state via a relaxation process. The relaxation of the excited electrons is a multi-step process. First, the excited electrons rapidly relax to the lowest vibrational state of the excited level ($S_{1(0)}$) by a radiationless transition. Second, the electrons undergo a transition from the excited level ($S_{1(0)}$) to a vibrational state of the ground state (S_0), with the energy lost in the transition given off in the form of light.

The fundamental difference between single-photon and two-photon excitation is in the ground-to-excited state transition. In the two-photon excitation process, two incident photons, each photon having an energy of $E/2$, need to be absorbed simultaneously to excite an electron from the ground state (S_0) to a higher energy state (S_1). There is a requirement that the photons be both spatially and temporally coincident for the transition to occur.

Typically single-photon excitation uses UV to visible laser sources for excitation, since the majority of biological fluorophores have absorption bands within these regions. Two-photon excitation typically uses wavelengths in the near infrared (NIR) region. The probability of two-photon excitation is significantly lower than that of single-photon excitation because two-photon excitation is a third-order process of polarisation, which leads to the requirement that excitation photons be spatially and temporally coincident. Therefore, to increase the efficiency of the two-photon excitation process a pulsed laser source with a typical pulse width of a few hundred femtoseconds is usually used for excitation.

The excitation volume for an objective lens employed for single-photon and two-photon excitation is illustrated in Figs. 2.3(a) and 2.3(b), respectively. It is seen from Fig. 2.3(a) that single-photon fluorescence is excited almost at all points within the focal volume, due to the linear excitation probability. For two-photon excitation, fluorescence is excited only in the focal region of the imaging objective, due to the quadratic

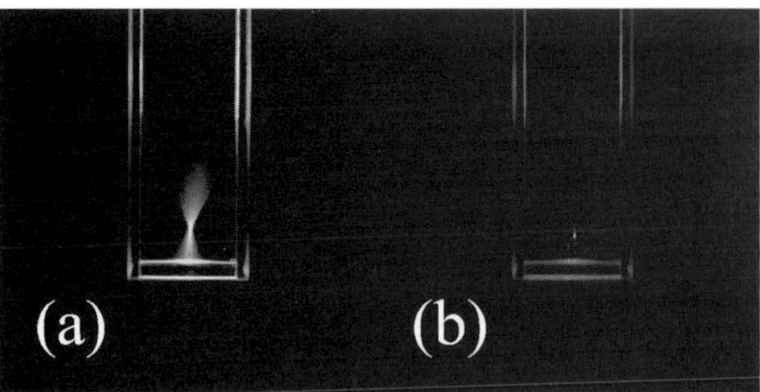

Fig. 2.3 Focal volume under single-photon (a) and two-photon (b) excitation. The laser beam is focused from the bottom upwards [28].

dependence on the excitation intensity. The result of this relationship is that two-photon excitation has an inherent optical sectioning property. Therefore it removes the requirement of inserting a pinhole in front of the detector to obtain the 3D imaging capability in confocal microscopy, although resolution can be improved further with the introduction of a finite sized pinhole [24–27]. Two-photon excitation can minimise photobleaching, phototoxicity and photodamage due to the confinement of the excitation volume. These factors are of primary importance in imaging through biological tissue, since it leads to an increased exposure time.

According to Mie scattering theory [29], a shorter wavelength has a larger scattering cross section which results in stronger scattering. This implies that two-photon excitation has an advantage over single-photon excitation in terms of the penetration depth in imaging through turbid media. Figure 2.4 demonstrates that, although the transverse resolution for single-photon (Fig. 2.4(a)) and two-photon (Fig. 2.4(b)) excitation is almost indistinguishable, the penetration depth for two-photon excitation (Fig. 2.4(d)) exceeds that for single-photon excitation (Fig. 2.4(c)) due to reduced scattering in two-photon excitation. The increase in the penetration depth demonstrated for two-photon excitation in Fig. 2.4(d) is between 50 and 100 μm. The effect of the scattering property on two-photon fluorescence microscopy will be studied in Chapter 3.

2.2.2   Reflection second harmonic generation microscopy through tissue

As indicated in the last section, imaging based on SHG is determined by electronic configurations, molecular symmetry, local morphology, orientation, molecular alignments and ultrastructures. In the case of imaging through biological tissue, it has been demonstrated from previous studies that the strength of SHG from tissue in a reflection-mode system satisfies the following expression [30]

$$I_{2\omega}(\lambda) \propto 16\pi \left(\frac{\omega^2}{n_\omega^2 n_{2\omega} c^2} \right) \kappa S_{2\omega}^2 \left| \chi_{\text{eff}}^{(2)}(\lambda) \right|^2 I_\omega^2, \tag{2.4}$$

where λ (or ω) is the excitation wavelength (or frequency). n_ω and $n_{2\omega}$ are the refractive indices of tissue at ω and 2ω, and can be assumed to be close to those of water if tissue

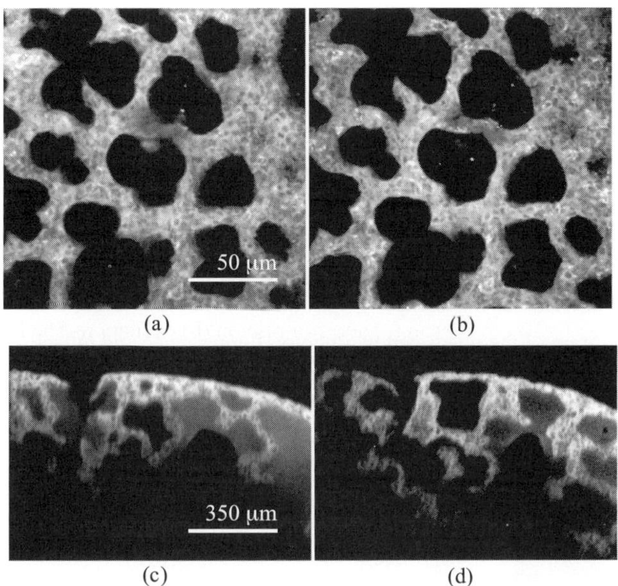

Fig. 2.4 Transverse ((a) and (b)) and vertical ((c) and (d)) image sections of rat lung tissue recorded from autofluorescence excited by single-photon ($\lambda = 488$ nm) with a 100 µm detection pinhole ((a) and (c)) and two-photon ($\lambda = 900$ nm) without a detection pinhole ((b) and (d)). The numerical aperture of the imaging objective is 0.9 (water immersion) [29].

is considered. The parameter κ and the backscattering coefficient $S_{2\omega}^2$ can be evaluated from Mie scattering theory [28]. Thus Eq. 2.4 gives the direct relation of SHG to the effective second-order susceptibility $\chi_{\text{eff}}^{(2)}$ which is sensitive to the local morphology of ultrastructures. Due to the quadratic dependence of the SHG strength on the excitation intensity, 3D imaging of SHG is possible [31]. This 3D imaging feature, together with the fact that SHG shows strong structural dependence, may offer a potential wavelength-sensitive non-invasive tool for exploring tissue components, tissue environments and ultimately disease diagnosis [32].

The experimental setup for reflection SHG microsocpy is shown in Fig. 2.5 [32]. It includes a femtosecond pulse laser with a pulse width between 70 fs and 100 fs and a wavelength tuning range from 730 nm to 870 nm. This laser was coupled to a confocal scanning microscope which could provide 3D imaging of SHG. The pinhole in front of the photomultiplier tube was removed. A water-immersion objective of numerical aperture (NA) 1.25 was used for microscopic imaging. A spectrograph was attached to the arc lamp port of the microscope for spectral measurement. Figure 2.6(a) shows the 3D reconstructed SHG image of the rat esophagus with a penetration depth of approximately 80 µm. Esophagus tissue was excised from the rat, and imaged within two hours. The excitation wavelength was 820 nm and the power on the sample was approximately 50 mW. The emission spectra of both SHG and two-photon excited fluorescence (TPEF) from the esophagus tissue in Fig. 2.6(b) exhibit a peak at wavelength 410 nm. A bandpass filter of 410/9 nm was placed before the PMT for the acquisition of SHG images.

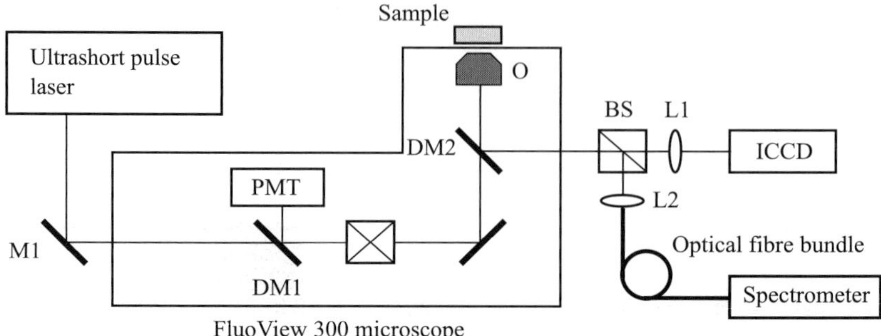

Fig. 2.5 Schematic diagram of the integrated two-photon fluorescence and reflection SHG microscopy setup. Ultrashort pulse laser: BS (beam splitter); M (mirror); PMT (photomultiplier); O (objective, NA = 1.25, 60×); DM (dichroic mirror). Spectrometer: ICCD (intensified coupled charge device) which will be discussed in Section 2.6.

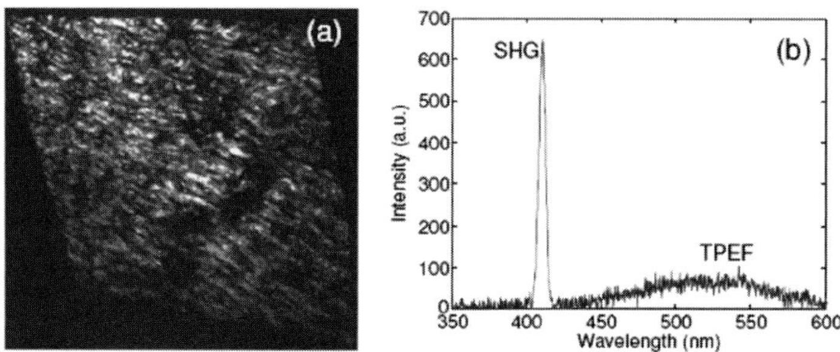

Fig. 2.6 (a) 3D reconstructed SHG image and (b) the emission spectra of rat esophagus.

2.3   Three-dimensional two-photon autofluorescence spectroscopy

For many years it has been known that all viable cells exhibit intrinsic fluorescence, i.e. autofluorescence, due to the natural presence of fluorescent molecules inside them [30, 33]. Until recently cellular autofluorescence was considered a source of unwanted interference; however, considerable attention has recently been paid to the cellular autofluorescence due to its potential as a marker to distinguish between healthy and cancerous tissue *in vivo* [34]. Biological tissue is also known to be highly diffusive, which makes it difficult to obtain the information required at a significant depth into tissue and to perform useful microscopic and/or spectroscopic analyses. One of the methods for solving this problem is to use two-photon excitation produced by a NIR ultrashort pulse laser beam [6]. In fact, due to the quadratic dependence of the excitation probability, it is possible to perform 3D two-photon autofluorescence spectroscopy [30, 35].

It has been known that cellular autofluorescence in mammalian cells is dominated by two distinct molecular fluorophores, with emission spectra around wavelengths of 440 nm and 520 nm under the excitation at wavelengths of 360 nm and 440 nm,

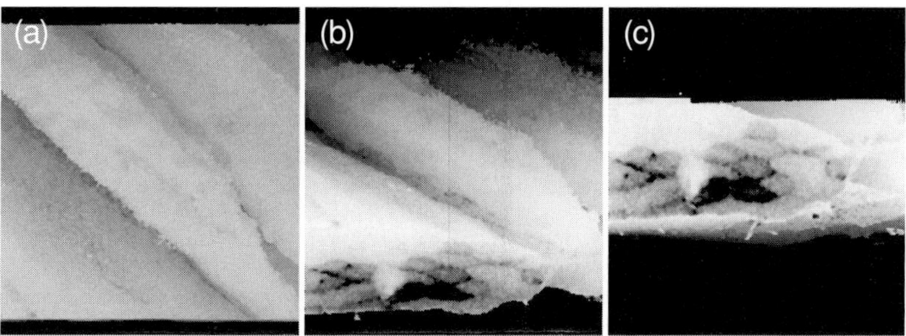

Fig. 2.7 Reconstructed 3D autofluorescence image of the rat skeletal muscle tissue at different rotation view angles: (a) 0°, (b) 45°, (c) 90°. Reprinted with permission from Ref. [35], S. Schilders and M. Gu, *Appl. Opt.* **38**, 720 (1999). © 1999, Optical Society of America.

respectively [30]. These spectra indicate the existence of the pyridine nucleotide (NADH) and flavin compounds bound in mitochondria and cytoplasm. In this section, we demonstrate that both fluorescence fluorophores in rat skeletal muscle tissue can be probed simultaneously by two-photon excitation at a wavelength of 800 nm. The experiment was performed on a system similar to Fig. 2.5 [35]. The specimen under inspection included fresh skeletal muscle tissue of thickness 60 μm from a single upper thigh of an adult rat. The sample was placed on a glass slide and fixed with a layer of wax for preservation. The average power incident upon the rat skeletal muscle tissue was 10 mW to avoid any photodamage.

Figure 2.7 illustrates the 3D image reconstructed from a series of image sections, demonstrating the structure of the skeletal muscle. It is clear that skeletal muscle fibres are grouped into fascicle structures [35,36]. As expected, no connective tissue or muscle fibre nuclei are visible in the reconstructed image.

Figure 2.8 shows autofluorescence spectra obtained at selected points within the skeletal muscle tissue. It is seen in Figs. 2.8(a), (b) and (c) that two peaks are present with emission maxima approximately at wavelengths 450 nm and 550 nm, respectively, while in Fig. 2.8(d) only one peak approximately at the wavelength 550 nm is present. The intensity at these peaks can be used to determine the local concentration of the two autofluorescence fluorophores within the muscle tissue. These autofluorescence peaks can most likely be attributed to NADH and riboflavin which are concentrated in the mitochondria and cytoplasm. It should also be noted that although the relative intensity of the two peaks changes from point to point within the skeletal muscle tissue, there is no change in the position of the fluorescence maxima. Therefore these fluorescence maxima could be used as a marker for healthy muscle tissue to discriminate against cancerous tissue.

The dependence of the fluorescence intensity on the incident power shows that the slope of the curves after regression analysis was 1.9 and 2.2 for the 450 nm and 550 nm peaks, respectively, which confirms the two-photon excitation process at these wavelengths. The degree of polarisation, $\gamma = (I_s - I_p)/(I_s + I_p)$, where I_p and I_s are the fluorescence intensities in the parallel and perpendicular directions and were 0.12 and 0.16 at wavelengths 450 nm and 550 nm, respectively. The lower γ value for the

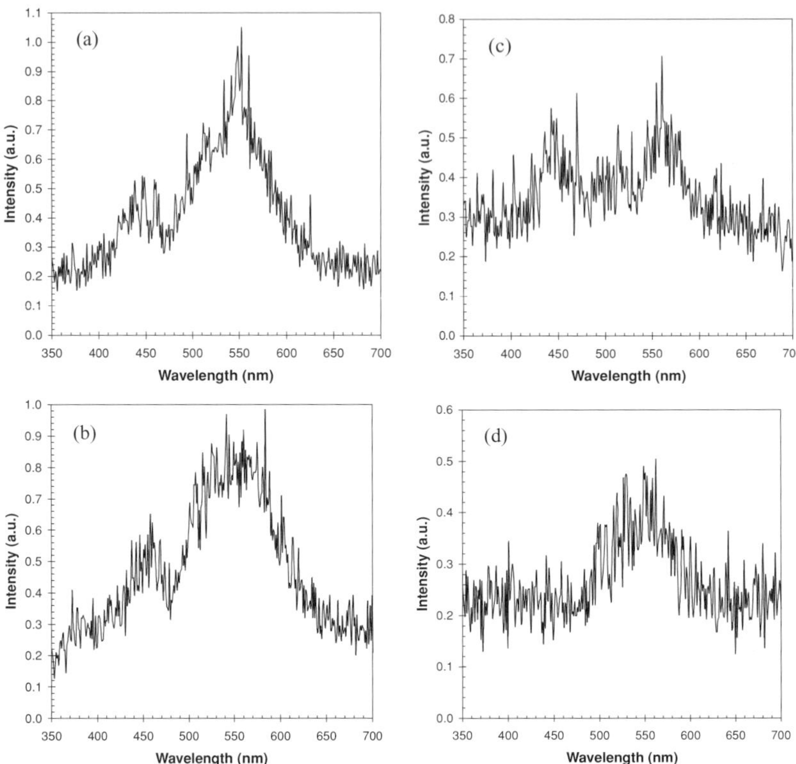

Fig. 2.8 Autofluorescence spectra from different points within the rat skeletal muscle tissue. Reprinted with permission from Ref. [35], S. Schilders and M. Gu, *Appl. Opt.* **38**, 720 (1999). © 1999, Optical Society of America.

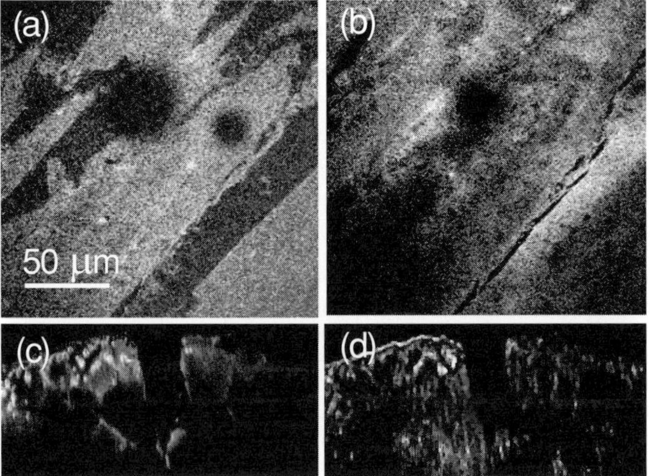

Fig. 2.9 Transverse ((a) and (b)) and vertical ((c) and (d)) image sections of the rat skeletal muscle tissue recorded at autofluorescence wavelengths 450 nm ((a) and (c)) and 550 nm ((b) and (d)). Reprinted with permission from Ref. [35], S. Schilders and M. Gu, *Appl. Opt.* **38**, 720 (1999). © 1999, Optical Society of America.

autofluorescence peak at wavelength 450 nm is due to the higher scattering of photons at this wavelength within the skeletal muscle tissue [9].

With the help of the appropriate bandpass and long pass filters, the two autofluorescence peaks can be separated and used to form 3D images respectively [9,35]. Figure 2.9 displays the two transverse image sections and two vertical image sections at the two autofluorescence peaks. It is seen that each autofluorescence fluorophore gives different 3D information about the skeletal muscle tissue, which means that the 3D spectroscopic images can be used to locate the chemical compounds. It is hoped that two autofluorescence peaks may be used as a marker to discriminate between healthy and cancerous muscle tissues, since their positions are fixed within the muscle tissue.

2.4 Effect of handling and fixation processes on two-photon autofluorescence spectroscopy

We have shown that the two distinct endogenous fluorophores, the NADH and flavin compounds bound in mitochondria and cytoplasm, can be excited simultaneously by two-photon absorption at wavelength 800 nm. This unique information offered by two-photon excitation provides a possibility to utilise the two endogenous fluorescence peaks as a marker to distinguish between healthy and abnormal tissues *in vivo*. As a key step towards *in vivo* non-invasive diagnosis by using two-photon technology, it is necessary to investigate the changes of the two autofluorescence peaks according to different handling and fixation processes that are routinely used in many pathological research and clinical diagnoses [37].

Skeletal muscle was from Balb/e mice euthanised by cervical dislocation [37]. The skeletal muscles were dissected free of surrounding tissues and prepared in two groups using two different handling processes. The first group (group 1) was prepared by block-cutting without pre-storage, which means that the muscle tissue was cut into bulky forms, and then directly placed in water, formalin and methanol for a certain period of time without pre-storing it in a freezer. The second group (group 2) was prepared by slice-sectioning with pre-storage, which means that the muscle tissue was stored in a freezer using OCT (optimal cutting temperature) compound for a period of time, then sliced into 50 μm thick section at $-20°C$ in a cryostat, and finally placed in water, formalin and methanol. Spectral measurements were performed following storage/fixation periods of 1, 2, up to 6 days. Spectral measurements from fresh muscle, which had not undergone any fixation, were also conducted for comparison.

The two spectral curves shown in Fig. 2.10(a) were measured on two separate spots of the tissue; both clearly show similar spectral characteristics with two emission peaks at around 470 nm and 540 nm. These two peaks are attributed to two types of fluorophore existing in the tissues, NADH (470 nm) and flavin (540 nm). The slight difference of magnitude between these two spectral curves results from the variation of fluorophore concentration at these measured areas. Note that the positions of the two fluorescence

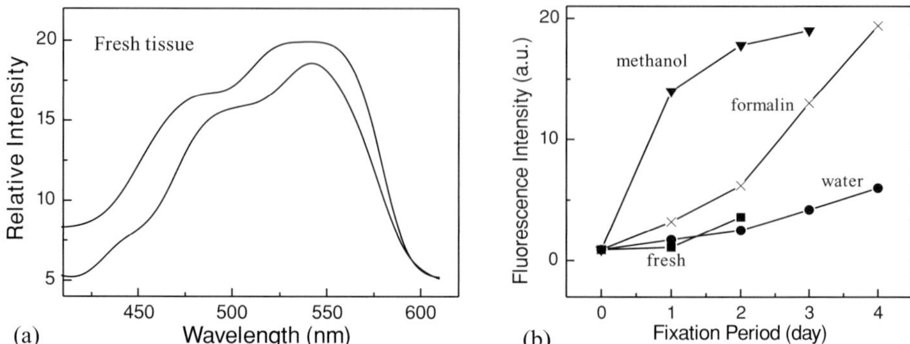

Fig. 2.10 (a) Two-photon autofluorescence spectra of fresh mouse skeletal muscle from two measured spots. (b) Measured two-photon fluorescence intensity as a function of storage or fixation period for the mouse skeletal tissues that were stored in water, or fixed in formalin and methanol, respectively. Reprinted with permission from Ref. [37], M. Xu, E. Williams, E. Thompson and M. Gu, *Appl. Opt.* **39**, 6312 (2000). © 2000, Optical Society of America.

peaks found in this work are slightly different from those shown in the last section in which the two fluorescence peaks were situated at 450 nm (NADH) and 550 nm (flavin). This difference is probably caused by the different sample preparation processes utilised.

Because moisture can affect fluorescence in tissue significantly, the measurements of fluorescence were performed immediately after the sample was taken out of the fixative to ensure that the sample was still moist with the fixative, and all measurements were carried out under the same conditions. Figure 2.10(b) shows the measured fluorescence intensity as a function of storage/fixation time for water and the two fixatives for group-1 samples and fresh muscles at the excitation wavelength of 800 nm. It is noteworthy that the fluorescence intensity is sensitive to the measurement conditions. The results presented in Fig. 2.10(b) are the relative average fluorescence intensity of an area of $50 \times 50 \ \mu m^2$. The initial point at day zero is the intensity for fresh muscle tissue. It is shown that the fluorescence intensity increases with storage/fixation period for water and the two fixatives; in particular methanol fixation results in a very rapid increase. The fluorescence intensity increases by one order of magnitude over the fixation period of two days for formalin and methanol fixatives while the effect of water is less pronounced. This phenomenon is not completely understood; probably due to the chemical or physical processes involved in the fixation that destroy other structures such as enzymes. It is also shown in Fig. 2.10(b) that the fresh sample exhibits an increase in fluorescence with time; this is probably because water has a quenching effect on fluorescence. As the sample was drying out, its fluorescence was increased.

The measured results of the two-photon fluorescence spectra for group-1 samples, which were stored in water or fixed either in formalin or methanol for 1, 2, 3 and 4 days, are shown in Fig. 2.11. The three spectra included in each set of curves were measured on three different spots within an area of $200 \times 200 \ \mu m^2$ and show similar spectral characteristics. The slight difference in magnitude among the three curves in each set

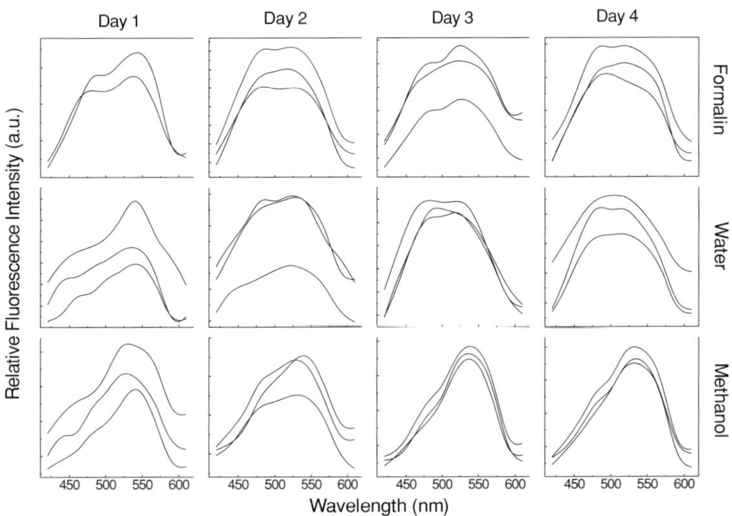

Fig. 2.11 Measured two-photon fluorescence spectra for group-1 samples stored in water, or fixed in formalin or methanol for 1, 2, 3 and 4 days. Reprinted with permission from Ref. [37], M. Xu, E. Williams, E. Thompson and M. Gu, *Appl. Opt.* **39**, 6312 (2000). © 2000, Optical Society of America.

was due to the slight difference of fluorophore concentration at those selected spots. Figure 2.11 shows that when samples were stored in water or fixed in formalin, their fluorescence spectra show little change with the periods, since two fluorescence peaks are persistent on the spectra. However, the results for those muscle tissues fixed in methanol are different. The spectra obtained from the methanol-fixed samples show only one distinguished fluorescence peak at wavelength 540 nm, the peak at wavelength 470 nm that was observed in fresh muscles becomes much less distinct and eventually disappears. Figure 2.12 displays the measured two-photon fluorescence spectra for group-2 samples, which were stored in water or fixed either in formalin or methanol for 1, 2, 3 and 4 days under the same measurement conditions as those shown in Fig 2.11. Similar spectral characteristics to Fig. 2.11 are observed.

The reason that the two-photon fluorescence spectra vary differently for water and the two fixatives is due to the different underlying fixation mechanisms [38]. Formalin is a cross-linking fixative that chemically forms covalent cross-links with proteins. It has good and rapid penetration into tissue and gives good tissue and protein preservation. Tissue can therefore be preserved in formalin for a relatively long time (up to two weeks) without incurring significant changes in the microenvironments of the native fluorophores and their autofluorescence spectra. Methanol is a coagulating fixative that rapidly changes the hydration state of the cellular component. It has good tissue penetration but is relatively less robust in preservation compared with formalin. In addition, methanol may extract water and proteins from the cell membrane in the tissue and damage the microenvironments, thus affecting the states of the endogenous fluorophores and their fluorescence spectra. In the methanol fixation process, the autofluorescence peak

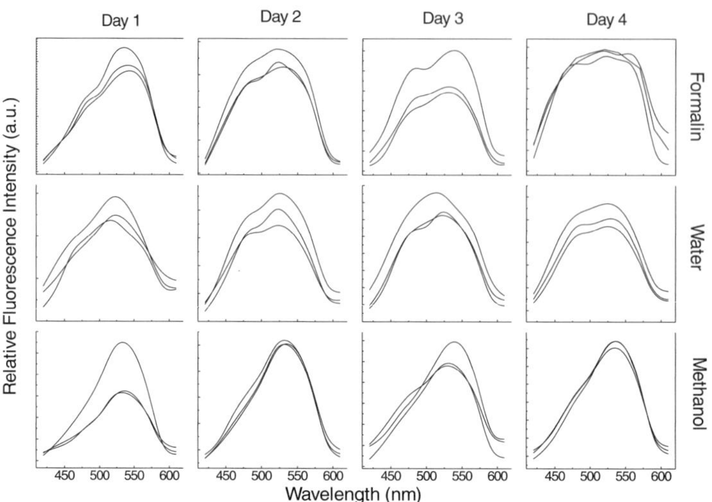

Fig. 2.12 Measured two-photon fluorescence spectra for group-2 samples stored in water, or fixed in formalin or methanol for 1, 2, 3 and 4 days. Reprinted with permission from Ref. [37], M. Xu, E. Williams, E. Thompson and M. Gu, *Appl. Opt.* **39**, 6312 (2000). © 2000, Optical Society of America.

at wavelength 470 nm varies more pronouncedly than that at wavelength 540 nm. This is because NADH is a hydrated protein, its state of microenvironment is more reactive to methanol than the state of flavins. Water is not usually treated as a fixative, because it does not fix tissue. It is a natural biological environment within tissue, and thus does not degrade tissue for a short period (up to four days), as indicated in Fig. 2.10. However, it was found that the spectra of mouse tissues changed significantly when left in water for longer than four days [37], more likely because of degradation of the tissue in the aqueous, non-fixing conditions.

The results obtained from the two groups of the sample prepared using two different handling processes indicate no significant difference in two-photon spectroscopy between these two handling processes. However, the slice-sectioning with OCT compound pre-storage method allows the sample to be kept in a fridge before using it and therefore offers an additional freedom to use the samples at a convenient time.

Handling and fixation are routine processes in most medical and biological research. It is required that the properties and structures of samples remain as close to their origin as possible after excision and fixation, which means that the state of the microenvironment should be kept intact during handling and fixation processes. Fluorescence spectroscopy is one of the methods that can be used to evaluate these processes because spectral information characterises the states of microenvironments within cells. The results shown above indicate that the spectra of mouse skeletal muscle were characterised by two autofluorescence peaks. The changes of the spectra caused by fixation and/or handling processes were identified by variations of these two fluorescence peaks. That the peak at wavelength 470 nm (NADH) varies more significantly than that at

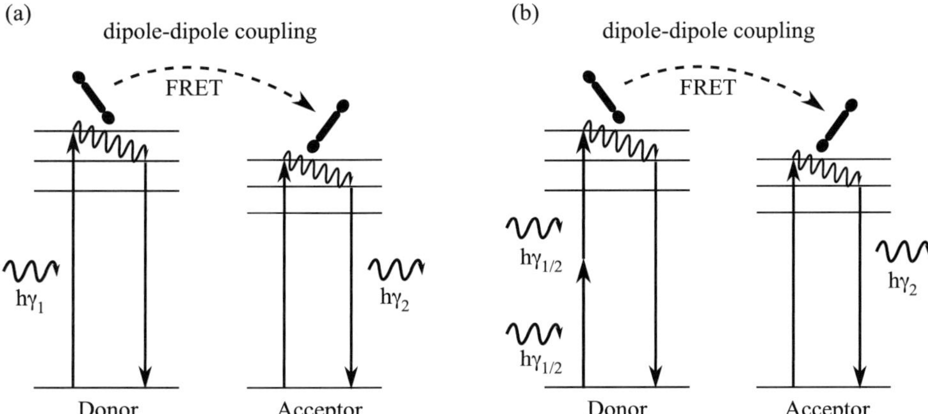

Fig. 2.13 Schematic diagram for FRET under single-photon (a) and two-photon (b) excitation.

wavelength 540 nm (flavins) in methanol fixation addresses the mechanism involved in the process. Such analysis becomes possible only under two-photon excitation, in which the two fluorescence peaks can be probed simultaneously.

2.5 Two-photon excitation fluorescence resonance energy transfer

Fluorescence resonance energy transfer (FRET) is a non-radiative energy transfer from an excited fluorescent donor molecule directly to an appropriately positioned acceptor molecule through the dipole–dipole coupling mechanism, i.e. the so-called Förster mechanism [39–42]. FRET is a technique that can be widely applied to probe biological and other complex systems for the determination of fluorophore separation and structures. Using fluorescence imaging microscopy, one can visualise the spatial and temporal distribution of FRET corresponding to cellular events.

The process of FRET is shown in Fig. 2.13. The FRET efficiency for a given pair of donors and acceptors varies inversely with the sixth power of the distance separating the donor and the acceptor; thus it can occur only over a distance limited to approximately 1–10 nm. Because of its strong distance dependence, FRET has been used to evaluate the proximity and interaction of molecules [39–42] and to visualise their dynamic behaviour in intact living cells [43]. Since FRET introduces an additional deactivation pathway for the excited donor molecule, it can be detected by monitoring the change in fluorescence emission at the spectral windows of the donor and the acceptor molecules.

Conventional techniques for imaging FRET in cell biology use confocal microscopes under single-photon excitation, as shown in Fig. 2.13(a) [41, 42]. In this case, most fluorophores in biological cells require UV or visible excitation sources, which is difficult to achieve due to the limited choice of lasers in this range. Furthermore, UV excitation can cause significant cell damage and involves complicated optical filters to separate the excitation and emission spectra. A technique for solving these problems is to use two-photon excitation, described in Section 2.1 [44].

By simultaneously absorbing two incident infrared photons, a fluorescent donor can be excited, as shown in Fig. 2.13(b). The excited donor then couples its energy to an acceptor through the Forster mechanism and visible fluorescence is then emitted from the acceptor. In particular, because of the quadratic dependence on the excitation intensity, two-photon excitation results in an inherent optical sectioning property that offers a possibility for 3D imaging, and less background fluorescence. This technique is particularly important in the detection of FRET as FRET pairs may be highly localised, making 3D imaging a necessity. Thus, 3D FRET imaging by two-photon excitation provides a better choice for biological and medical applications. In this section, we demonstrate 3D FRET imaging of living cells that were encoded with genetic vectors for two-photon excitation [44]. The experimental system is similar to that in Fig. 2.5. In this case, the pulsed laser has a wavelength tuning range from 690 nm to 1020 nm, which provides a source for two-photon excitation.

The samples used were Chinese hamster ovary tumour cell lines ectopically expressing fusion proteins of the human cell surface protein, CD46 and either enhanced blue fluorescence protein (EBFP) or enhanced green fluorescence protein (EGFP). CD46 is a ubiquitously expressed glycoprotein that protects cells from destruction by blood complement, and also acts as a measles and herpes virus receptor. The fluorochromes, EBFP and EGFP, are derived from Aequorea victoria, and have been used extensively for analysis of FRET in single-photon systems. The signals detected in these cells are a combination of low level autofluorescence and fluorescence from the EGFP or EBFP which resides at the cell membrane as well as in intracellular vesicles and in the Golgi apparatus of the cell.

Under single-photon excitation, the peak excitation wavelength is 375 nm for EBFP and 488 nm for EGFP [45]. The peak emission wavelength is 440 nm for EBFP and 510 nm for EGFP. When these two vectors are encoded in the same cell and properly positioned, EBFP acts as a donor and EGFP as an acceptor in a FRET event. We first analysed the excitation and emission spectra for EGFP and EBFP under two-photon excitation, the peak excitation wavelength was found to be approximately 750 nm for EBFP, and 960–980 nm for EGFP. However, because the objectives are optimised for the visible range and have limited transmission in the wavelength range longer than 900 nm, the wavelength used for two-photon excitation was 770 nm for EBFP and 890 nm for EGFP. The fluorescence emission of four transfected cell lines, expressing (i) no fluorochrome, (ii) EGFP-CD46, (iii) EBFP-CD46 and (iv) both EGFP-CD46 and EBFP CD46, was measured under excitation at 770 nm and 890 nm. To increase the proximity of EGFP and EBFP in the double-labelled cells, cells were treated with an antibody to CD46. For separation of the blue (EBFP) and green (EGFP) emissions, two bandpass filters with a 30 nm transmission bandwidth were used. The centre wavelengths of the filters were 440 nm and 510 nm, respectively. The excitation power was approximately 2 mW for both the excitation wavelengths, 770 nm and 890 nm, and was kept constant for excitation of different cells. The voltage and gain of the detector were also kept constant.

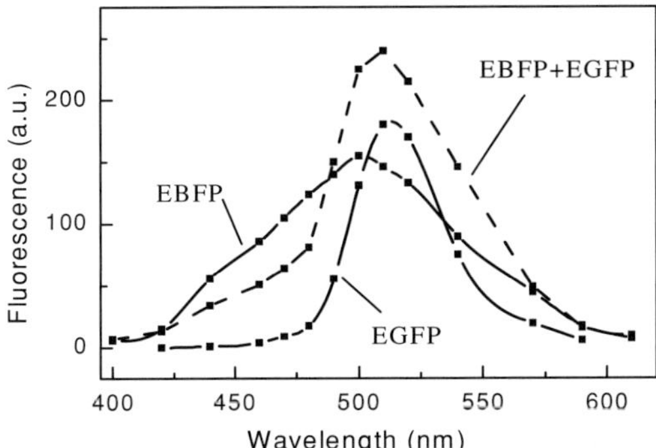

Fig. 2.14 Measured fluorescence spectra from cells with single- and double-labelled EBFP and EGFP, respectively. Excitation wavelength for EBFP and EBFP+EGFP cells is 770 nm, and for EGFP cells is 890 nm. Reprinted with permission from Ref. [44], M. Xu, B. Crimen, M. Ludford-Menting, X. Gan, S. Russell and M. Gu, *Scanning* **23**, 9 (2001). © 2001, John Wiley & Sons.

The measured emission spectra for EBFP and EGFP at excitation wavelengths of 770 nm and 890 nm are shown in Fig. 2.14. The results indicate that the EBFP emission spectrum shifts to the wavelength range peaked at 490 nm while the EGFP emission spectrum stays the same, when they are compared with those under single-photon excitation at 375 nm for EBFP and 488 nm for EGFP. This is probably due to different transition rules involved under two-photon excitation. The dashed line in Fig. 2.14 is the measured spectrum for double-labelled EBFP and EGFP cells under 770 nm excitation. It has an emission peak at 510 nm and an emission tail in the blue fluorescence region, which indicates the combination of the emission from EBFP and EGFP and confirms the existence of FRET in the double-labelled cells. To verify that the FRET event is caused by two-photon excitation, we measured the dependence of the fluorescence intensity at wavelength 510 nm on the excitation power for the doubled-labelled cells [44]. The slope of the log–log plot line is 1.90 ± 0.1, indicating the quadratic dependence of the fluorescence signal on the excitation power, and thus confirming the fluorescence excitation by two-photon absorption [44].

The two-photon fluorescence images ($100 \times 100 \ \mu m^2$) at two excitation wavelengths, 770 nm and 890 nm, are given in Fig. 2.15. The cells that do not express EBFP and EGFP show little emission of the background fluorescence at both excitation wavelengths (not shown here). Figures 2.15(a) and 2.15(b) are images of cells with single-labelled EBFP, and Figs. 2.15(c) and 2.15(d) are images of cells with EGFP. These images show that cells with EBFP fluoresce strongly at 770 nm excitation but weakly at 890 nm excitation (the intensity differs by a factor of five), while cells with EGFP fluoresce weakly at 770 nm excitation but strongly at 890 nm excitation (the intensity differs by a factor of six). These results indicate a relatively low interference of excitation between EBFP and EGFP, providing a method for verification of the existence of EBFP and EGFP.

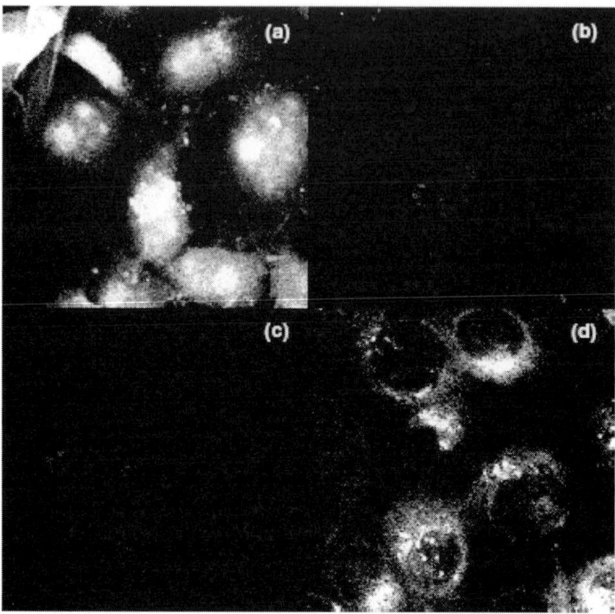

Fig. 2.15 Two-photon fluorescence images at the two excitation wavelengths, 770 nm ((a) and (c)) and 890 nm ((b) and (d)). (a) and (b) are cells with single-labelled EBFP. (c) and (d) are cells with single-labelled EGFP. The image size is 100×100 μm^2. Reprinted with permission from Ref. [44], M. Xu, B. Crimen, M. Ludford-Menting, X. Gan, S. Russell and M. Gu, *Scanning* **23**, 9 (2001). © 2001, John Wiley & Sons.

Figure 2.16 displays the 3D images of the same area (50×50 μm^2) in the double-labelled cells at the two excitation wavelengths. Each set of images includes ten slices separated by 1 μm along the depth of the sample. Figures 2.16(a) and 2.16(b) were excited at 770 nm and Fig. 2.16(c) at 890 nm. Figures 2.16(a) and 2.16(c) were obtained at the green fluorescence region and Fig. 2.16(b) at the blue fluorescence region. For 890 nm excitation, the signal detected in the blue window (not shown here) is more than a factor of six weaker than that detected in the green region, implying the proper excitation and emission of EGFP. A bright fluorescent area is observed in all slices in Figs. 2.16(a) and 2.16(c), as expected. However, it appears only in the first five slices in Fig. 2.16(b). The quenching of blue fluorescence in the remaining five slices in Fig. 2.16(b) indicates the presence of FRET within this region. This is because, in the absence of FRET, images in Fig. 2.16(b) should have strong fluorescence emission when excited at 770 nm. In the presence of FRET, the energy is transferred from the donor (EBFP) to the acceptor (EGFP), and therefore less fluorescence is emitted from the donor but more fluorescence from the acceptor (see the last five slices in Fig. 2.16(a)). The 3D localisation property of FRET within cells under two-photon excitation may prove advantageous over single-photon excitation because of the wavelength tunability of a pulsed laser and the intrinsic optical sectioning effect resulting from the quadratic dependence on the excitation power in the former case.

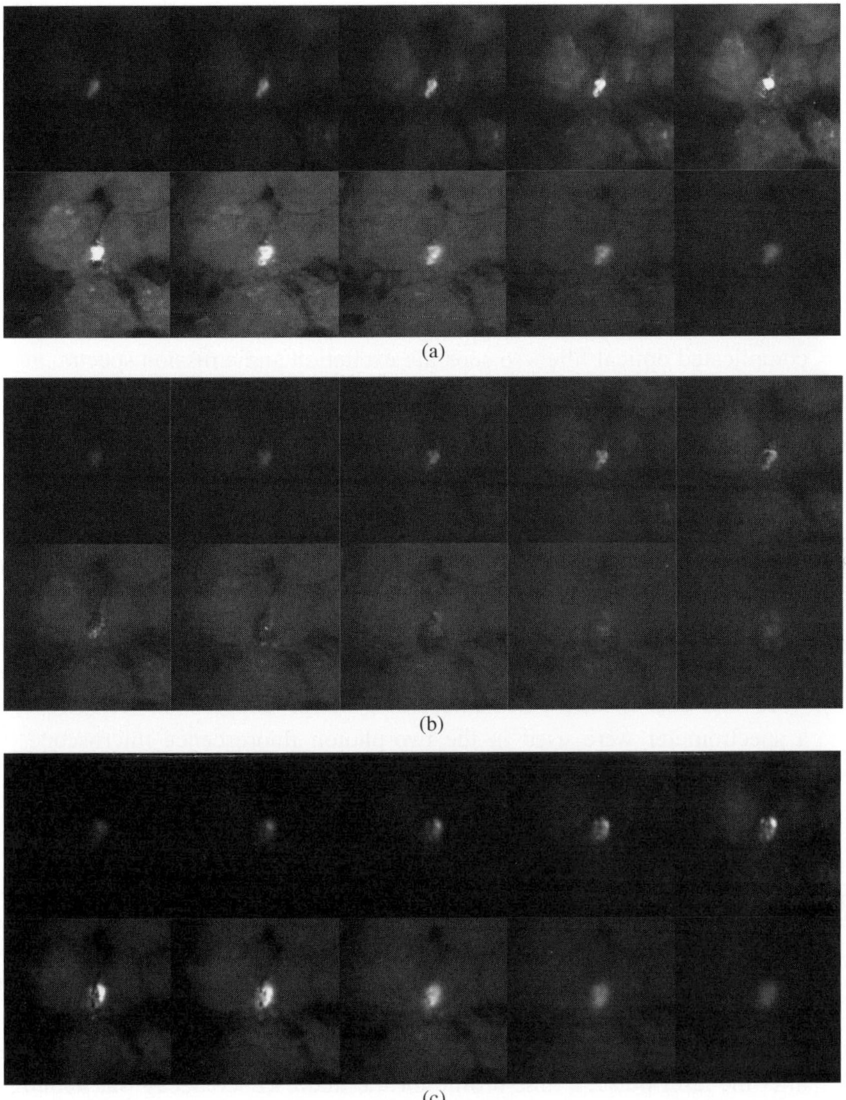

Fig. 2.16 Two-photon fluorescence 3D images of the same area ($50 \times 50\ \mu m^2$) in the double-labelled cells at the two excitation wavelengths, 770 nm ((a) and (b)) and 890 nm (c). (a) and (c) were obtained at the green fluorescence region while (b) was at the blue fluorescence region. Reprinted with permission from Ref. [44], M. Xu, B. Crimen, M. Ludford-Menting, X. Gan, S. Russell and M. Gu, *Scanning* **23**, 9 (2001). © 2001, John Wiley & Sons.

2.6 Two-photon fluorescence lifetime imaging

Intermolecular and intramolecular FRET between two fluorescent variants fused to two different host proteins, or at two different sites within the same protein, offers a unique opportunity to monitor real-time protein–protein interactions or protein conformational changes. The observation of such dynamic molecular events in living cells provides a vital insight into the action of biological molecules. To measure the FRET efficiency one can use fluorescence ratios of the donor and acceptor with and without energy transfer. This is typically achieved by producing a sample, measuring the donor/acceptor ratio, then bleaching the acceptor and repeating the measurement [46]. An alternative approach is the use of donor FLIM measurements. In the presence of FRET the donor lifetime is reduced [42]. FLIM measurements require the use of intensity modulated or pulsed light sources. Combining FLIM imaging devices with ultrashort pulse laser sources for two-photon excitation provides a means of performing lifetime and spectral measurements of FRET with the advantages that two-photon imaging offers. It removes the need for complicated optical filters to separate excitation and emission spectra, and for live cell imaging two-photon excitation has advantages over single-photon excitation such as reduced fluorescence bleaching effects and photodamage.

The combination of these techniques, two-photon microscopy, FLIM and FRET, offers a means of examining protein interaction in a fixed or live cell. In this section, we describe the FLIM technique to measure the two-photon fluorescence lifetimes of fluorophores that can be applied to two-photon FRET experiments: the two-photon spectra and lifetimes of extracellular Alexa350, EGFP, EGFP-CD46 fusion protein and Cy3 labelled antibody as well as cellular EGFP expressed in Cos7 cells [47].

To this end, a time gated intensified charge coupled device (ICCD) camera and a spectrometer were used in the two-photon fluorescence microscope, as shown in Fig. 2.5 [47]. The excitation wavelengths used ranged from 740 nm to 870 nm and the excitation power delivered to the sample was limited to approximately 6 mW to avoid photodamage.

To acquire spectra and lifetime measurements required the dichroic DM2 in the laser beam path (see Fig. 2.5). With DM2 in position the fluorescence was coupled out to the gated ICCD via the lens L1 and to the spectrometer via the beam-splitter BS and lens L2. The ICCD camera had a temporal resolution of 200 ps and the capture rate was synchronised to the pulse repetition rate (80 MHz) of the pulsed laser. This allows fluorescence from each pulse to be captured. By acquiring images at increasing delays after the laser pulse, a time profile for the fluorescence decay was acquired. The delay was successively increased by a fixed interval (200 or 400 ps) after each exposure (or integration period). In order to avoid temporal artefacts, the ICCD camera exposure time or integration period for a single point measurement was matched to the time required for a single frame scan of the confocal system. The result was then averaged over two frames before being recorded. The time required to collect data for each plot was approximately 1 to 2 minutes. For the pixel by pixel lifetime image, two temporal measurements were averaged over four frames, with a total acquisition time of approximately 8 seconds [47].

To observe the fluorescence spectra a beam-splitter BS and a lens L2 were used to couple the fluorescence to a spectrometer via an optic fibre bundle. The spectra were recorded with a spectral resolution of 3 nm over a range of 300 nm. The integration time used was between 1 and 4 seconds.

Table 2.2 Two-photon excitation wavelengths, together with observed fluorescence peaks and lifetimes.

	Alexa350	EGFP	EGFP-CD64	Cy3	Cellular EGFP
Excitation (nm)	740	870	870	870	870
Peak fluorescence (nm)	460	511	511	575	511
Lifetime (ns)	3.62	2.49	2.38	0.02, 0.8	1.69

The EGFP coding sequence was fused to the C-terminus of the extracellular domain of CD46 [48] so that the four short consensus repeats of CD46 were followed in frame by EGFP. EGFP and EGFP-CD46 were expressed in Cos7 cells using Lipofectamine Plus. Two days post transfection cells were mounted in AntiFade. For extracellular examination of Alexa350, EGFP, EGFP-CD46 and Cy3 labelled antibody the samples were prepared by placing a single drop of each, at a low concentration, onto a microscope slide. All samples were imaged at room temperature ($\sim$22°C) and a pH of about 7. Images were taken using electronic magnifications ranging from $1\times$ to $7\times$, depending on the fluorescence level. The scanning rate was typically 1050 ms/frame, image size 512×512.

The fluorescence spectra for Alexa350, EGFP, EGFP-CD46 and Cy3 under two-photon excitation were first measured for maximising the signal. Table 2.2 shows the wavelengths at which maximum signal was obtained. The two-photon fluorescence spectra of Alexa350, EGFP and Cy3 were similar to that of single-photon spectra with shifts in the peak fluorescence of approximately 10 nm to the red for Alexa350 and Cy3 [48].

The FLIM system was characterised by measuring the two-photon fluorescence lifetime of a 1 mM solution of rhodamine 6G (R6G). The result of our lifetime measurement for 1 mM R6G was 3.66 ns, which is in good agreement with results previously reported [49]. The two-photon fluorescence lifetime of Alexa350 was found to be 3.6 ns. For the protein variants EGFP and fused EGFP-CD46, a small variation was seen in two-photon lifetimes, EGFP (2.49 ns) and EGFP-CD46 (2.38 ns). These results correlate well with a previous report suggesting a shorter lifetime for fused EGFP [50]. Cy3 was found to have a bi-exponential decay with lifetime components of 0.02 ns and 0.8 ns. EGFP expressed in cells was predominantly cytoplasmic, and so fluoresced only weakly in the centre of the cell.

The observed two-photon lifetime for cellular EGFP of 1.69 ns (Fig. 2.17) averaged from three spectra at different spots was much shorter than the two-photon extracellular lifetimes (Fig. 2.17(a)). The two-photon lifetimes of the fluorophores examined are summarised in Table 2.2. Figure 2.17(b) shows the mapping of the lifetimes, indicating the spatial dependence of the lifetime within a cell. The observed two-photon lifetime of cellular EGFP (1.6 ns) was much shorter than the two-photon extracellular lifetimes. It is known that EGFP is highly susceptible to forming dimers. It is suspected that it is the formation of these dimers that causes the shortened lifetime we observe for cellular EGFP.

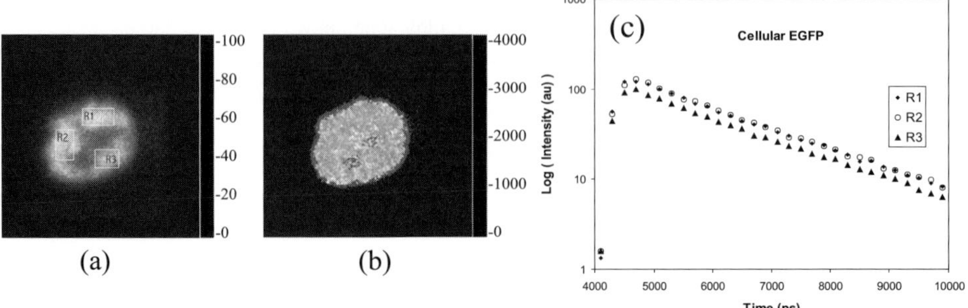

Fig. 2.17 Two-photon fluorescent lifetime of cellular EGFP. Image (a) is an intensity image of EGFP fluorescence captured with the ICCD camera. Image (b) is a pixel by pixel lifetime image. The lifetimes from regions R1, R2 and R3 marked in image (a) are shown in (c). Reprinted with permission from Ref. [47], R. Ashman, B. Crimen, M. Ludford-Menting, S. Russell and M. Gu, *Scanning* **25**, 116 (2003). © 2003, John Wiley & Sons.

References

[1] Y. R. Shen. *The Principles of Nonlinear Optics*. New York, John Wiley, 1984.

[2] L. Fu. *Fibre-optical Nonlinear Endoscopy*. Ph.D. thesis, Faculty of Engineering and Industrial Research, Swinburne University of Technology, Australia, 2006.

[3] M. Göppert-Mayer. Uber Elementarakte mit zwei Quantensprungen. *Ann. Phys. (Leipzig)*, **5**:273–294, 1931.

[4] W. Kaiser and C. G. Garrett. Two-photon excitation in caf$_2$:eu^{2+}. *Phys. Rev. Lett.*, **7**:229–231, 1961.

[5] S. Singh and L. T. Brandley. Three-photon absorption in napthalene crystals by laser excitation. *Phys. Rev. Lett.*, **12**:612–614, 1964.

[6] W. Denk, J. H. Strickler and W. W. Webb. Two-photon laser scanning fluorescence microscopy. *Science*, **248**:73–75, 1990.

[7] S. Hell, S. K. Bahlmann, M. Schrader, A. Soini, H. Malak, I. Gryczynski and J. R. Lakowicz. Three-photon excitation in fluorescence microscopy. *J. Biomed. Opt.*, **1**:71–74, 1996.

[8] F. Helmchen and W. Denk. Deep tissue two-photon microscopy. *Nat. Methods*, **2**:932–940, 2005.

[9] S. Schilders. *Microscopic Imaging in Turbid Media*. Ph.D. thesis, Department of Physics, Victoria University of Technology, Australia, 1999.

[10] D. R. Larson, W. R. Zipfel, R. M. Williams, S. W. Clark, M. P. Bruchez, F. W. Wise and W. W. Webb. Water-soluble quantum dots for multiphoton fluorescence imaging *in vivo*. *Science*, **300**:1434–1436, 2003.

[11] H. Wang, T. B. Huff, D. A. Zweifel, W. He, P. S. Low, A. Wei and J. X. Cheng. *In vitro* and *in vivo* two-photon luminescence imaging of single gold nanorods. *Proc. Natl. Acad. Sci. USA*, **102**:15752–15756, 2005.

[12] J. Li, D. Day and Min Gu. Ultra-low energy threshold for cancer photothermal therapy using transferrin-conjugated gold nanorods. *Adv. Mater.*, **20**:1–6, 2008.

[13] R. Hellwarth and P. Christensen. Nonlinear optical microscopic examination of structure in polycrystalline ZnSe. *Opt. Com.*, **12**:318–322, 1974.

[14] J. N. Gannaway and C. J. R. Sheppard. Second-harmonic imaging in the optical scanning microscope. *Opt. Quant. Elect.*, **10**:435–440, 1978.

[15] V. E. Deer, E. Klien and S. Fine. General biological effects of laser radiation. *Federation Proceedings*, **24**:99–103, 1965.

[16] M. M. Zaret. Analysis of factors of laser radiation producing retinal damage. *Federation Proceedings*, **24**:62–64, 1965.

[17] I. Freund and M. Deutsch. Second-harmonic microscopy of biological tissue. *Opt. Lett.*, **11**:94–96, 1986.

[18] P. Campagnola and L. Loew. Second harmonic imaging microscopy for visualizing biomolecular arrays in cells, tissues and organisms. *Nat. Biotechnol.*, **21**:1356–1360, 2003.

[19] Y. Barad, H. Eizenberg, M. Horowitz and Y. Silberberg. Nonlinear scanning laser microscopy by third-harmonic generation. *App. Phys. Lett.*, **70**:922–924, 1997.

[20] P. Stoller, K. M. Reiser, P. M. Celliers and A. M. Rubenchik. Polarization-modulated second harmonic generation in collagen. *Biophys. J.*, **82**:3330–3342, 2002.

[21] P. D. Maker and R. W. Tehunek. Study of optical effects due to an induced polarization third order in the electric field strength. *Phys. Rev.*, **137**:A801–A818, 1965.

[22] M. D. Duncan, J. Reintjes and T. J. Manuccia. Scanning coherent anti-Stokes Raman microscope. *Opt. Lett.*, **7**:350–352, 1982.

[23] C. L. Evans, E. O. Potma, M. Puoris'haag, D. Côté, C. P. Lin and X. S. Xie. Chemical imaging of tissue *in vivo* with video-rate coherent anti-Stokes Raman scattering microscopy. *Proc. Natl. Acad. Sci. USA*, **102**:16807–16812, 2005.

[24] C. J. R. Sheppard and M. Gu. Image formation in two-photon fluorescence microscopy. *Optik*, **86**:104–106, 1990.

[25] M. Gu and C. J. R. Sheppard. Effects of a finite-sized pinhole on 3d image formation in confocal two-photon fluorescence microscopy. *J. Mod. Opt.*, **40**:2009–2024, 1993.

[26] M. Gu and C. J. R. Sheppard. Comparison of three-dimensional imaging properties between two-photon and single-photon fluorescence microscopy. *J. Microscopy*, **177**:128–137, 1995.

[27] M. Gu. *Principles of Three-Dimensional Imaging in Confocal Microscopes*. Singapore, World Scientific, 1996.

[28] D. Morrish. *Morphology Dependent Resonance of a Microsphere and its Application in Near-field Scanning Optical Microscopy*. Ph.D. thesis, Faculty of Engineering and Industrial Research, Swinburne University of Technology, Australia, 2005.

[29] C. F. Bohren and D. R. Huffman. *Absorption and Scattering of Light by Small Particles*. New York, John Wiley, 1983.

[30] Y. Guo, Q. Z. Wang, N. Zhadin *et al.* Two-photon excitation of fluorescence from chicken tissue. *Opt. Lett.*, **36**:968–970, 1997.

[31] Y. Guo, P. P. Ho, H. Savage *et al.* Second-harmonic tomography of tissues. *Opt. Lett.*, **22**:1323–1325, 1997.

[32] X. Deng, E. D. Williams, E. W. Thompson, X. Gan and M. Gu. Second-harmonic generation from biological tissues: effect of excitation wavelength. *Scanning*, **24**:175–178, 2002.

[33] B. Chance, P. Cohen, F. Jobsis and B. Schoener. Intracellular oxidation states *in vivo*. *Science*, **137**:499–502, 1962.

[34] R. R. Alfano, D. B. Tata, J. Cordero, P. Tomashefsky, F. W. Longo and M. A. Alfano. Laser induced fluorescence spectroscopy from native cancerous and normal tissue. *IEEE J. Quant Elect.*, **QE-20**:1507–1511, 1984.

[35] S. P. Schilders and M. Gu. Three-dimensional autofluorescence spectroscopy of rat skeletal muscle tissue under two-photon excitation. *Appl. Opt.*, **38**:720–723, 1999.

[36] W. J. Krause. *Essentials of Human Histology*. Boston, Little, Brown and Company (inc.), 1994.

[37] M. Xu, E. D. Williams, E. W. Thompson and M. Gu. Effect of handling and fixation processes on fluorescence spectroscopy of rat skeletal muscles under two-photon excitation. *Appl. Opt.*, **39**:6312–6317, 2000.

[38] J. B. Pawley. *Handbook of Biological Confocal Microscopy*. New York, Plenum Press, 1994.

[39] Th. Forster. Intermolecular energy migration and fluorescence. *Ann. Phys.*, **2**:55–75, 1948.

[40] G. Coker, B. Meer and S. Chen. *Resonance Energy Transfer, Theory and Data*. New York, Wiley-VCH, 1994.

[41] A. K. Kirsch, V. Subramaniam, A. Jenei and T. M. Jovin. Fluorescence resonance energy transfer detected by scanning near-field optical microscopy. *J. Microscopy*, **194**:448–454, 1999.

[42] E. Kohen and J. G. Hirschberg. *Cell Structure and Function by Microspectrofluorometry*. San Diego, Academic Press, 1989.

[43] N. Mahajan, K. Linder, G. Berry, G. Gordon, R. Heim and B Herman. Bcl-2 and bax interactions in mitochondria probed with green fluorescent protein and fluorescence resonance energy transfer. *Nat. Biotechnol.*, **16**:547–552, 1998.

[44] M. G. Xu, B. Crimeen, M. J. Ludford-Menting, X. Gan, S. M. Russell and M. Gu. Three-dimensional localisation of fluorescence resonance energy transfer in living cells under two-photon excitation. *Scanning*, **23**:9–13, 2001.

[45] B. A. Pollok and R. Heim. Using GFP in FRET-based applications. *Trends in Cell Biology*, **9**:57–60, 1999.

[46] W. Becker, K. Benndorf, A. Bergmann, C. Biskup, K. König, U. Tirplapur and T. Zimmer. Fret measurements by TCSPC laser scanning microscopy. *Proceedings of SPIE*, **4431**:94–98, 2001.

[47] R. Ashman, M. J. Ludford-Menting, S. Russell and M. Gu. Spectra and lifetime of fluorescence resonance energy transfer fluorophores under two-photon excitation. *Scanning*, **25**:116–120, 2003.

[48] S. M. Russell, R. L. Sparrow, I. F. C. McKenzie and D. F. J. Purchell. Tissue-specific and allelic expression of the complement regulator cd46 is controlled by alternative splicing. *European Journal of Immunology*, **22**:1513–1518, 1992.

[49] M. Straub and S. W. Hell. Fluorescence lifetime three-dimensional microscopy with picosecond precision using a multifocal multiphoton microscope. *App. Phys. Lett.*, **73**:1769–1771, 1998.

[50] R. Pepperkok, A. Squire, S. Geley and P. I. H. Bastiaens. Simultaneous detection of multiple green fluorescent proteins in live cells by fluorescence lifetime imaging microscopy. *Current Biology*, **9**:269–272, 1999.

3 Two-photon fluorescence microscopy through turbid media

As discussed in Chapters 1 and 2, biological tissue is a highly scattering medium which will affect image resolution, contrast and signal level. This chapter discusses the effect of multiple scattering in a tissue-like turbid medium on two-photon fluorescence microscopy. Section 3.1 discusses a model based on imaging of microspheres embedded in a turbid medium. A quantitative study of the limiting factors on image quality is given in Section 3.2. In particular, the limitation on the penetration depth in turbid media, revealed from Monte-Carlo simulation and experimental measurements, is presented in Section 3.3.

3.1 Two-photon fluorescence microscopy of microspheres embedded in turbid media

Two-photon fluorescence microscopy has been extensively used due to its significant advantages over single-photon fluorescence microscopy. This technology has been used for *in vivo* imaging of thick biological samples [1–3]. Since the required image information is taken at a large depth within a biological specimen, optical multiple scattering within tissue may result in a severe distortion on images obtained in this situation. Thus, the effect of optical multiple scattering on fluorescence image quality should be understood if high quality images are to be obtained at significant depths into a biological specimen. In this section, we present measured images of small fluorescent microspheres embedded in a turbid medium which has different scattering characteristics under single-photon and two-photon excitation [4]. Imaging of small spheres embedded in a turbid medium has practical importance since it can be considered to be an approximate model of imaging small tumours embedded in biological tissue. It should be pointed out that this model relies on the tumours being the only fluorescent objects within turbid media, which means that the surrounding tissue does not fluoresce. The measured images are also compared with the Monte-Carlo simulation [5–8].

3.1.1 Measurement of two-photon fluorescence images

The fluorescence imaging measurements were performed on a system similar to Fig. 2.5 [1, 9]. For single-photon excitation, an Ar ion laser at wavelength 488 nm was used. A femtosecond pulse laser was employed for two-photon excitation at a

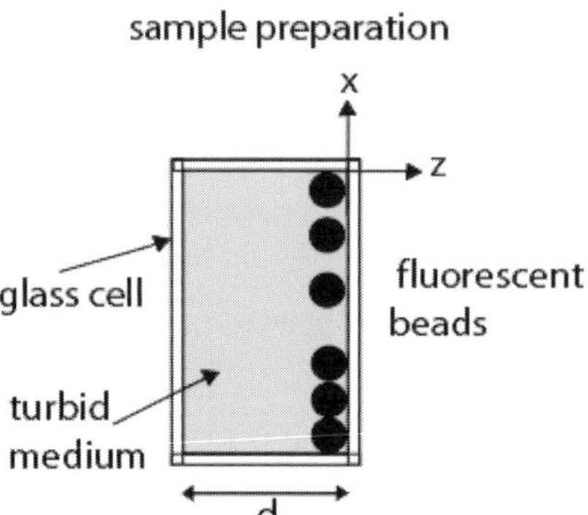

Fig. 3.1 Schematic diagram of a sample cell including microspheres and a turbid medium. Reprinted with permission from Ref. [5], M. Gu, S. Schilders and X. Gan, *J. Mod. Opt.* **47**, 959 (2000). © 2000, Taylor & Francis, www.informaworld.com.

wavelength of 800 nm. The numerical aperture of the imaging objective used for all imaging was 0.75. The incident power for both single-photon and two-photon excitation was approximately 3 mW at the back aperture of the imaging objective. In the case of single-photon excitation, a pinhole of 300 μm in diameter was placed in front of the detector to produce an optical sectioning effect similar to that under two-photon excitation without using a pinhole [10, 11].

The object embedded in a turbid medium was a cluster layer of 10 μm (diameter) fluorescent polystyrene microspheres (standard deviation 1.028 μm) dried on the glass slide of a cell (Fig. 3.1). The peak fluorescence wavelength from microspheres was measured to be approximately 520 nm at excitation wavelengths, λ, of 488 nm (single-photon) and 800 nm (two-photon). The cell had lateral dimensions 2 cm $\times$ 1 cm and thickness d. It was filled with polystyrene microspheres suspended in water. Two types of polystyrene microsphere of diameters 0.107 μm and 0.202 μm were used as turbid media. According to Mie scattering theory [12], the corresponding scattering-mean-free-path (SMFP) length l_s and the anisotropy values, g, were calculated for a given particle weight concentration of 2.5% and listed in Table 3.1. The optical thickness n is defined by the cell thickness d divided by l_s.

In Fig. 3.2, images of a 10 μm polystyrene microsphere cluster embedded in a turbid medium consisting of scattering beads of diameter 0.107 μm at $d = 200$ μm are illustrated. A comparison of the fluorescence images recorded under single-photon (Figs. 3.2(a) and (b)) and two-photon (Fig. 3.2(c)) excitation demonstrates the effectiveness of two-photon excitation over single-photon in imaging through turbid media. The substantial improvement in image quality observed under two-photon excitation (Fig. 3.2(c)) is due to two factors.

The first factor results from a longer wavelength used under two-photon excitation, which results in fewer scattering events (i.e. smaller scattering cross section) experienced by the illumination beam. The second factor results from the quadratic dependence of two-photon excitation on the incident intensity. This feature implies that two-photon excitation generates less out-of-focus fluorescence which reduces the structural detail

Table 3.1 Calculated values of the scattering-mean-free-path length, l_s, and the anisotropy value, g, for four sizes of polystyrene microsphere at wavelengths of 488 nm, 800 nm and 520 nm.

Sphere diameter κ (μm)	Geometric cross section σ_g (μm^2)	Concent. ρ (Part./μm^3)	Relative particle size, A $(\frac{a}{\lambda})$	Scattering efficiency $Q_s = \frac{\sigma_s}{\sigma_g}$	Scattering cross section, σ_s (μm^2)	smfp length, l_s (μm)	Anisotropy value, g
			(a) Single-photon excitation, $\lambda = 488$ nm				
0.107	0.009	37.11	0.1096	0.0239	2.16×10^{-4}	124.9	0.146
0.202	0.032	5.52	0.2069	0.1586	5.07×10^{-3}	35.7	0.54
0.48	0.181	0.411	0.4918	1.1896	2.15×10^{-1}	11.3	0.86
1.056	0.876	0.038	1.0819	3.4931	3.06	8.6	0.93
			(b) Fluorescence, $\lambda = 520$ nm				
0.107	0.009	37.11	0.1029	0.0198	1.78×10^{-4}	151.3	0.131
0.202	0.032	5.52	0.1942	0.1356	4.34×10^{-3}	41.8	0.482
0.48	0.181	0.411	0.4615	1.0566	1.91×10^{-1}	12.7	0.851
1.056	0.876	0.038	1.0154	3.368	2.95	8.9	0.92
			(c) Two-photon excitation, $\lambda = 800$ nm				
0.107	0.009	37.11	0.0669	0.0038	6.02×10^{-3}	786.3	0.054
0.202	0.032	5.52	0.1263	0.0389	1.24×10^{-3}	145.38	0.20
0.48	0.181	0.411	0.3	0.4197	7.59×10^{-2}	32.05	0.73
1.056	0.879	0.038	0.66	1.9937	1.75	15.16	0.90

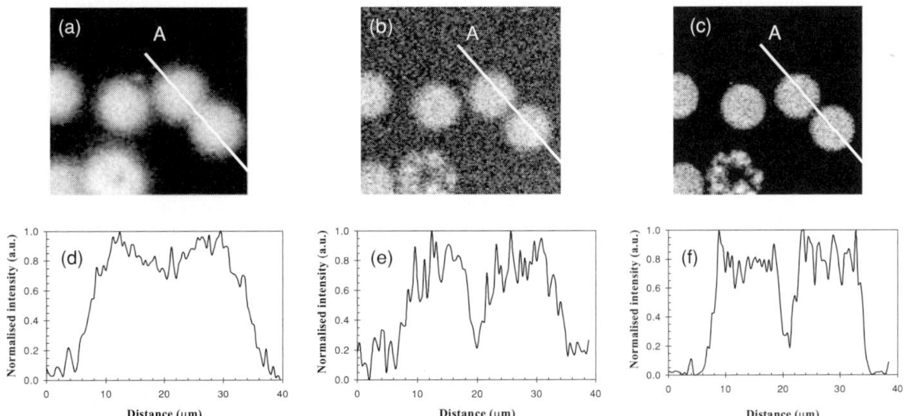

Fig. 3.2 Measured images of a 10 μm polystyrene microsphere cluster embedded within a turbid medium consisting of scattering beads of diameter 0.107 μm. The intensity of the images was normalised to unity. (a) Single-photon excitation without a detection pinhole; (b) single-photon excitation with a 300 μm detection pinhole; (c) two-photon excitation without a detection pinhole; (d), (e) and (f) intensity cross sections at position A marked in (a), (b) and (c), respectively. Reprinted with permission from Ref. [5], M. Gu, S. Schilders and X. Gan, *J. Mod. Opt.* **47**, 959 (2000). © 2000, Taylor & Francis, www.informaworld.com.

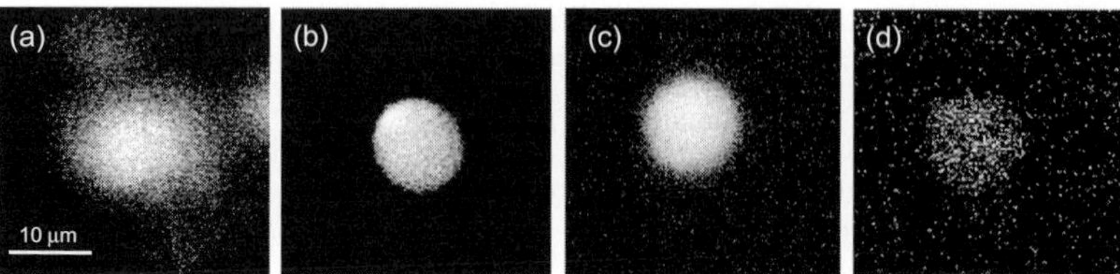

Fig. 3.3 Measured images of 10 μm polystyrene microspheres embedded within a turbid medium consisting of scattering beads of diameter 0.202 μm. (a) Single-photon excitation with a 300 μm detection pinhole ($d = 100$ μm); (b) two-photon excitation without detection pinhole ($d = 100$ μm); (c) two-photon excitation without detection pinhole ($d = 150$ μm); (d) two-photon excitation without detection pinhole ($d = 300$ μm). The intensity of the images was normalised to unity. Reprinted with permission from Ref. [5], M. Gu, S. Schilders and X. Gan, *J. Mod. Opt.* **47**, 959 (2000). © 2000, Taylor & Francis, www.informaworld.com.

and contributes more scattered photons to the image. A comparison of Figs. 3.2(a) and 3.2(b) demonstrates that the inclusion of a detection pinhole of 300 μm in diameter under single-photon excitation reduces the resolution deterioration caused by the out-of-focus information as well as the number of collected scattered photons. It should be noted that, although the image in Fig. 3.2(b) was obtained with a weak confocal arrangement, the corresponding signal-to-noise ratio is significantly poorer than that under two-photon excitation for the given detection sensitivity. Figure 3.2(b) primarily demonstrates that even when the out-of-focus fluorescence is suppressed under single-photon excitation, scattering of the illumination light results in fewer unscattered photons reaching the detector, which may limit image quality when the sample becomes optically thick.

This property is confirmed in a comparison of the images under single-photon (Fig. 3.3(a)) and two-photon (Fig. 3.3(b)) excitation in an optically thick turbid medium consisting of 0.202 μm beads. In this case, incident and fluorescence beams approximately experience three and two scattering events respectively under single-photon excitation. It can be concluded from Fig. 3.3(a) that image quality under single-photon excitation is primarily limited by the fast degradation of resolution due to the scattering of the illumination beam.

Figures 3.3(b) to 3.3(d) demonstrate the effect of the optical thickness n on image quality under two-photon excitation. In Fig. 3.3(d), the incident and fluorescence beams approximately experience two and eight scattering events respectively. It can be clearly seen from this image that the limiting factor on image quality under two-photon excitation is the signal-to-noise ratio of the images because resolution in Figs. 3.3(b) to 3.3(d) does not change appreciably. Therefore the fast decrease in fluorescence signal under two-photon excitation may lead to a limitation on the penetration depth d.

3.1.2 Comparison with Monte-Carlo simulation

In order to demonstrate further the difference between single-photon and two-photon fluorescence imaging, we use the Monte-Carlo method [5, 13] to simulate a thin circular

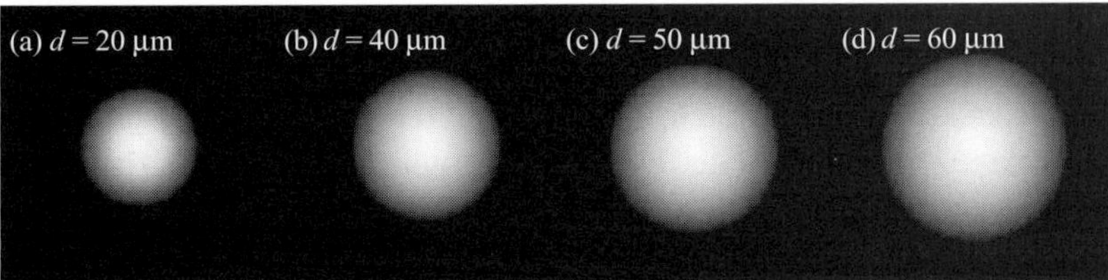

Fig. 3.4 Simulated fluorescence images of a 20 μm circular disk under single-photon excitation. (a) $d = 20$ μm, (b) $d = 40$ μm, (c) $d = 50$ μm, (d) $d = 60$ μm. Reprinted with permission from Ref. [5], M. Gu, S. Schilders and X. Gan, *J. Mod. Opt.* **47**, 959 (2000). © 2000, Taylor & Francis, www.informaworld.com.

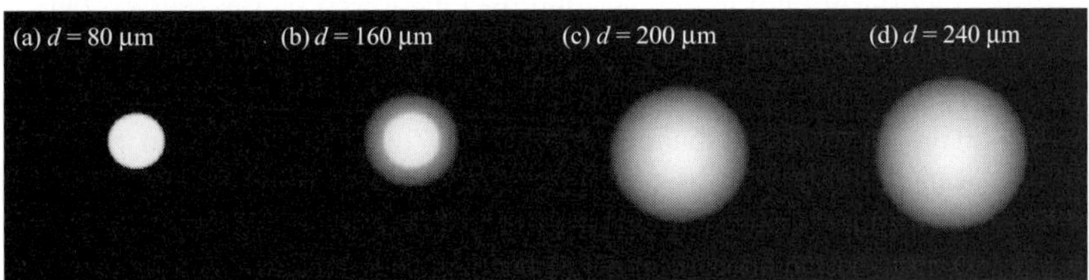

Fig. 3.5 Simulated fluorescence images of a 20 μm circular disk under two-photon excitation. (a) $d = 80$ μm, (b) $d = 160$ μm, (c) $d = 200$ μm, (d) $d = 240$ μm. Reprinted with permission from Ref. [5], M. Gu, S. Schilders and X. Gan, *J. Mod. Opt.* **47**, 959 (2000). © 2000, Taylor & Francis, www.informaworld.com.

disk (20 μm in diameter) embedded in a turbid medium. It can be assumed that the turbid medium consists of 0.202 μm scattering particles, and has a concentration of 22 particles/μm^3. According to Mie scattering theory [12], the corresponding SMFP length for wavelengths 488 nm, 520 nm and 800 nm is 8.9 μm, 10.4 μm and 36.2 μm, respectively.

In Fig. 3.4, single-photon fluorescence images of a thin circular disk for different depths are shown (the image size is 100 μm by 100 μm). In Fig. 3.4(a), the incident and fluorescence beams approximately experience two scattering events individually. The corresponding image shows a sharp part, which has a similar size to that of the original object, and a blurred part. The former is caused by unscattered photons whereas the latter results from scattered photons. As the depth d increases, the image becomes blurred due to the rapid reduction of the number of unscattered photons. The signal level is approximately decreased by one order of magnitude from Fig. 3.4(a) to Fig. 3.4(d). These features are consistent with those demonstrated by Fig. 3.2(b) and Fig. 3.3(a).

However, the situation may be different under two-photon excitation, as shown in Fig. 3.5. It can be noticed that at $d = 80$ μm (Fig. 3.5(a)) in which case the incident and fluorescence beams experience two and eight scattering events, respectively, the corresponding image is nearly diffraction limited; no significant blurring can be seen from the image. The theoretical calculation shows that the signal level in Fig. 3.5(a) is approximately three orders of magnitude lower than that for $d = 0$. These results confirm the experimental observation in Fig. 3.3(d).

It is further suggested from Figs. 3.5(a) to 3.5(d) that in the range of 80 μm $< d <$ 200 μm, the near diffraction-limited image of the disk becomes weak and a blurred part of the image starts to build up. These two parts of the images are contributed by two types of fluorescence light, that excited by ballistic photons and that excited by scattered photons. Ballistic photons form a diffraction-limited spot on the focal plane and contribute to a diffraction-limited sharp image [5, 14], while scattered photons excite fluorescence in a much broader region and produce a blurred image. As the depth d increases, the size of the blurred image increases and the blurred image eventually overshadows the sharp part of the image, which means that the scattered light becomes dominant in image formation. This feature also exists in single-photon fluorescence imaging. Since the fluorescence light excited by scattered photons quickly becomes dominant if $d > l_s$, the near diffraction-limited image is often overshadowed by the blurred part of the image. However, because of the quadratic dependence of fluorescence light on the excitation intensity under two-photon excitation, the effect of ballistic photons on the sharp part of an image is greatly enhanced. Therefore the two overlapped parts can be clearly observed in Figs. 3.5(b) and 3.5(c) for an optically thick turbid medium. Finally, when $d = 240$ μm (Fig. 3.5(d)), only the blurred image can be seen. The calculated signal level in Fig. 3.5(d) is approximately decreased by three orders of magnitude, compared with that in Fig. 3.5(a).

Imaging of microspheres embedded in turbid media qualitatively reveals that in the case of two-photon fluorescence imaging through a turbid medium, the decrease in signal level, i.e. signal-to-noise ratio, is faster than the degradation in resolution, while in the case of single-photon excitation, the degradation in image resolution poses the limitation in depth penetration. This phenomenon will be quantitatively understood in the following sections.

3.2 Limiting factors on image quality in imaging through turbid media

In the last section, attention has been paid to the qualitative effect of multiple scattering on two-photon fluorescence microscopy. In fact, an important issue is how multiple scattering affects image resolution and contrast as well as signal level under single-photon and two-photon excitation. In this section, image quality is quantitatively characterised by the sharpness of images of a bar embedded in turbid media consisting of different types of scattering particles [15].

A 1 mm wide fluorescent polymer bar was embedded within a turbid medium (Fig. 3.6). The peak fluorescence wavelength of the bar was measured to be approximately 520 nm under single-photon and two-photon excitation. Image resolution α is defined as the distance between the 10% and 90% intensity points, measured from the edge responses after they are fitted.

Four types of polystyrene microsphere suspended in water were used as turbid media. The diameters κ of the four types of microsphere are shown in Table 3.1. The range of the microspheres' sizes chosen was consistent with that of the scattering particle size in real biological samples which do not include large scattering particles (>1 μm) [16].

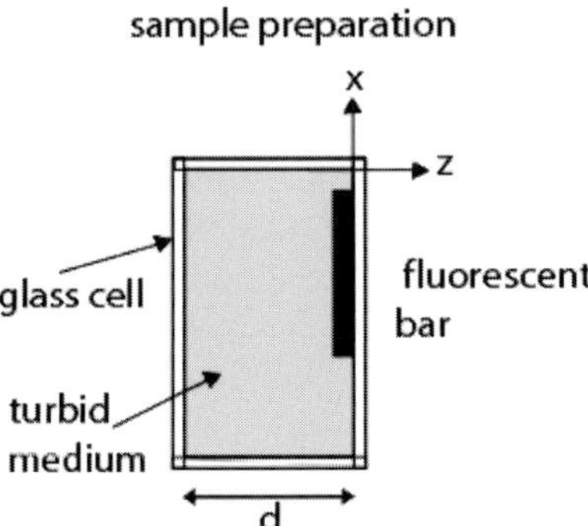

Fig. 3.6 Schematic diagram of a sample cell including a fluorescent bar and a turbid medium [9].

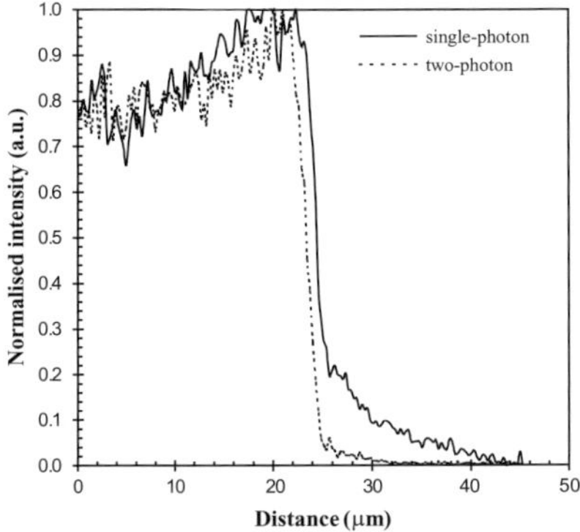

Fig. 3.7 Cross section of the image intensity of a bar embedded in a cell ($d = 100$ μm) consisting of water under single-photon and two-photon excitation. Reprinted with permission from Ref. [15], S. Schilders and M. Gu, *Microsc. Microanal.* **6**, 156 (2000). © 2000, Cambridge University Press.

The scattering parameters of the four sizes of polystyrene microsphere are shown in Table 3.1. Each type of polystyrene microsphere was placed in a glass cell with lateral dimensions of 2 cm × 1 cm. The glass cell thickness d was varied from 25 μm up to 300 μm.

For a fluorescent bar embedded in a cell consisting of distilled water, images in this case are constructed purely by unscattered photons obeying the prediction by diffraction theory and therefore show the maximum resolution achievable with the imaging system used [11]. The measured value of α in this case is approximately 9.5 μm and 2.8 μm under single-photon and two-photon excitation, respectively (Fig. 3.7). The lower resolution under single-photon excitation is caused by the fact that spherical aberration resulting from the refractive index mismatch between the cover glass of the cell and water under single-photon excitation is stronger than that under two-photon excitation [17]. When the distilled water is replaced by polystyrene microspheres of diameter 0.202 μm, image resolution under single-photon and two-photon excitation becomes approximately

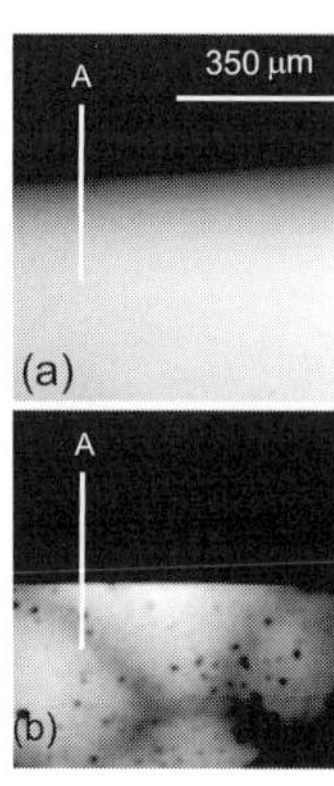
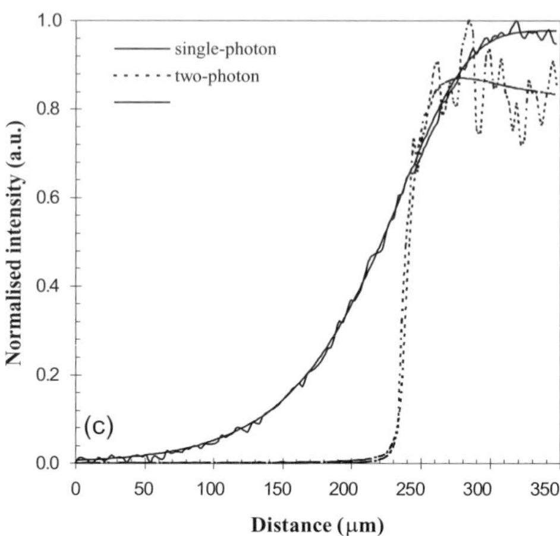

Fig. 3.8 Images of a bar embedded in a cell ($d = 100$ μm) consisting of polystyrene microspheres of diameter 0.202 μm under (a) single-photon excitation and (b) two-photon excitation. The intensity cross sections at position A and the corresponding fitted curves are shown in (c). Thin solid curves are the fitted curves. Reprinted with permission from Ref. [15], S. Schilders and M. Gu, *Microsc. Microanal.* **6**, 156 (2000). © 2000, Cambridge University Press.

147 μm and 17 μm, respectively (Fig. 3.8). This result shows a reduction of resolution by 15 and 6 times compared with that in Fig. 3.7.

To further demonstrate the effect of multiple scattering on image resolution under single-photon and two-photon excitation, we show in Fig. 3.9 the image resolution α and the optical thickness n (n is defined as the cell thickness divided by the SMFP length) as a function of the cell thickness d for four types of scattering particle. It can be clearly seen from Fig. 3.9 that the image resolution under single-photon excitation deteriorates significantly faster than that under two-photon excitation as d increases. For example, compared with those at $d = 0$, the image resolutions under single-photon and two-photon excitation are decreased by approximately 12 and 5 times at $d = 300$ μm, 24 and 8 times at $d = 150$ μm, 26 and 12 times at $d = 50$ μm, and 8 and 3 times at $d = 25$ μm in Figs. 3.9(a), (b), (c) and (d), respectively. This result is caused by the fact that the SMFP length under single-photon excitation is smaller than that under two-photon excitation. A smaller particle size leads to a larger difference of the SMFP length between single-photon and two-photon excitation. It is therefore concluded that two-photon excitation can significantly improve image resolution by reducing the number of scattering events experienced by the excitation beam, in particular when scattering particles are small.

In the case of single-photon excitation, the α value can be measured only when the cell thickness d is increased up to 150 μm, 50 μm and 25 μm, as shown in Figs. 3.9(b), 3.9(c) and (d), respectively. This phenomenon happens because the corresponding images significantly blur although signal strength is still detectable. However, the cell thickness d under two-photon excitation is limited to 75 μm and 50 μm in Figs. 3.9(c) and 3.9(d), respectively, mainly because the corresponding two-photon fluorescence signal is too

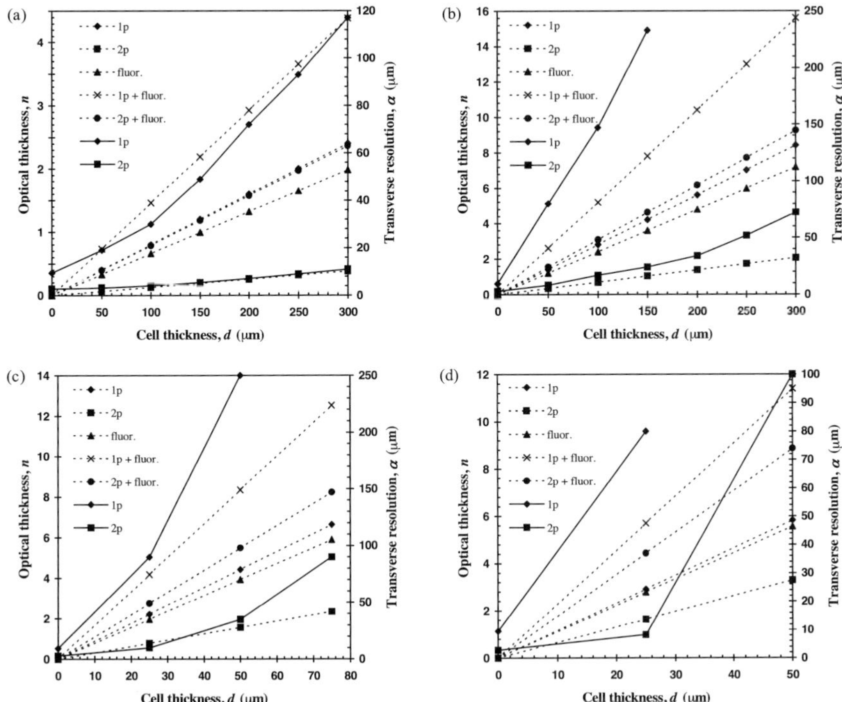

Fig. 3.9 Optical thickness *n* under single-photon (1p) and two-photon (2p) excitation (dashed) and image resolution α (solid) as a function of the cell thickness *d*: (a) turbid medium including beads of diameter 0.107 μm; (b) turbid medium including beads of diameter 0.202 μm; (c) turbid medium including beads of diameter 0.48 μm; (d) turbid medium including beads of diameter 1.056 μm. Reprinted with permission from Ref. [15], S. Schilders and M. Gu, *Microsc. Microanal.* **6**, 156 (2000). © 2000, Cambridge University Press.

weak to be detectable as the SMFP length decreases with the size of scattering particles (see Table 3.1). Thus, image quality under single-photon excitation is mainly limited by the degradation of image resolution caused by multiple scattering, while image quality under two-photon excitation is mainly limited by the degradation of signal strength/contrast.

This conclusion is further demonstrated in Figs. 3.8 and 3.10. Figs. 3.8(a) and 3.8(b) are the single-photon and two-photon fluorescence images of a bar embedded in a given turbid sample, which show the stronger degradation of the image resolution in the former case. When the thickness of the turbid sample is increased by two, images under single-photon excitation exhibit no sharpness at the edge of the bar (not shown here). But the edge of the bar can be still seen with a low contrast under two-photon excitation (Fig. 3.10).

The image resolution achievable under two-photon excitation is determined by the number of scattering events experienced by the excitation beam, which leads to the three regions shown in Fig. 3.9. The first region is demonstrated by Fig. 3.9(a) in which multiple scattering is not so strong because the optical thickness for the excitation beam

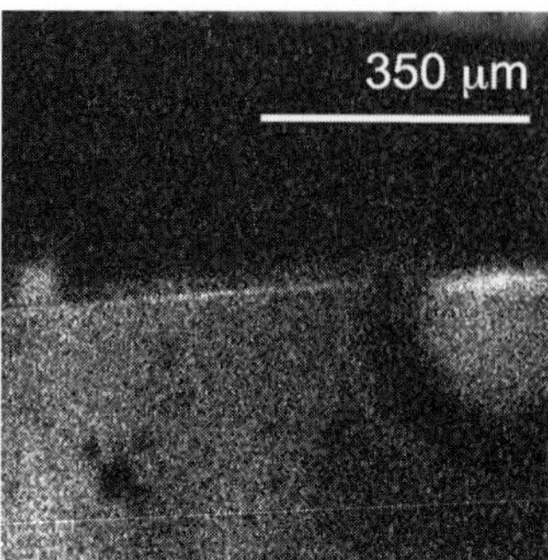

Fig. 3.10 Image of a bar embedded within a turbid medium including polystyrene microspheres of diameter 0.202 μm under two-photon excitation (thickness $d = 300$ μm). Reprinted with permission from Ref. [15], S. Schilders and M. Gu, *Microsc. Microanal.* **6**, 156 (2000). © 2000, Cambridge University Press.

is less than 1. In this case, two-photon excitation is produced mainly by unscattered photons of the illumination beam. As a result, the image resolution achievable is close to that measured without turbid media. As the number of scattering events experienced by the excitation beam increases, the unscattered component of the illumination beam significantly reduces. Therefore the value of α deteriorates quickly due to the quadratic dependence of the two-photon fluorescence on the incident power. This feature is illustrated in Fig. 3.9(b) which shows that α increases at a steady rate up to about 200 μm. After a depth of 200 μm, α increases quickly, indicating that multiple scattering of the illumination beam is strong in this region (the second region). A similar trend is seen in Fig. 3.9(c). If the excitation beam experiences more scattering events, scattered photons of the illumination beam become dominant, leading to a significant reduction in image resolution. This third region can be seen from Fig. 3.9(d) when d is larger than 25 μm.

Another major physical difference between Figs. 3.9(a), (b), (c) and (d) is the scattering anisotropy of the turbid media under inspection (see Table 3.1). As the scattering particle size increases a larger proportion of scattered photons propagate in the forward (illumination) direction due to their larger g value. Thus the image quality achievable for a given optical thickness n of the illumination beam should be improved as the size of scattering particle size increases. It is seen from Fig. 3.9 that at a given number of scattering events of the illumination beam, the image resolution α achievable under two-photon excitation is improved when a turbid medium consisting of larger scattering particles (Fig. 3.9(d)) is used. However, such an improvement under single-photon excitation is not pronounced because of the poorer resolution which leads to less accurate measurements.

It should be pointed out that spherical aberration caused by the mismatch of refractive indices between the cover glass of the cell and water can also lead to a reduction in image resolution. This type of aberration can broaden the point spread function of the imaging objective by a factor of two to three and reduce the intensity at the diffraction focus by approximately 90% for a water layer of thickness 200 μm [17].

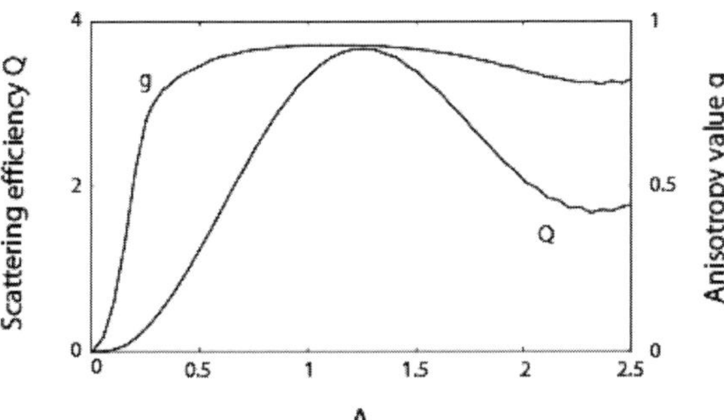

Fig. 3.11 Scattering efficiency Q and the anisotropy value g in Mie scattering as a function of the scattering parameter A. The refractive indices of scatterers (spherical particles) and the immersion medium are 1.59 and 1.33, respectively. Reprinted with permission from Ref. [20], M. Gu, X. Gan, A. Kisteman and M. Xu, *App. Phys. Lett.* **77**, 1551 (2000). © 2000, American Institute of Physics.

However, these degradations are negligible compared with those caused by multiple scattering.

3.3 Quantitative comparison of penetration depth between two-photon excitation and single-photon excitation

Image quality under two-photon excitation may be appreciably degraded when the scattered photons in the excitation beam become dominant. The image quality achieved by a given detector is limited by the two-photon fluorescence strength rather than by the image resolution. However, the degradation in image resolution is a main limiting factor in obtaining high quality images in the case of single-photon excitation. This effect on imaging through tissue media, which is a topic related to photodynamic therapy and early detection of small tumours [1–3, 18], is further studied using Monte-Carlo simulation [8, 19, 20] in this section.

Biological tissue is usually composed of small scatterers such as bacteria, viruses, malignant cells and so on. The size of these scatterers varies from 0.1 μm to a few micrometres [12, 16]. Therefore, the dominant scattering effect caused by these scatterers is Mie scattering rather than Rayleigh scattering. The physical difference between these two types of scattering is that the former is anisotropic scattering whereas the latter is isotropic scattering. The strength of Rayleigh scattering is inversely proportional to the fourth power of the illumination wavelength. However, Mie scattering exhibits a more complicated nature, as shown in Fig. 3.11.

Figure 3.11 shows the scattering efficiency Q, which is defined as the ratio of the scattering cross section σ_s to the geometrical cross section, and the anisotropy value g in Mie scattering as a function of the scattering parameter A (A is defined as the ratio of the size of a scattering particle a to the light wavelength λ) [12]. It is seen from this example that for a given scatterer size, Q and g decrease with the illumination wavelength if A is approximately less than one. The decrease in the scattering efficiency implies the reduction of multiple scattering events, which leads to high image resolution and signal level in microscopy imaging, illustrated in the last two sections. However the decrease in the anisotropy value results in a broad distribution of scattered photons in the focal region [7], which may reduce image resolution and signal level. The competition of these two processes, together with the quadratic dependence of the two-photon fluorescence intensity, determines the limiting factor on penetration depth.

To demonstrate the limiting factor of multiple scattering on the penetration depth under two-photon and single-photon excitation, we consider a turbid medium consisting of scattering particles (diameter 0.202 μm) suspended in water. The turbid medium was placed in a glass cell with lateral dimensions of 2 cm × 1 cm, as shown in Fig. 3.6. The thickness of the glass cell, d, was varied from 25 μm up to 250 μm which is the maximum working distance of the objective used in experiments. Considering a uniform fluorescent polymer bar embedded at the bottom of the glass cell and excited under single-photon ($\lambda = 488$ nm) and two-photon ($\lambda = 800$ nm) excitation with a peak fluorescence wavelength approximately at 520 nm, we have the scattering parameters as summarised in Table 3.1.

To avoid the effect of the refractive-index mismatching between the turbid medium and the cover glass of the cell, a water-immersion objective was used. In the case of single-photon excitation, a pinhole of 300 μm (optical unit ~ 3) in diameter was placed in front of the detector to produce an optical sectioning effect with strength similar to that under two-photon excitation without using a pinhole [10, 11]. This arrangement implies that the ability of rejecting scattered photons caused by the optical sectioning effect is the same in the two cases.

Figure 3.12(a) shows the dependence of the measured fluorescence signal level on the sample thickness d under single-photon and two-photon excitation. The signal level has been normalised by the fluorescence signal measured without the turbid medium ($d = 0$). Such normalisation is equivalent to the assumption that the fluorescence intensity at $d = 0$ is the same under single-photon and two-photon excitation and provides a benchmark for comparison between the two cases. It is seen that the two-photon fluorescence signal level is better than the single-photon fluorescence signal level when d is less than the single-photon SMFP length. After that depth, the two-photon fluorescence signal level drops more quickly than the single-photon fluorescence signal level. For example, at $d = 225$ μm, the former is approximately two orders of magnitude lower than the latter. To confirm this measured dependence, we used the Monte-Carlo simulation method [8, 19, 20] to calculate the signal levels. The theoretical prediction shown in Fig. 3.12(a) agrees with the experimental results in the sense that the two-photon fluorescence signal level is lower than the single-photon fluorescence signal level.

To understand the phenomenon in Fig. 3.12(a), we should point out that both unscattered and scattered illumination photons can contribute to fluorescence emission. However, the fluorescence contributed from these two groups of photons behaves in different ways between two-photon and single-photon excitation.

We first consider the case when the thickness d is less than the single-photon SMFP length. In this case, the number of unscattered illumination photons is considerable in the total number of photons but decreases exponentially with d/l_{s}. In other words, the

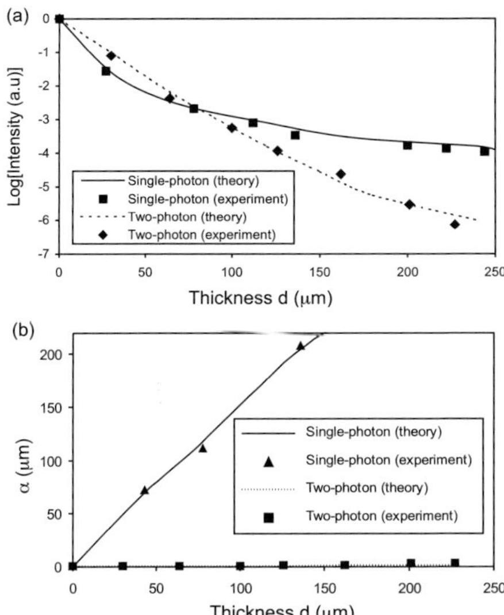

Fig. 3.12 (a) Signal level under two-photon and single-photon excitation as a function of the penetration depth in a turbid medium. (b) Image resolution under two-photon and single-photon excitation as a function of the penetration depth of a turbid medium. Reprinted with permission from Ref. [20], M. Gu, X. Gan, A. Kisteman and M. Xu, *App. Phys. Lett.* **77**, 1551 (2000). © 2000, American Institute of Physics.

logarithm of the emitted fluorescence signal contributed by unscattered illumination photons decreases linearly with d, as observed in Fig. 3.12(a). Due to the quadratic dependence of two-photon fluorescence on the illumination intensity, the number of two-photon fluorescence photons produced by the unscattered illumination photons decreases exponentially with $2d/l_s$. As a result, for a given depth d, the two-photon fluorescence signal excited by the unscattered illumination photons is stronger, because the two-photon SMFP length is approximately four times as large as that for single-photon excitation, see Table 3.1.

When d becomes larger than the single-photon SMFP length, the two-photon fluorescence signal still decreases exponentially with $2d/l_s$ but scattered illumination photons make a significant contribution to the generation of single-photon fluorescence. As a result, the fall of single-photon fluorescence becomes slower than that of two-photon fluorescence.

Once the thickness d is larger than the two-photon SMFP length, scattered photons play a dominant role in the excitation and emission processes. In general, scattered photons distribute in a broad region near the geometric focus [8]. Because of the lower anisotropy value g and the larger l_s value, the scattered illumination photons under two-photon excitation distribute in a broader region than those under single-photon excitation. This feature results in a lower photon density near the geometric focus. Therefore, the two-photon fluorescence emission excited by scattered photons is less efficient than the single-photon fluorescence emission. Further, due to the quadratic dependence under two-photon excitation, two-photon fluorescence emission excited by

Fig. 3.13 Measured transverse (x–y) and axial (x–z) images through a thick muscle tissue sample under two-photon and single-photon excitation. (a) and (b) Transverse two-photon fluorescence images at depths of 10 and 70 μm. (c) Axial two-photon fluorescence image. (d) and (e) Transverse single-photon fluorescence images at depths of 10 and 70 μm. (f) Axial single-photon fluorescence image. Image size: 200 μm. The intensity is normalised by the value at $d = 0$. Reprinted with permission from Ref. [20], M. Gu, X. Gan, A. Kisteman and M. Xu, *App. Phys. Lett.* **77**, 1551 (2000). © 2000, American Institute of Physics.

scattered illumination photons becomes even less efficient. In addition, two-photon fluorescence photons excited by scattered illumination photons have a lower possibility of reaching the detector because they originate from a broader region. As a result, two-photon fluorescence produced by the scattered illumination photons exhibits a lower signal level than single-photon fluorescence.

However the stronger suppression of the contributions from scattered illumination and fluorescence photons under two-photon excitation may be advantageous in terms of image resolution. To confirm this feature, we measured the transverse edge image of the fluorescence polymer bar. Image resolution α is defined as in Section 3.2. The dependence of the resolution α on the sample thickness d is depicted in Fig. 3.12(b) which also includes the Monte-Carlo simulation results corresponding to the experimental condition. A good agreement between experiments and theoretical predictions is observed. For the turbid medium we used, the image resolution under two-photon excitation is two orders of magnitude higher than that under single-photon excitation.

Experiments similar to the conditions in Fig. 3.12 were also carried out for different sizes of scatterers using the measure detailed in Section 3.2. In general, the smaller the scatterer size (the smaller the anisotropy value and the larger the SMFP length), the quicker the reduction of the two-photon fluorescence signal. Based on this property, we can conclude that for a real tissue medium consisting of different sizes of Mie scatterers, two-photon fluorescence signal level is lower than single-photon signal level at a large depth. This conclusion is demonstrated in Fig. 3.13.

Figure 3.13 shows the transverse $(x\text{–}y)$ and axial $(x\text{–}z)$ autofluorescence images of muscle tissue under two-photon ($\lambda = 800$ nm) and single-photon ($\lambda = 488$ nm) excitation. The image intensity has been normalised by the maximum intensity at the surface of the sample. The muscle tissue has an average single-photon SMFP length of approximately 20 μm [21]. The two-photon transverse images (Figs. 3.13(a) and 3.13(b)) were recorded at depths of 10 and 70 μm, showing a higher resolution but becoming significantly weaker at a large depth than the single-photon images (Figs. 3.13(d) and 3.13(e)). It can be seen from the axial images (Figs. 3.13(c) and 3.13(f)) that although the degradation of two-photon resolution is not pronounced, the two-photon fluorescence signal (Fig. 3.13(c)) decreases faster than the single-photon fluorescence signal (Fig. 3.13(f)).

As a general conclusion of this chapter, the penetration depth of two-photon excitation in imaging through a tissue medium is smaller than that of single-photon excitation because of the lower signal level in the former case. However, within the depth of detectable signal, two-photon excitation leads to image resolution approximately two orders of magnitude higher than that under single-photon excitation.

References

[1] S. P. Schilders and M. Gu. Three-dimensional autofluorescence spectroscopy of rat skeletal muscle tissue under two-photon excitation. *Appl. Opt.*, **38**:720–723, 1999.

[2] B. R. Masters, P. T. C. So and E. Gratton. Multi-photon excitation fluorescence microscopy and spectroscopy of *in vivo* human skin. *Biophys. J.*, **72**:2405–2412, 1997.

[3] Y. Guo, Q. Z. Wang, N. Zhadin *et al.* Two-photon excitation of fluorescence from chicken tissue. *Opt. Lett.*, **36**:968–970, 1997.

[4] V. Daria, C. Blanca, O. Nakamura, S. Kawata and C. Saloma. Image contrast enhancement for two-photon fluorescence microscopy in a turbid medium. *Appl. Opt.*, **37**:7960–7967, 1998.

[5] M. Gu, S. Schilders and X. Gan. Two-photon fluorescence imaging of microspheres embedded in turbid media. *J. Mod. Opt.*, **47**:959–965, 2000.

[6] C. M. Blanca and C. Saloma. Monte Carlo analysis of two-photon fluorescence imaging through a scattering medium. *Appl. Opt.*, **37**:8092–8102, 1998.

[7] X. Gan and Min Gu. Spatial distribution of single-photon and two-photon fluorescence light in scattering media: Monte Carlo simulation. *Appl. Opt.*, **39**:1575–1579, 2000.

[8] X. S. Gan and M. Gu. Fluorescence microscopic imaging through tissue-like turbid media. *J. App. Phy.*, **87**:3214–3221, 2000.

[9] S. Schilders. *Microscopic Imaging in Turbid Media*. Ph.D. thesis, Department of Physics, Victoria University of Technology, Australia, 1999.

[10] M. Gu and C. J. R. Sheppard. Effects of a finite-sized pinhole on 3d image formation in confocal two-photon fluorescence microscopy. *J. Mod. Opt.*, **40**:2009–2024, 1993.

[11] Min Gu. *Principles of Three-Dimensional Imaging in Confocal Microscopes*. Singapore, World Scientific, 1996.

[12] C. F. Bohren and D. R. Huffman. *Absorption and Scattering of Light by Small Particles*. New York, John Wiley & Sons, 1983.

[13] X. Gan and M. Gu. Effective point-spread function for fast image modeling and processing in microscopic imaging through turbid media. *Opt. Lett.*, **24**:741–743, 1999.

[14] F. Liu, J. Ying and R. R. Alfano. Spatial distribution of two-photon-excited fluorescence in scattering media. *Appl. Opt.*, **38**:224–229, 1999.

[15] S. Schilders and M. Gu. Limiting factors on image quality in imaging through turbid media under single-photon and two-photon excitation. *Microsc. Microanal.*, **6**:156–160, 2000.

[16] S. P. Morgan, M. P. Khong and M. G. Somekh. Effects of polarization state and scatterer concentration on optical imaging through scattering media. *Appl. Opt.*, **36**:1560–1565, 1997.

[17] P. C. Ke and M. Gu. Characterisation of trapping force in the presence of spherical aberration. *J. Mod. Opt.*, **45**:2159–2168, 1998.

[18] V. E. Centonze and J. G. White. Multiphoton excitation provides optical sections from deeper within scattering specimens than confocal imaging. *Biophys. J.*, **75**:2015–2024, 1998.

[19] X. Deng, X. Gan and M. Gu. Multiphoton fluorescence microscopic imaging through double-layer turbid tissue media. *J. App. Phy.*, **91**:4659–4665, 2002.

[20] M. Gu, X. Gan, A. Kisteman and M. Xu. Comparison of penetration depth between single-photon excitation and two-photon excitation in imaging through turbid tissue media. *App. Phys. Lett.*, **77**:1551–1553, 2000.

[21] W. F. Cheong, S. A. Prahl and A. J. Welch. A review of the optical properties of biological tissues. *IEEE J. Quant Elect.*, **26**:2166–2185, 1990.

4   Fibre-optical nonlinear microscopy

The techniques introduced in Chapters 1 and 2 are emerging technologies that offer significant promise as tools for diagnostic imaging at the cellular level. Using devices founded on well established techniques such as confocal microscopy [1] and confocal fluorescence microscopy [2, 3], instruments capable of providing point-of-care pathological analysis of malignant and cancer causing tissues are becoming practical realities. Through examination of the physical properties of inherent autofluorescence or fluorescent dyes that are used as markers in conjugation with biological samples, very good detection of cellular processes can be achieved. Tagging of target biological cells makes it possible to examine cells *in vivo* and achieve real time three-dimensional (3D) visualisation for diagnosis of the pathological state.

However, the inherent nature of these devices is such that the conditions under which these techniques can be applied is fundamentally limited. In most cases (for definitive analysis) a surgical biopsy is performed on the patient and the sample is extensively prepared for observation by the pathologist on bulk, bench-top imaging apparatus. Ideally, examination of whole, intact specimens within internal cavities of the body would be the preferred method that may decrease patient trauma and eliminate diagnosis lag time.

One of the recent developments in confocal fluorescence microscopy is the introduction of optical fibres and fibre-optical components into the microscope geometry [4]. Optical fibre couplers in particular offer the most compact and cost effective solution. Since the core diameter of a single-mode fibre is comparable to the diameter of a conventional confocal pinhole [5–8], a fibre coupler based system combines the unique characteristics of confocal imaging with the ability to deliver illumination to a remote sample. This technology has enormous potential to lead to the practical realisation of *in vivo* diagnosis of diseased tissues.

This chapter is intended to serve as a practical guide to fibre-optic nonlinear microscopy. It aims to provide the reader with the knowledge necessary for practical implementation of such systems and to highlight key considerations that need to be taken into account to ensure effective deployment of the instruments. Section 4.1 presents an introduction into fibre-optical two-photon confocal microscopy. The detail of fibre-optical two-photon and second harmonic generation (SHG) microscopy based on a single mode coupler is described in Section 4.2 and Section 4.3, respectively. Section 4.4 describes the feasibility of fibre-optical two-photon micro-endoscopy.

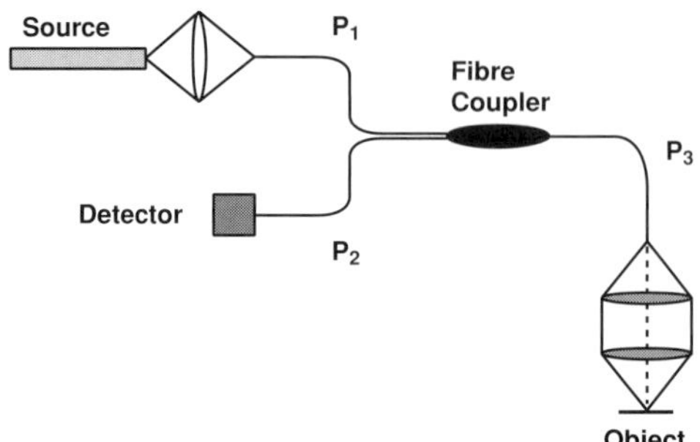

Fig. 4.1 Schematic diagram of a reflection-mode fibre-optical confocal scanning microscope using a single-mode fibre coupler. Port 3 of the coupler (P_3) serves as the illumination source and detection device simultaneously. The coupler itself replaces the beamsplitter. P_1, P_2 and P_3 are the ports of the coupler.

4.1   Fibre-optical confocal microscopy

4.1.1   Image formation

Before giving a complete description of the complexities of nonlinear fibre-optic imaging, it is useful to prime the reader with a simplistic introduction to imaging with optical fibres. In the context of this chapter, this discussion is based around confocal microscopy [9].

A schematic diagram of a fibre-optical confocal reflection-mode microscope based on an optical fibre coupler is shown in Fig. 4.1. Key to this discussion is the fact that port 3 of the coupler is used as an illumination and collection device simultaneously, and the coupler junction itself replaces the conventional bulk-optic beamsplitter. This has the advantage that the system is automatically aligned for detection and fewer components need to be manually aligned compared with conventional confocal microscopy (recall that the diameter of a single mode fibre is of the order of approximately 4 μm, similar to that of a conventional pinhole). This arrangement does not require the laser to be suspended near the input port of the microscope as is frequently done in commercial confocal systems [9], and vibrations in the system can be reduced because the laser and cooling fans can be isolated from the microscope.

While the small aperture of the fibre core can effectively replace the conventional confocal pinhole, image formation for a non-fluorescent object in a confocal reflection-mode system using optical fibres is fundamentally different from a confocal system using pinhole masks [5, 10, 11]. Imaging in the former case is purely coherent even for a finite value of the fibre spot size [10], while imaging in the latter is partially coherent due to using a finite-sized detector.

The physical reason for this important difference is that a finite-sized detector responds incoherently to the light intensity impinging upon the detector, whereas an optical fibre is a waveguide and has a set of eigenfield modes, which can respond coherently to the amplitude and phase of light [12]. Because of this fundamental difference the effect of the fibre spot size on confocal images differs from that of the detector of finite size [13, 14]. The purely coherent nature of imaging in the fibre-optical confocal mode may prove advantageous when the device is used for quantitative microscopy or for spectroscopic imaging.

In a fibre-optical confocal microscope the tip of the optical fibre has a finite-sized cross section, which accordingly affects the properties of image formation. A single-mode fibre emits a beam which has an approximate Gaussian cross section [13]. As a result, the geometry of the microscope can be designed to fill the objective pupil. If the pupil is over-filled then maximum resolution is obtained at the expense of loss of light; however, if the pupil is under-filled then the resolution is degraded [9]. In the case of single mode fibres with Gaussian profiles, theoretical analyses [9–11] have shown that the fibre-optical confocal microscope can be described by the normalised fibre spot size given by

$$A_j = \left(\frac{2\pi a a_j}{\lambda_j d_1} \right)^2 \quad j = 1, 2, \tag{4.1}$$

where d_1 is the distance between the fibre and the objective lens, a is the radius of the pupil function of the objective and a_j is the fibre spot size.

Under the Gaussian approximation two system parameters, A_1 and A_2, representing the illumination and collection fibres respectively, can be derived from Eq. 4.1 and are defined as [9]

$$A_1 \approx \frac{V_i^2}{2\ln V_i} \left[\frac{N_i}{N_{f1}} \right]^2 \tag{4.2}$$

and

$$A_2 \approx \frac{V_d^2}{2\ln V_d} \left[\frac{N_d}{N_{f2}} \right]^2, \tag{4.3}$$

where N_i and N_d are the numerical apertures of the objective on the illumination and detection sides, under the paraxial approximation. N_{f1} and N_{f2} are the numerical apertures of the illumination and collection fibres. V_i and V_d are the normalised fibre frequency parameters, which are related to the illumination and collection wavelengths, respectively. In the case of a single fibre being used for simultaneous illumination and collection, $N_i = N_d$ and $N_{f1} = N_{f2}$. Therefore, if the normalised fibre frequency parameters are equal $A_1 = A_2 = A$. It is clear that the values of the normalised spot sizes, A_1 and A_2, are determined by the ratios of the numerical apertures of the system on illumination and detection sides to those of the optical fibres. However, in the case of two-photon excitation, V_i and V_d are not equal.

If the numerical apertures of the fibres are much larger than those of the system, then only a small part of the source light is coupled into the fibre-optical confocal system (Fig. 4.2(a)). In this case, $A_1 = 0$ and $A_2 = 0$, so that the tips of the fibres can be considered a point source and a point detector, and the system behaves as a true confocal system. However, when A_1 and A_2 have finite size, the numerical aperture

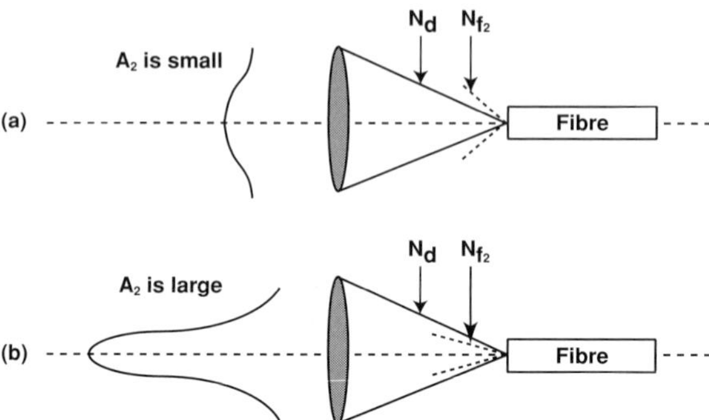

Fig. 4.2 Coupling of a Gaussian beam into a single mode optical fibre: (a) the numerical aperture of the fibre is larger than that of the illumination; (b) the numerical aperture of the fibre is smaller than that of the illumination.

of the fibres is smaller than those of the system (Fig. 4.2(b)) and the finite-sized cross sections of the fibres play an important role in the microscope. In this case, the pupil of the objective is under-filled and its effective pupil function is now shaded with a function that depends on the fibre profiles. Hence, the fibre-optical confocal microscope is now determined by the properties of the optical fibre [12]. This suggests that a fibre-optical confocal microscope can be operated in a set of novel modes by changing the fibre parameters.

It should be pointed out that if the sample under investigation with a fibre-optical confocal scanning microscope is labelled with fluorescent materials the purely coherent imaging nature of the system is completely destroyed due to the interaction of the light with the sample. The emitted fluorescence propagates in all directions and has no phase relation. In this case the collection fibre combined with the finite-sized detector acts effectively as a shaded incoherent detector. Under these conditions the image formation in fibre-optical confocal scanning fluorescence microscopy is fundamentally different from that described and reverts to an incoherent process.

4.1.2   Milestones in fibre-optical confocal microscopy

It has been demonstrated that the resolution performance of a reflection-mode fibre-optical confocal microscope implementing a single-mode optical fibre coupler to replace the beamsplitter and conventional confocal pinholes has the same fundamental characteristics as a conventional confocal microscope [8, 14]. Furthermore, rigorous theoretical analyses [5, 10] have revealed that imaging of a non-fluorescent sample with optical fibre is purely coherent even for a finite value of the fibre spot size [11]. This feature is of fundamental importance and was the catalyst for exploring new applications in which the instrument could be employed.

Confocal interference microscopy [15, 16] is a useful method for determining both the phase and amplitude information, with which it is possible to measure the aberration caused by the optical system [17]. The information can also be used to reconstruct a 3D phase object and for characterisation of thin films and rough surfaces. A fibre-optical 3D microscope with high depth sensitivity has been reported where a pair of single-mode optical fibre couplers were employed to form a six port all-fibre Michelson interferometer [18]. It is well known that the degree of coherence determines the visibility of the interference fringes. In this respect, the fibre-optical confocal scanning microscope gives better results than a pinhole imaging system because of the purely coherent imaging feature. A confocal interference microscope has been developed in which a reference beam was superimposed on the normal confocal signal [12, 14]. However, in this work, the system was constructed using bulk optical components and two lengths of single-mode fibre. As a consequence, the measured signal was quite susceptible to air currents and vibrations in the environment. To overcome this, a single four-port optical fibre coupler has been implemented [19, 20] since, in this geometry, the light in most paths of the device is shielded from the external environment by the fibre. The coherent imaging property also results in an improved signal strength because good matching of the fibre mode with the incoming diffraction pattern can be achieved [21].

In 1980 it was proposed that it may be possible to further improve the resolution of a confocal microscope by using a higher number of lenses in a transmission-mode confocal system [22]. Double-pass confocal microscopy arose after it was found that the spot size of a point object was much improved for two lenses, but further traversing did not result in further significant improvement [22]. Practical realisation of the double-pass confocal microscope is achieved by placing a fixed mirror after the collector lens, which ensures the illumination beam traverses the object twice. An optical fibre coupler has been adapted to double-pass confocal microscopy and results in a compact geometry that can be easily switched between reflection, transmission and double-pass modes [23].

Fibre-based confocal microscopes have also been used for fluorescence imaging [4], although these systems are not purely coherent in nature. Nevertheless, it has been shown that an important property resulting from using fibres in confocal fluorescence microscopy is that there is no missing cone of spatial frequencies in the 3D optical transfer function (OTF) that occurs with a pinhole detector. This feature results in an improvement in the 3D fluorescence imaging performance of the system [9].

Optical fibre was implemented in scanning differential interference contrast optical microscopy in the 1980s [24] and was demonstrated to have high depth resolution. Confocal differential microscopy is fundamentally reliant on the positioning of the focus of an objective. Placing the sample exactly at the focal point is not the most advantageous for achieving high depth resolution; the zero derivative of the response curve with respect to the sample position means least sensitivity to sample height variation. In contrast, if one places the sample slightly away from the focal point, so that its position is at the slope of the response curve, the sensitivity becomes greatest. At the slopes, sample height variation causes a differential change of signal. The sensitivity of this effect can be utilised to image surface structures with great depth resolution. Fibre optical confocal differential microscopy [25] was realised in the early 1990s, demonstrating that the purely coherent imaging nature still exists for non-fluorescent samples. Furthermore, in that study, multi-mode fibres were employed.

4.2   Two-photon fluorescence imaging systems using a single-mode optical fibre coupler

As discussed in Chapter 2, the process of two-photon excitation of fluorescence and SHG is highly nonlinear and requires a high photon density at the focus of an imaging objective. Therefore, the use of ultrashort pulse lasers is usually required. These sources produce extremely high peak-power bursts of light of very short temporal duration. While effective in conventional microscopy, these sources are not ideal for nonlinear fibre based instruments because (among other things) the material dispersion effect of the fused-silica significantly temporally broadens the ultrashort pulse. As a consequence, the probability of two-photon absorption or harmonic generation is reduced because the process is highly dependent on confining a large number of photons in a confined space and in an extremely short time.

Although there are obvious physical difficulties, a nonlinear fibre-optical based microscope would be a superior instrument. Previous studies have demonstrated that a single-mode optical fibre can be used for the delivery of an ultrashort pulse laser beam, however, the optical system designs used in experiments are still based largely on bulk optical components [26, 27]. This fact limits the functionality of the instrument in terms of its practicability. To achieve a cost-effective and compact arrangement for two-photon fluorescence and SHG microscopy, one can use a multi-port fibre coupler to replace major bulk optical components for illumination delivery and signal collection.

In this section, the design and characterisation of a compact two-photon fluorescence microscope that implements a three-port single-mode fibre coupler for an infrared wavelength are presented [28]. In addition to the compactness, the new geometry exhibits a self-aligning nature because delivery of incident illumination and signal collection are achieved via the same fibre port. Since the small fibre aperture acts as a confocal pinhole, image resolution in this new system is higher than that in conventional two-photon fluorescence microscopy without a pinhole [29]. The extension of the design into a SHG microscope is given in Section 4.3.

4.2.1   Fibre-optical two-photon fluorescence microscopy

An experimental arrangement of a two-photon fluorescence microscope using a single-mode optical fibre coupler is given in Fig. 4.3 [29, 30]. An 800 nm 70 fs pulse laser beam was coupled into a port (port 3) of a three-port fused silica single-mode fibre coupler via a 0.25 NA 10× objective O_1. The coupler is designed for operation at wavelength 785 nm with an equal splitting ratio of 50:50 and has a mode field diameter of approximately 5.4 μm. The core/cladding ratio and NA of the coupler are approximately 5/125 and 0.17, respectively. Each arm of the coupler is approximately 1 metre in length.

Rotation of a neutral density filter, ND, placed before the objective O_1 allowed variation of the input power. The output beam from port 1 of the coupler was collimated by a second 0.25 NA 10× objective O_2 and then passed through a variable aperture, AP,

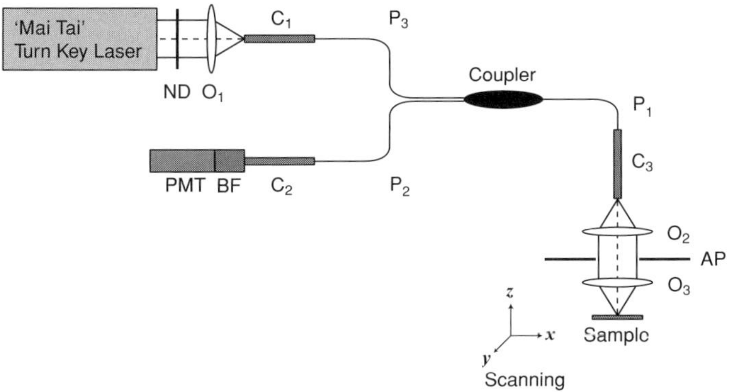

Fig. 4.3 Schematic diagram of the experimental two-photon fluorescence microscope using a single-mode optical fibre coupler. O_1 and O_2: 0.25 NA 10× microscope objectives, O_3 (0.85 NA 40×, ∞/170) imaging objective, ND: neutral density filter, BF: bandpass filter, AP: variable aperture, C_1, C_2, C_3: fibre chucks.

to fill the back aperture of an imaging objective O_3. The emitted fluorescence from a sample was collected by the objective O_3 and returned via the same optical path used for pulse delivery. The signal was delivered via port 2 of the coupler into a photomultiplier tube (PMT). A bandpass filter, BF, operating at wavelength 550 ± 20 nm, was placed in the beam path to ensure that only the fluorescence signal was detected. Each port of the fibre coupler was placed in a chuck holder in an x–y–z micropositioner to allow precise positioning of the fibre tip at the focus of an objective.

4.2.2　Coupling efficiency and splitting ratio

With the introduction of a fibre coupler to a confocal system, both delivery of the incident illumination and collection of the fluorescence signal are achieved through the fibre tip of port 1 (Fig. 4.3). As pointed out in Section 4.2.1, the fibre coupler used in experiments splits light input to port 1 to ports 2 and 3 with a ratio of 50:50 at the operating wavelength of 785 nm. This property requires extensive investigation given the operating conditions of the instrument.

Typically, light coupled to ports 2 and 3 and emerging from port 1 of the fibre coupler can be achieved with an efficiency of between approximately 20–38% in the wavelength range between 770 and 870 nm. This implies that a laser power of up to 80–150 mW in this wavelength range can be delivered to a microscope objective. Furthermore, in this wavelength range the measured transmission efficiency of the imaging objective, O_3, is approximately 50–67%. Therefore, this coupler is capable of delivering a laser beam of sufficient power for two-photon excitation of a sample [28, 30, 31].

Since the collection of the fluorescence emitted from a sample is through port 1 of the fibre coupler and two-photon fluorescence falls in the visible range, the coupling efficiency of the coupler may be reduced and an equal splitting ratio between ports 2 and 3 may not necessarily be maintained at this wavelength. To confirm these features one

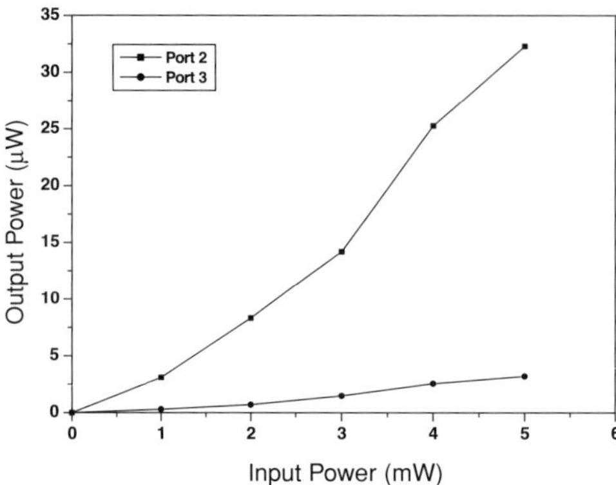

Fig. 4.4 Splitting ratio at ports 2 and 3 of a 50:50 785 nm single-mode optical fibre coupler for input illumination of wavelength 532 nm to port 1. At this wavelength the coupler splits with a 90:10 ratio. Reprinted with permission from Ref. [31], D. Bird and M. Gu, *J. Micros.* **208**, 35 (2002). © 2002, Blackwell Publishing.

should simply monitor the mode profile and the coupling efficiency at ports 2 and 3 while port 1 is illuminated by a continuous wave beam at the 532 nm (see Fig. 4.4).

The field distribution from ports 2 and 3 should be of a single-mode profile, which is consistent with the estimation based on the core size and numerical aperture of the coupler in the visible range. The coupling efficiency at port 2 in this case is approximately 1% with a splitting ratio of 90:10 between ports 2 and 3. As a result, using port 2 for the signal collection of two-photon fluorescence and port 3 for delivery of the pulsed beam means that the coupler acts as a low-pass filter and that the strength of two-photon fluorescence signal can be maximised.

4.2.3   Spectral and temporal broadening

An experimental investigation into the magnitude of the temporal pulse broadening through the fibre coupler is also required in order to fully characterise the abilities of the two-photon system. In an optical fibre, the most significant form of broadening referred to as linear dispersion occurs as a consequence of group velocity dispersion (GVD). Since different spectral components of a pulsed beam travel at slightly different speeds within the core of an optical fibre, the pulse width is broadened over its initial width. However, in the presence of intense electromagnetic fields (such as those produced by an ultrashort pulse beam) nonlinear broadening may be observed. To investigate this feature, temporal profiles of the ultrashort pulse illumination emerging from port 1 of the single-mode fibre coupler for a range of input power up to 400 mW to fibre port 3 were recorded using a streak camera. The increase in the temporal full width at half maximum (FWHM) with increasing input power to the coupler is shown in Fig. 4.5. In this chart, an increase in the temporal FWHM of the ultrashort pulse is clearly visible for increased power coupled to the fibre. The effect of linear dispersion can be considered through observation of the low input power condition, i.e. the 100 mW data point.

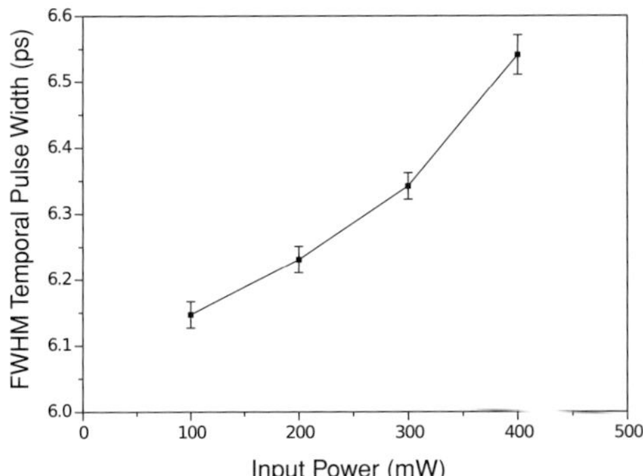

Fig. 4.5 Temporal FWHM of a pulsed beam as a function of input power to port 3 of the 2 × 1 fibre coupler. Reprinted with permission from Ref. [31], D. Bird and M. Gu, *J. Micros.* **208**, 35 (2002). © 2002, Blackwell Publishing.

As would be expected, there is a slight increase in the temporal FWHM of the pulse as a result of increasing the optical input power to fibre port 3 to 400 mW. At this point (the last point in Fig. 4.5) the pulse has broadened to approximately 6.55 ps. This broadening arises as a consequence of nonlinear spectral broadening through self-phase modulation (SPM), which is a manifestation of an intensity dependent refractive index. It should be pointed out that SPM-induced temporal broadening results in a slightly broader temporal pulse width than if the effects of GVD alone are considered. The maximum temporal pulse width measured corresponds to a broadening factor of approximately 93 assuming an input transform limited pulse width of 70 fs.

The effect of the optical fibre on ultrashort pulse propagation has been investigated in terms of the spectral distribution in a previous work [26]. This phenomena also requires close attention in the case of an optical fibre coupler. To characterise the spectral distribution of an ultrashort pulse beam after propagation through the single-mode fibre coupler, a spectrum analyser was used to capture the spectral profile of the emerging illumination. Spectral profiles over a range of input power to the fibre coupler of up to 400 mW were recorded. The FWHM for each recorded spectral profile allows for an analysis of the effect of SPM on the spectral broadening of the ultrashort-pulsed beam.

Figure 4.6 depicts the captured spectra from port 1 of the fibre coupler for input illumination to port 3. The initial pulse spectrum produced by the illumination source is included on the same axes for comparison and has a spectral FWHM of approximately 10.8 ± 0.1 nm. The measured FWHM for an input power of 400 mW is approximately 43.2 nm, demonstrating a broadening factor of approximately 4.1 relative to the input pulse.

The spectral FWHM as a function of the input power is plotted in Fig. 4.7 where it is indicative that the degree of broadening of the pulse spectrum is greater with an increase of the input power to the fibre coupler. It is noted from Fig. 4.6 that the generation of new frequency components can be observed in the pulse wings as it propagates through the fibre coupler. This result is a direct result of SPM.

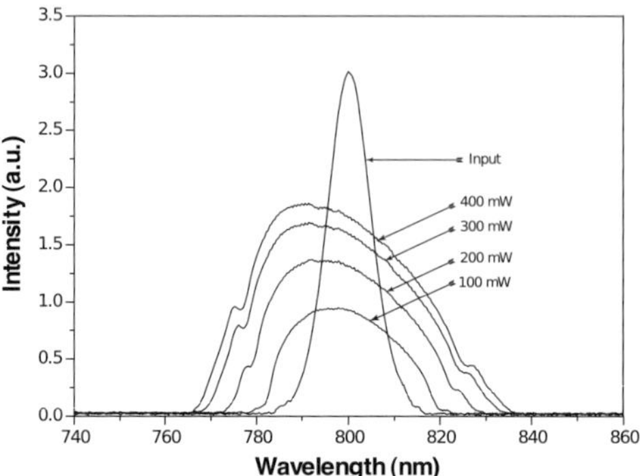

Fig. 4.6 Spectral profiles captured at port 1 of a single-mode optical fibre coupler for input illumination to port 3 of the fibre coupler. Reprinted with permission from Ref. [31], D. Bird and M. Gu, *J. Micros.* **208**, 35 (2002). © 2002, Blackwell Publishing.

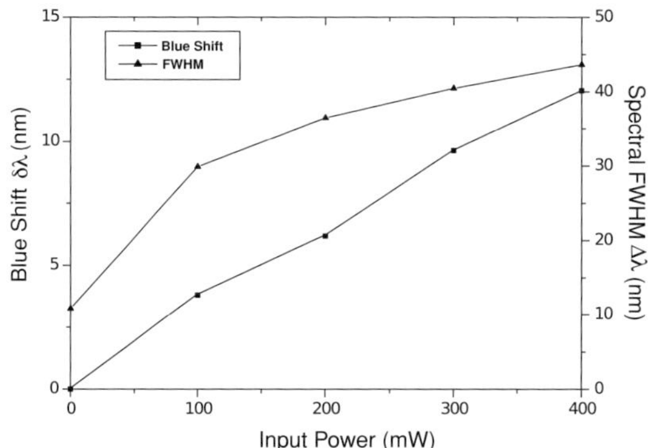

Fig. 4.7 Spectral FWHM $\Delta\lambda$ and spectral blue shift $\delta\lambda$ as a function of input power to the 2×1 single-mode optical fibre coupler. Reprinted with permission from Ref. [31], D. Bird and M. Gu, *J. Micros.* **208**, 35 (2002). © 2002, Blackwell Publishing.

The magnitude of the peak blue shift, $\delta\lambda$, is measured from Fig. 4.6 and is plotted as a function of the input power to the coupler in Fig. 4.7 together with the FWHM. It is clear that the magnitude of the blue shift is dependent on the magnitude of the input power. For an input power of 400 mW, a blue shift of 12.3 ± 0.1 nm is observable. The impact of this broadening and shifting on the resolution performance of a fibre coupler based imaging system is investigated in Section 4.2.4. The peak shift of the ultrashort pulse towards the blue region is due to the effects of self-steepening, a high order nonlinear effect that results as a consequence of an intensity dependent group velocity [32].

4.2.4   Fluorescence axial response

The previous section showed that an improvement in resolution could be achieved in two-photon fluorescence microscopy through the use of a single-mode optical fibre [26]. This

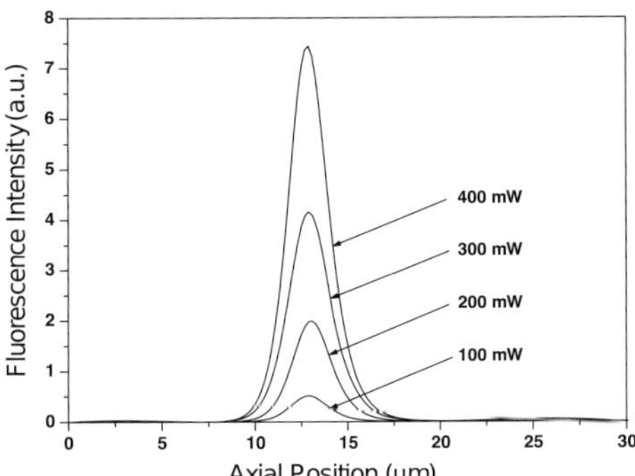

Fig. 4.8 Axial response to a thin fluorescent sample as a function of input power to port 3 of the optical fibre coupler.

feature was demonstrated through measurement of the axial responses to a fluorescent polymer block [33] and it was concluded that this improvement was a result of the effects of SPM-induced spectral broadening and self-steepening. Since short wavelength components of an ultrashort pulse beam are more significantly weighted in the diffraction integral, which results in a smaller spot size at the focus of an objective, axial resolution improvement can be achieved. Since the effects of SPM and self-steepening have also been shown to arise using a fibre coupler (Section 4.2.3), an analysis into the resolution performance of the fibre coupler based system is required.

To investigate this property, the axial response of the system to a thin fluorescent layer sample was obtained by continuous scanning in the z-direction (Fig. 4.3). The fluorescent layer was produced by evaporating a few drops of fluorescent AF-50 dye dissolved in isopropanol alcohol on a cover slide. The sample was excited via two-photon absorption at a wavelength of 800 nm. By varying the input power to port 3 of the fibre coupler in the range 100–400 mW, a set of axial response curves can be obtained, which allows for analysis of the resolution performance of the instrument.

Typical axial responses to the thin fluorescent layer are shown in Fig. 4.8 where the fluorescence intensity detected by the photomultiplier tube has not been normalised. This method allows for simultaneous monitoring of the two-photon efficiency of the system, where it can be seen that the peaks of the response curves are increasing with the square of the input power, confirming two-photon excitation. Careful analysis of the axial responses when normalised to unity reveals a slight improvement in axial resolution of the system with increasing input power to the fibre coupler. This is shown in Fig. 4.9, where the dependence of the FWHM Δz on the incident power to port 3 is depicted. This feature is caused by the fact that the spectrum of the pulsed beam is broadened and blue-shifted due to self-phase modulation [32] and self-steepening [32], respectively, in a fibre [26].

As highlighted in Section 4.1.1, the fibre tips of port 1 and port 3 of the fibre coupler are analogous to conventional confocal pinholes. The introduction of a confocal pinhole to a conventional two-photon microscope leads to a significant improvement in the optical sectioning ability of the system. Under this experimental condition, it

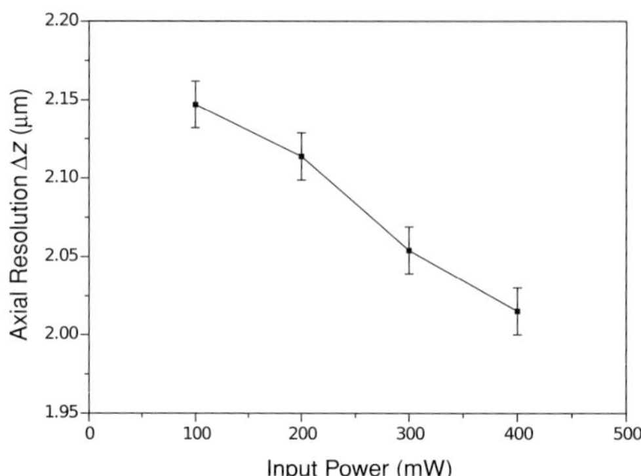

Fig. 4.9 Axial resolution, Δz, as a function of input power to port 3 of the optical fibre coupler. Reprinted with permission from Ref. [30], D. Bird and M. Gu, *Opt. Lett.* **27**, 1031 (2002). © 2002, Optical Society of America.

is observed that the FWHM of the axial response measured in the new two-photon fluorescence microscope, Δz, is approximately 2.1 µm for an input power of 200 mW, which is reduced by approximately 30% compared with that measured by a large area detector without a pinhole [26]. The decrease in the FWHM indicates the enhancement of the optical sectioning effect and therefore the improvement in axial resolution. In this case, the resolution improvement results from the aperture of port 1, which acts as an effective confocal aperture [9].

4.2.5   Three-dimensional optical transfer function analysis

Optical transfer function analysis is a powerful tool that gives a complete description of the imaging process. The confocal fluorescence scanning microscopy imaging mode has received a great deal of attention [34–44] including investigations into the effects of the detector size [37–43] and the effect of the central obstruction radius of the annular lens [41, 43]. From these studies it is known that the cutoff spatial frequency of the OTF in both the axial and transverse directions for a confocal fluorescent system with a point source and a point detector is twice as large as those in a conventional incoherent microscope. Furthermore, in the case of finite-sized detectors, the OTF has a missing cone of spatial frequencies and a tail of negative values [42, 43] that results in poor transmission of axial information and the introduction of image artefacts, respectively. All of these studies, however, are based on a single-photon process and therefore consideration needs to be given to the fibre-optical two-photon fluorescence case.

An incoherent imaging system can be analysed in terms of the 3D OTF that is given by the 3D Fourier transform of the effective intensity point spread function [9, 45]. The OTF gives the transmission efficiency of each spatial frequency of the object through an imaging system. The fibre-optical two-photon fluorescence microscope can thus be described by the 3D OTF $C(\mathbf{m})$ given by [46]

$$C(\mathbf{m}) = C_1(\mathbf{m}) \otimes_3 C_2(\mathbf{m}), \tag{4.4}$$

where

$$C_1(\mathbf{m}) = \mathbf{F}_3\left\{ \left| f_1(\mathbf{M}_1 x, \mathbf{M}_1 y) \otimes_2 h_1(\mathbf{M}_1 \mathbf{r}) \right|^4 \right\} \tag{4.5}$$

and

$$C_2(\mathbf{m}) = \mathbf{F}_3\left\{ \left| f_2^*(\mathbf{M}_1 x, \mathbf{M}_1 y) \otimes_2 h_2(\mathbf{r}) \right|^2 \right\}. \tag{4.6}$$

Here $\mathbf{F}_3$ is the 3D Fourier transform with respect to $\mathbf{r}_s$ and $\mathbf{m}$ represents the spatial frequency vector with two transverse components m and n and one axial component s. Compared with Eq. (8) of reference [44], Eq. 4.5 is the 3D OTF for a fibre-optical single-photon fluorescence microscope with identical illumination and collection fibres and equal illumination and fluorescence wavelengths. Thus Eq. 4.5 can be rewritten as an auto-convolution operation

$$C_1(\mathbf{m}) = C_1'(\mathbf{m}) \otimes_3 C_1'(\mathbf{m}), \tag{4.7}$$

where

$$C_1'(\mathbf{m}) = \mathbf{F}_3\left\{ \left| f_1(\mathbf{M}_1 x, \mathbf{M}_1 y) \otimes_2 h_1(\mathbf{M}_1 \mathbf{r}) \right|^2 \right\}. \tag{4.8}$$

Suppose that both illumination and collection fibres are single-mode fibres of mode spot radii a_1 and a_2 and that the Gaussian approximation [13] holds for circularly polarised fibres. Further assume that the paraxial approximation [9] is used and the objective and collector lenses, P_1 and P_2 are of identical circular aperture of radius a. Therefore, the analytical expressions for $C(\mathbf{m})$, $C_1'(\mathbf{m})$ and $C_2(\mathbf{m})$ can be derived, in a cylindrical coordinate system as used previously [9, 44], as

$$C(l, s) = (C_1'(l, s) \otimes_3 C_1'(l, s)) \otimes_3 C_2(l, s), \tag{4.9}$$

$$C_1'(l, s) = \frac{\sqrt{\pi}\exp\left\{ -A_1\left[\frac{\beta^2 l^2}{4} + \left(\frac{s}{l}\right)^2 \right] \right\}}{\sqrt{A_1}\beta l}$$
$$\times \operatorname{erf}\left[\sqrt{A_1}\operatorname{Re}\left(\sqrt{1 - \left(\frac{|s|}{l} + \frac{\beta l}{2}\right)^2} \right) \right] \tag{4.10}$$

and

$$C_2(l, s) = \frac{\sqrt{\pi}\exp\left\{ -A_2\left[\frac{l^2}{4} + \left(\frac{s}{l}\right)^2 \right] \right\}}{\sqrt{A_2}l}$$
$$\times \operatorname{erf}\left[\sqrt{A_2}\operatorname{Re}\left(\sqrt{1 - \left(\frac{|s|}{l} + \frac{l}{2}\right)^2} \right) \right]. \tag{4.11}$$

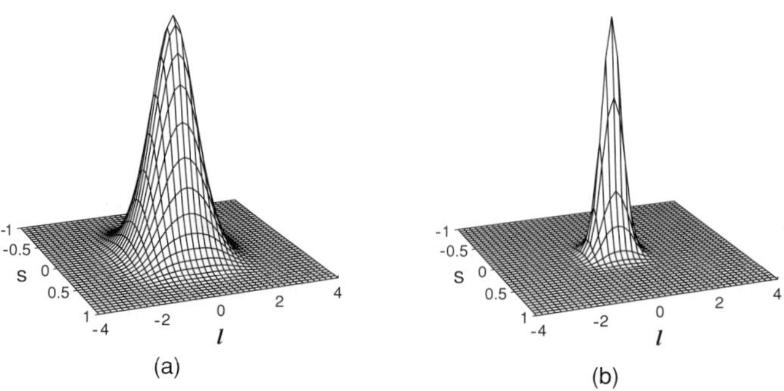

Fig. 4.10 3D OTFs for fibre-optical two-photon fluorescence microscopy using a fibre coupler ($A_2 \approx 4A_1$), for (a) $A_1 = 2$ and (b) $A_1 = 10$. Reprinted with permission from Ref. [46], M. Gu and D. Bird, *J. Opt. Soc. Am. A.* **20**, 941 (2003). © 2003, Optical Society of America.

Here A_j is the normalised fibre spot size for illumination and collection fibres defined by Eq. 4.1 in Section 4.1.1 and d_1 is the distance between the fibre and the lens P_j. Here, the variables l and s are the radial and axial spatial frequencies normalised by $\sin(\alpha)/\lambda_2$ and $4\sin^2(\alpha/2)/\lambda_2$, respectively, where $\sin(\alpha)$ is the numerical aperture of the objective lens. β in Eq. 4.10 is the wavelength ratio of λ_1 to λ_2. Re{} denotes the real part of its argument, which becomes zero for negative values of the argument.

The function erf in Eqs 4.10 and 4.11 is defined as

$$\mathrm{erf}(x) = \frac{2}{\sqrt{\pi}} \int_0^x \exp(-\tau^2)\, \mathrm{d}t, \tag{4.12}$$

which is called the error function [47].

In order to reveal the dependence of the 3D OTF on the normalised fibre parameters, the 3D OTF has been numerically calculated in order to understand the experimental results. A comparison between the performance of the OTFs for varying normalised fibre parameters can be made through observation of Fig. 4.10, which depicts the calculated 3D OTF for the two-photon fibre coupler based imaging system.

Since the two-photon excited fluorescence wavelength is approximately half that of the excitation wavelength, in this case the fibre parameters are related via $A_2 \approx 4A_1$, assuming that the fibre spot size is the same for both the illumination and collection process. Figure 4.10(a) is the 3D OTF for $A_1 = 2$. A direct comparison with Fig. 4.10(b) ($A_1 = 10$) reveals that both the transverse and axial responses of the 3D OTF in the former case are stronger and that the cutoff spatial frequencies in both the axial and transverse directions are increased. This feature means that more information pertaining to the finer features and details in an object can be imaged by the smaller fibre parameter value, leading to higher image resolution.

4.2.6   Discussion

The 3D imaging capability of a two-photon fibre-coupler-based system has been demonstrated through various imaging experiments. A series of image sections taken at

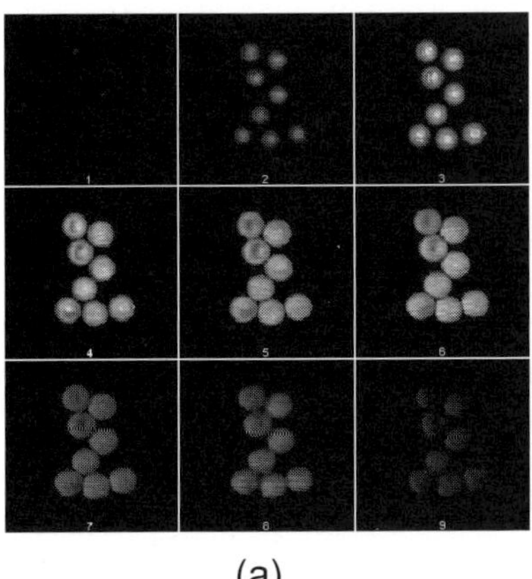

(a)

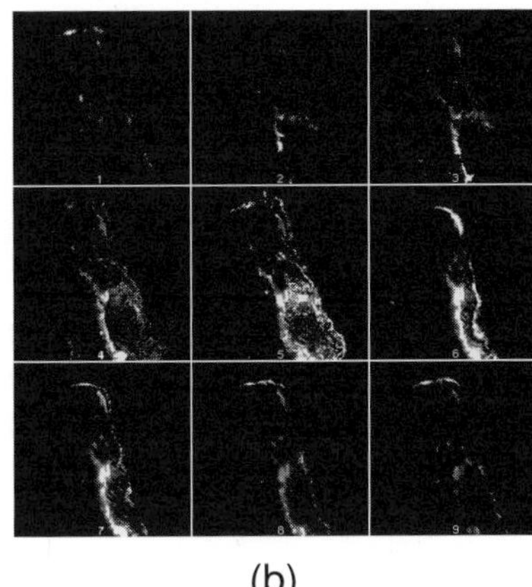

(b)

Fig. 4.11 A series of x–y images of (a) fluorescent polymer microspheres and (b) *Griffithsia* sea algae obtained by two-photon fluorescence microscopy using a 2×1 single-mode optical fibre coupler. The lateral size of each slice is 60 μm × 60 μm for the beads and 150 μm × 150 μm for the *Griffithsia*. Each section was recorded at an axial depth of 1 μm into the sample. The excitation power is 35 mW at the focus. Reprinted with permission from Ref. [30], D. Bird and M. Gu, *Opt. Lett.* **27**, 1031 (2002). © 2002, Optical Society of America.

1 μm steps into an ensemble of 10 μm fluorescent polymer microspheres is shown in Fig. 4.11(a). The applicability of the instrument for biological study is shown in Fig. 4.11(b), which displays a series of image sections of the sea algae, *Griffithsia*. Both image sets exhibit high contrast and the pronounced optical sectioning property of the system as a result of the confocal aperture of the fibre tip.

The results presented in this section lead to the possibility of fibre-optical two-photon fluorescence endoscopy if an ultrashort pulse fibre laser and gradient index (GRIN) lenses are used in conjunction with a single-mode optical fibre coupler. This topic is explored further in Section 4.4. Such an endoscopic device may be an important complementary instrument of optical coherence tomography (OCT) [48, 49]. Both OCT and two-photon fluorescence microscopy have been demonstrated to be effective techniques for imaging scattering media such as biological tissues. The contrast mechanisms for these two techniques, however, are intrinsically different in that the former is based on the detection of backscattered illumination, whereas the latter is based on the detection of fluorescence light. Consequently, OCT imaging usually provides information on sample morphology, whereas two-photon fluorescence microscopy has the ability to provide information on sample functionality. It should be possible therefore, based on these findings, to achieve a simultaneous operation of two-photon fluorescence microscopy and optical coherence tomography using a fibre coupler. Simultaneous imaging by OCT and two-photon fluorescence microscopy has recently been demonstrated [50]; however, the system used in those experiments was based entirely on bulk optical components. It should be pointed out that although the use of a fibre coupler is convenient from a compact/miniaturisation point of view and can lead to a significant improvement in the axial resolution of a two-photon system, these features come at the expense of signal level. In addition, the use of the fibre coupler will make the system unsuitable for non-descanned applications.

4.3   Fibre-optical second harmonic generation microscopy

Although SHG was demonstrated almost as soon as the first laser had been built, only more recently has it emerged rapidly as a powerful contrast mechanism in nonlinear microscopy [51]. Like two-photon fluorescence microscopy, SHG microscopy possesses intrinsic benefits including an optical sectioning property, reduced out-of-plane photobleaching and phototoxicity, and deeper penetration depth. In particular, due to its coherent scattering nature, SHG enables direct imaging of highly polarisable and ordered non-centrosymmetric structures without exogenous molecular probes, and polarisation anisotropy to extract nonlinear organisation of samples. Therefore, SHG imaging of endogenous proteins such as collagen, microtubules and actomyosin in live tissue may provide a new angle for tissue morphology, cell–cell interaction and disease diagnosis [52, 53]. As is the case with two-photon fluorescence microscopy, at present the majority of SHG studies of biological tissue are obtained on the bench top with bulk optics, that preclude *in vivo* applications on living animals. In the case of SHG imaging, these studies are (in general) limited even further by virtue of commonly employed transmission detection geometries. Although the effectiveness of a fibre coupler has been demonstrated in two-photon fluorescence microscopy to overcome some of the more obvious practical restraints (Section 4.2), the difficulty with adopting fibre-optic components in SHG microscopy is due to the fact that the wavelength range where SHG signal falls is further away from the designed wavelength of the fibre-optic components compared with that of normal two-photon fluorescence. Moreover, polarisation preservation of the laser illumination beam and the collected SHG signal are essential in order to probe the orientation of (for example) structural proteins in tissues. In this section, fibre-optic SHG microscopy using a three-port single-mode fibre coupler is described, and it is revealed that such an instrument is capable of collecting SHG signal and two-photon fluorescence simultaneously and efficiently [54–57].

4.3.1   Coupling efficiency and splitting ratio

Experimentally, the arrangement of a SHG microscope based on a single-mode optical fibre coupler is shown in Fig 4.12. The backscattered SHG signal is collected by the imaging objective and delivered to the signal arm of the fibre coupler to either a photomultiplier tube for imaging or a CCD-based fibre-coupled spectrograph to record spectra. The bandpass filter placed in front of the PMT is replaced with a narrow notch filter to ensure that only the SHG signal is detected. Switching between an imaging and spectral acquisition mode is achieved in practice simply by changing the direction of a flip mirror arranged such that the collected signal can be delivered to either detector input [54, 55].

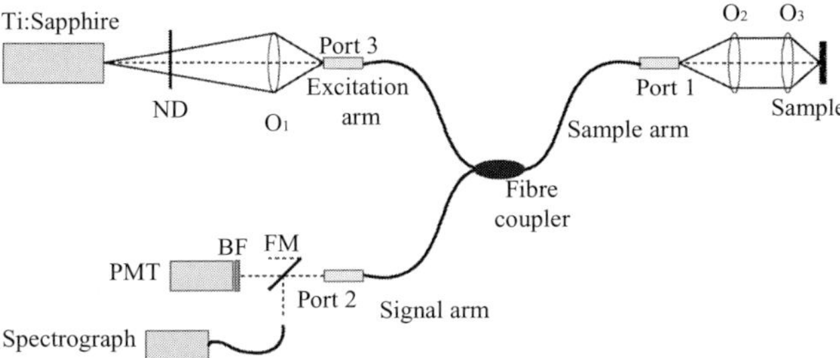

Fig. 4.12 Schematic diagram of a SHG microscope based on a single-mode fibre coupler for polarisation anisotropy measurement. ND: neutral density filter, GTP: Glan Thompson polariser, O_1 and O_2: microscope objectives, O_3: Olympus $40\times$ 0.85 NA imaging objective.

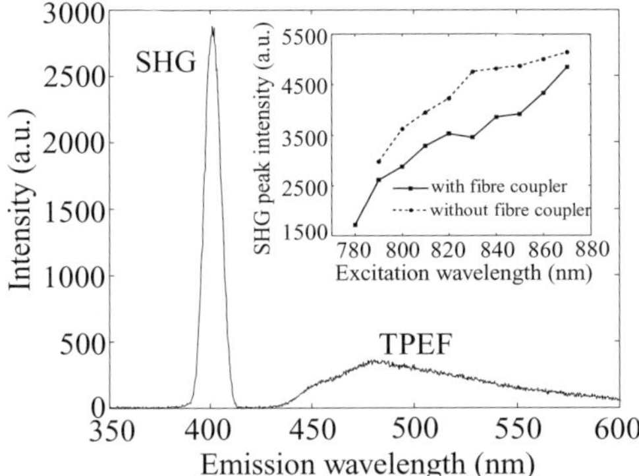

Fig. 4.13 Emission spectra of AF-50 dye at excitation wavelength of 800 nm. Inset shows the SHG peak intensity as a function of the excitation wavelength. The solid curve in the inset is derived from the spectra collected by focusing the output beam from port 2 of the fibre coupler into the spectrograph, while the dashed curve is that from the spectra collected directly from the sample without the use of the fibre coupler. The excitation power is approximately 4 mW at the sample and the excitation polarisation is parallel to the fundamental laser polarisation. Reprinted with permission from Ref. [55], L. Fu, X. Gan and M. Gu, *Opt. Lett.* **30**, 385 (2005). © 2005, Optical Society of America.

The typical coupling efficiency from port 3 to port 1 varies between 20 and 41% in the wavelength range of 770–870 nm and 29–41% in the range 435–532 nm from port 2 to 1 (shown in inset A of Fig. 4.13). The 435 nm laser beam is obtained by the combination of a Ti:sapphire laser and a frequency doubler while the 532 nm laser beam is a continuous beam. Consequently, a laser power of up to approximately 80–150 mW in the near-infrared wavelength range can be expected to be delivered to the imaging objective. Although the ultrashort pulse laser beam is broadened to a few picoseconds due to group velocity dispersion and self-phase modulation during the delivery, sufficient power can still be achieved after the propagation of the laser beam through the coupler. To understand the propagation properties of the SHG signal (which is in the visible wavelength range) through the fibre coupler, the mode profile and the coupling efficiency need to be measured at ports 2 and 3 when a beam at wavelength 435 nm and at wavelength 532 nm illuminates port 1, respectively. Ideally it should be found that in both cases the field distribution from port 2 remains a single-mode profile whereas the field distribution from port 3 presents a multi-mode pattern, which would be consistent with the estimation based on the V parameter of such a fibre in the visible range [53]. Indeed, in this case, the coupling efficiency at port 2 for wavelength 435 nm illumination is approximately 29% with a splitting ratio of 99.6/0.4 between ports 2 and 3. The corresponding measurements for wavelength 532 nm are 41% and 99.7/0.3, respectively. The reduction in the coupling efficiency at wavelength 435 nm may result from the greater loss of a higher mode at the coupler junction. The coupling measurements from port 1 to ports 2 and 3 show that the strength of the visible beams guided by the fibre coupler in the signal arm is two orders of magnitude higher than that in the excitation arm. As a result, the fibre coupler acts as a short-pass filter at a visible wavelength, which may optimise the delivery of pulsed laser beam and the collected SHG signal.

The effectiveness of the fibre-optic SHG microscope is demonstrated by the SHG spectra recorded from a thin layer of AF-50 dye. The layer is produced in a similar manner to that described in Section 4.2.4 by evaporation of a mixture of AF-50 dye and isopropyl alcohol on a coverslip with an averaged thickness of approximately 250 nm measured by atomic force microscopy. This dye has extended conjugated pi networks and aromatic heteroatom substitution possesses large second-order and third-order nonlinear susceptibilities [58]. Therefore, SHG and two-photon fluorescence can be produced simultaneously. In this case, the femtosecond laser is scanned between 780 and 870 nm in wavelength and the excitation polarisation is parallel to the polarisation of the incident laser. The emission spectra of AF-50 dye are collected by spectrograph. As shown in Fig. 4.13, the emission spectrum with an excitation wavelength of 800 nm reveals a sharp SHG peak at 400 nm and a two-photon fluorescence lobe from wavelength 430 nm to 600 nm. For all the excitation wavelengths used, the SHG spectra peak at exactly half the excitation wavelengths and have the same bandwidth of approximately 10 nm, which is consistent with the excitation laser spectral width. The SHG peak intensity as a function of excitation wavelength, obtained with and without the fibre coupler, is depicted in the inset of Fig. 4.13. The increased SHG peak intensity propagated through the fibre coupler and detected by the spectrograph may arise from the increased SHG cross section of AF-50 and the improved coupling efficiency of the fibre coupler in the visible range, which has been discussed in the last paragraph. Fig. 4.13 also implies that the simultaneous collection of SHG signal and two-photon fluorescence can be achieved efficiently through the fibre coupler, although second harmonic and two-photon fluorescence occur at a much shorter wavelength region than the designed operating wavelength of the fibre coupler.

4.3.2   Second-harmonic generated axial response

To investigate the optical sectioning ability of the fibre-optic SHG microscope, the axial response of the system to a thin layer as described above is measured. The dependence of

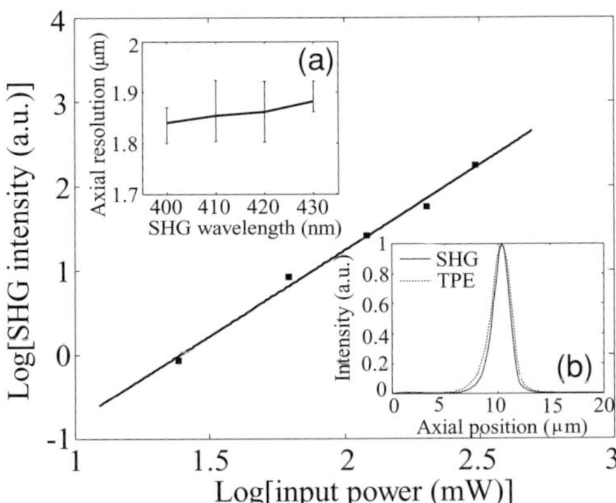

Fig. 4.14 Quadratic dependence of the SHG peak intensity on the excitation power. Inset (a) shows the FWHM of axial responses as a function of the SHG wavelength. Inset (b) gives the axial response of the system for SHG signal and two-photon fluorescence from a thin layer of AF-50 dye. The excitation wavelength is 800 nm. Reprinted with permission from Ref. [55], L. Fu, X. Gan and M. Gu, *Opt. Lett.* **30**, 385 (2005). © 2005, Optical Society of America.

the SHG peak intensity on the input power to port 3 of the coupler is shown in Fig. 4.14. It can be seen that the gradient of the log–log plot is approximately two, confirming the second-order nonlinear frequency conversion process. Inset (a) of Fig. 4.14 shows the FWHM of the SHG axial responses as a function of the detected SHG wavelength. Bandpass filters having appropriate central wavelengths with a bandwidth of approximately 9 nm are placed before the photomultiplier tube to collect the SHG signal. It is shown that the FWHM of the SHG axial responses varies between 1.8 and 1.9 as the excitation wavelength tunes from 800 to 860 nm. In Fig. 4.14 inset (b), the axial responses obtained with the SHG and two-photon fluorescence signals at an excitation wavelength of 800 nm are depicted. To record the two-photon fluorescence axial response, a 510/20 bandpass filter is placed before the PMT to eliminate the second harmonic and reflected fundamental wavelengths. Figure 4.14 inset (b) reveals that the FWHM of these curves is approximately 1.8 μm and 2.1 μm, respectively, demonstrating that a slight improvement in resolution of approximately 14% for SHG signal collection is obtained if compared with the two-photon fluorescence axial response. This feature is due to the fact that the SHG wavelength is shorter than the two-photon fluorescence wavelength. To demonstrate the imaging capability and polarisation sensitivity of the instrument, we measure the SHG and two-photon fluorescence images of a triangle-shaped sample excited by two beams of parallel and perpendicular polarisation states. Usually SHG polarisation anisotropy can be extracted by rotating an analyser before the detector or the input laser polarisation. As demonstrated elsewhere [30] the fibre coupler shows the feature of linear polarisation preservation along the birefringent axis from near infrared to visible wavelength regions.

The sample consists of a paper sheet (75 g/m^2) with a mixture of AF-50 dye and isopropyl alcohol. The SHG and two-photon fluorescence images are obtained from the

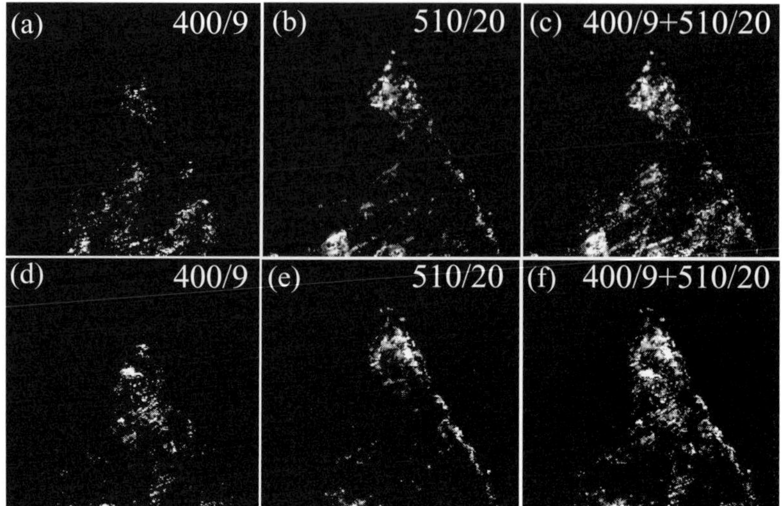

Fig. 4.15 SHG and two-photon fluorescence images of a triangle-shaped paper sheet excited by parallel polarisation (upper row) and perpendicular polarisation (lower row) beams at wavelength 800 nm. The excitation power is approximately 4 mW on the sample. SHG images (a) and (d) are obtained with a 400/9 nm bandpass filter and two-photon fluorescence images (b) and (e) with a 510/20 nm bandpass filter. Images in (e) and (f) are obtained by overlaying the SHG and two-photon fluorescence image. Scale bar is 50 μm. Reprinted with permission from Ref. [55], L. Fu, X. Gan and M. Gu, *Opt. Lett.* **30**, 385 (2005). © 2005, Optical Society of America.

same sample site at the same focal plane by using a 400/9 nm and a 510/20 nm bandpass filter, respectively, as shown in Fig. 4.15. It can be seen that SHG and two-photon fluorescence microscopy do not produce exactly the same pattern of contrast. This feature is caused by the different microscopic contrast mechanisms that can provide the complementary information shown in the overlay of the SHG and two-photon fluorescence images. Further observation from the SHG images shows that SHG images with orthogonal-polarisation beam excitations display a distinguishing appearance which is due to the fact that the SHG signal arises from dipole interaction. The fibre-optic SHG microscope reveals an axial resolution of approximately 1.8 μm, demonstrating an improvement in resolution compared with the case under two-photon excitation. Both SHG and two-photon fluorescence images can be collected simultaneously through the new instrument under parallel and perpendicular excitation polarisations of laser beams. Such multiple imaging modalities may prove advantageous in nonlinear optical endoscopy.

4.3.3   Three-dimensional coherent transfer function analysis

It is now clear that SHG microscopy differs physically from two-photon fluorescence microscopy since it is a coherent imaging process. Therefore, the 3D optical transfer function derived in Section 4.2.5 to describe the imaging performance in fibre-optical

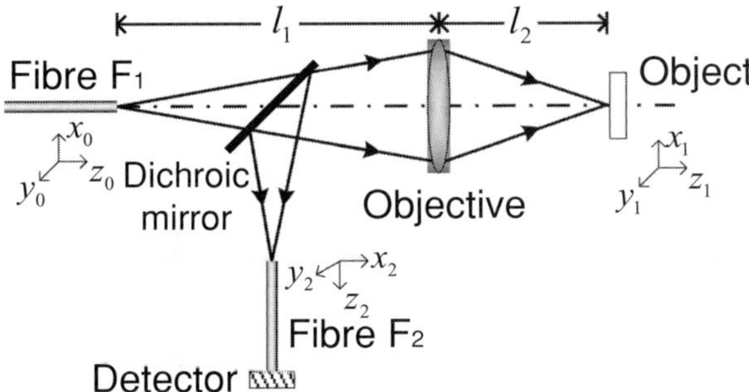

Fig. 4.16 Schematic diagram of a fibre-optical SHG scanning microscope. Reprinted with permission from Ref. [57], M. Gu and L. Fu, *Opt. Express* **14**, 1175 (2006). © 2006, Optical Society of America.

two-photon fluorescence microscopy is not applicable to fibre-optical SHG microscopy because the system behaves fully coherently even for finite values of fibre spot size [5,10]. Whereas fibre-optical two-photon fluorescence microscopy can be understood by an optical transfer function, fibre-optical SHG microscopy can be described by a 3D coherent transfer function (CTF), which can be expressed by an analytical expression [11]. This section describes 3D image formation in fibre-optical SHG microscopy using the concept of the 3D CTF to reveal the dependence of axial resolution on fibre coupling parameters [57].

To investigate the effect of optical fibres on illumination and collection separately, refer to the schematic diagram of a fibre-optical SHG microscope as shown in Fig. 4.16. Note that for the purpose of this discussion, two optical single-mode fibres F_1 and F_2 are used to deliver illumination at wavelength $2\lambda_0$ and collect SHG signal at wavelength λ_0, respectively. When the illumination optical fibre F_1 and the collection optical fibre F_2 are identical, the performance of the system is equivalent to that using a fibre coupler.

As SHG is a coherent process, the analysis of image formation in fibre-optical SHG microscopy is similar to that in fibre-optical non-fluorescence microscopy described elsewhere [11]. Considering that the electric field of the SHG emission from a sample is proportional to the square of the field of the illumination field impinging on the sample, the image intensity from a scan point $\mathbf{r}_s = (x_s, y_s, z_s)$ in the fibre-optical SHG microscope can be expressed (if the optical axis is assumed along the z direction) as

$$
I(\mathbf{r}_s) = \Big| \int_\infty^\infty U_2^*(x_2, y_2)\delta(z_2) \Big[\int_\infty^\infty U_1(x_0, y_0)\delta(z_0)
$$

$$
\times \exp[ik(z_0 - z_1)]h_1(\mathbf{r}_0 + \mathbf{M}_1\mathbf{r}_1)\delta\mathbf{r}_0 \Big]^2
$$

$$
\times \varepsilon(\mathbf{r}_s - \mathbf{r}_1)\exp[ik(\pm z_1 - z_2)]h_2(\mathbf{r}_1 + \mathbf{M}_2\mathbf{r}_2)\delta\mathbf{r}_2\mathbf{r}_1 \Big|^2, \qquad (4.13)
$$

where the letters $\mathbf{r}_i$ ($i = 0,1,2$) represent vectors with components x_i, y_i and z_i. The functions $U_1(x, y)$ and $U_2(x, y)$ are the amplitude mode profile on the output end of fibre F_1 and on the input end of fibre F_2, respectively. $*$ denotes the conjugate operation and the parameters $\mathbf{M}_1$ and $\mathbf{M}_2$ are diagonal matrices of the magnification factors of the illumination and collection lenses, respectively. The term in the first square brackets is a result of the quadratic dependence of SHG. $h_1(r)$ and $h_2(r)$ are the 3D amplitude point spread functions (PSFs) for the objective in illumination and collection paths, respectively. $\varepsilon(\mathbf{r})$ is the object function representing the SHG strength of the object. Using the 3D convolution relation, one can simplify Eq. 4.13 as

$$I(\mathbf{r}_s) = \left| h_{\mathrm{eff}}(\mathbf{r}_s) \otimes_3 \varepsilon(\mathbf{r}_s) \right|^2, \tag{4.14}$$

where $\otimes_3$ denotes the 3D convolution operation. h_{eff} is the 3D effective PSF for fibre-optical SHG microscopy and is given by

$$h_{\mathrm{eff}}(\mathbf{r}) = \left[U_1(\mathbf{M}_1 x, \mathbf{M}_1 y) \otimes_2 h_1(\mathbf{M}_1 \mathbf{r}) \right]^2 \left[U_1^*(\mathbf{M}_1 x, \mathbf{M}_1 y) \otimes_2 h_2(\mathbf{r}) \right], \tag{4.15}$$

where $\otimes_2$ denotes the 2D convolution operation.

Note that Eqs. 4.13 and 4.15 represent a superposition of the light amplitude from a sample and therefore imply that (like fibre-optical non-fluorescence microscopy [5, 10, 11]) fibre-optical SHG microscopy is purely coherent. This fact is particularly important when performing SHG interferometric microscopy and/or tomography.

Therefore, fibre-optical SHG microscopy can be analysed in terms of the 3D CTF that is given by the 3D Fourier transform of the effective PSF [5, 11]. The 3D CTF, $c(\mathbf{m})$, for fibre-optical SHG microscope can thus be described by

$$c(\mathbf{m}) = c_1(\mathbf{m}) \otimes_3 c_2(\mathbf{m}), \tag{4.16}$$

where

$$c_1(\mathbf{m}) = \mathbf{F}_3 \left\{ \left[U_1(\mathbf{M}_1 x, \mathbf{M}_1 y) \otimes_2 h_1(\mathbf{M}_1 \mathbf{r}) \right]^2 \right\} \tag{4.17}$$

and

$$c_2(\mathbf{m}) = \mathbf{F}_3 \left\{ U_2^*(\mathbf{M}_1 x, \mathbf{M}_1 y) \otimes_2 h_2(\mathbf{r}) \right\}. \tag{4.18}$$

Here $\mathbf{F}_3$ is the 3D Fourier transform with respect to $\mathbf{r}_s$ and $\mathbf{m}$ represents the spatial frequency vector with two transverse components m and n and one axial component s.

For a system using a circular lens, $c_1(\mathbf{m})$ is the 3D CTF for a fibre-optical reflection-mode non-fluorescence microscope with wavelength $2\lambda_0$ and can be analytically expressed as

$$c_1(l, s) = \exp(-2A_1 s) \begin{cases} 1, & \frac{l^2}{2} \leq s \leq \frac{1}{2} - l(1 - l) \\[2mm] \frac{2}{\pi} \sin^{-1} \left\{ \frac{1 - 2s}{2l(2s - l^2)^{\frac{1}{2}}} \right\} & \frac{1}{2} - l(1 - l) \leq s \leq \frac{1}{2} \\[2mm] 0, & \text{otherwise.} \end{cases} \tag{4.19}$$

Similarly, Eq. 4.18 represents the 3D CTF for a single circular lens with wavelength λ_0 and weighted by the Fourier transform of the fibre mode profile $U_2^*(x, y)$, given by

$$c_2(l, s) = \exp(-A_2 l^2/2)\delta(s - l^2/2). \tag{4.20}$$

In Eqs. 4.19 and 4.20, $A_j = [2\pi a a_j/(\lambda_j d_j)]^2$ $(j = 1, 2)$ is the normalised fibre spot size for illumination and collection fibres. d_j is the distance between the fibre ends and the objective. The variables $l(l = \sqrt{m^2 + n^2})$ and s denote the radial and axial spatial frequencies normalised by $\sin\alpha/\lambda_0$ and $4\sin^2(\alpha/2)/\lambda_0$, respectively, where $\sin\alpha$ is the numerical aperture of the objective of radius a. Here it is assumed that both illumination and collection fibres are single-mode of mode spot radii a_1 and a_2. It has been shown that A_j is proportional to the square of the ratio of the numerical aperture of the objective in the illumination and collection paths to the numerical aperture of the fibres F_1 and F_2 [9]. According to Eq. 4.16, the 3D CTF can be numerically evaluated, if the delta function in Eq. 4.20 is taken into account, by

$$c(l, s) = \iint_\sigma \exp\left[\frac{-A_2(m^2 + n^2)}{2}\right]$$
$$\times c_1\left(\sqrt{(m - l)^2 + n^2}, s - \frac{m^2 + n^2}{2}\right) dm\, dn, \tag{4.21}$$

where σ represents the area overlapped by $m^2 + n^2 = 1$ and $(m - l)^2 + n^2 = 1$. Finally, the 3D CTF for fibre-optical SHG microscopy can be explicitly written as

$$c(l, s) = \int_0^{\sqrt{1-(l/2)^2}} \left[\int_{1-\sqrt{1-n^2}}^{\sqrt{1-n^2}} \exp\left[\frac{-A_2(m^2 + n^2)}{2}\right]\right.$$
$$\left.\times c_1\left(\sqrt{(m - l)^2 + n^2}, s - \frac{m^2 + n^2}{2}\right) dm\right] dn. \tag{4.22}$$

It should be pointed out that 3D CTF for fibre-optical SHG microscopy has a spatial frequency passband of $l^2/4 < (s + s_0) < 1$ with axial and transverse cutoffs 1 and 2, respectively. Here $s_0 = 1/[\sin^2(\alpha/2)]$ is a constant axial spatial-frequency shift resulting from the reflection imaging geometry [10]. This feature is the same as fibre-optical reflection-mode non-fluorescence microscopy.

When $A_j \to 0$, which corresponds to the case when the numerical aperture of the fibre is much larger than that of the objective in the illumination and collection paths, the 3D CTF describes confocal SHG microscopy of a point source and a point detector. If either A_1 or A_2 becomes infinity, $c(l, s)$ becomes Eq. 4.19 or Eq. 4.20. No image is formed because $c(l, s)$ is zero. For a SHG microscope based on a fibre coupler (Section 4.3), $A_2 = 4A_1$ and the corresponding 3D CTFs for $A_1 = 0, 1, 5$ are shown in Figs. 4.17(a)–(c). Figure 4.17(a) represents the 3D CTF for confocal SHG microscopy with a point source and a point detection. For a finite value of A_1, the strength of the 3D CTF is reduced in particular along the axial direction, as shown in Figs. 4.17(b) and (c). Figure 4.17(d) also denotes the 3D CTF when a point source is used ($A_1 = 0$) and a fibre is used for collection. In this case, the collection function of the objective becomes

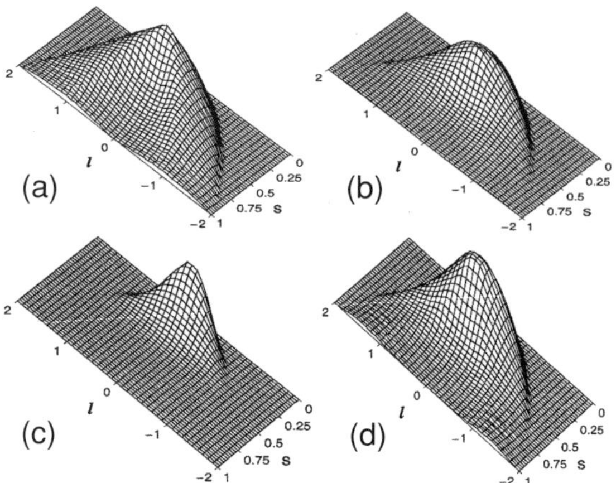

Fig. 4.17 3D CTF for fibre-optical SHG microscopy. (a) $A_1 = 0$, $A_2 = 4A_1$, (b) $A_1 = 1$, $A_2 = 4A_1$, (c) $A_1 = 5$, $A_2 = 4A_1$, (d) $A_1 = 0$, $A_2 = 5$. Reprinted with permission from Ref. [57], M. Gu and L. Fu, *Opt. Express* **14**, 1175 (2006). © 2006, Optical Society of America.

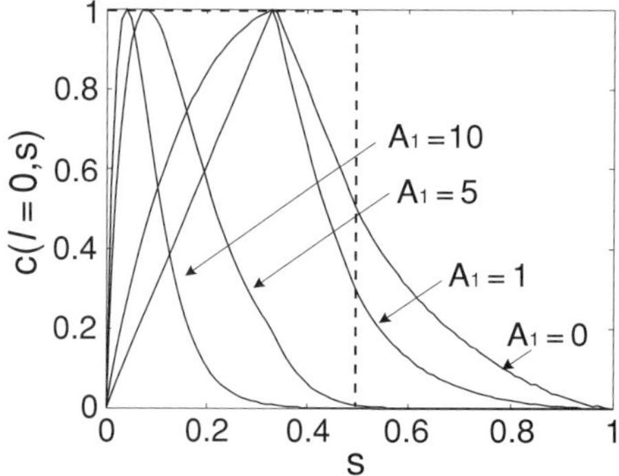

Fig. 4.18 Axial cross section of the 3D CTF for fibre-optical SHG microscopy using a fibre coupler ($A_2 = 4A_1$) for different values of the normalised optical spot size parameter A_1. The dashed curve represents the case $A_1 = 0$ and $A_2 \to \infty$. Reprinted with permission from Ref. [57], M. Gu and L. Fu, *Opt. Express* **14**, 1175 (2006). © 2006, Optical Society of America.

weak. Ultimately, when $A_2 \to \infty$, $c_2(l, s)$ approaches a delta function at $l = 0$ and thus the 3D CTF for SHG microscopy is given by Eq. 4.19.

It is important to investigate the axial cross section, $c(l = 0, s)$ of the 3D CTF further as it gives the axial imaging performance. The solid curves in Fig. 4.18 represent the normalised axial cross section of the 3D CTF for SHG microscopy using a fibre coupler ($4A_1 = A_2$), while the dashed line depicts the condition for $A_1 = 0$ and $A_2 \to \infty$. For $4A_1 = A_2 = 0$, the CTF increases linearly up to $s = 1/3$, which is contributed by the constant region in Eq. 4.19. After the maximum value at $s = 1/3$, the CTF decreases and finally cuts off at s $= 1$. When $4A_1 = A_2$ and $A_1 \neq 0$, the strength of the CTF for $s < 1/3$

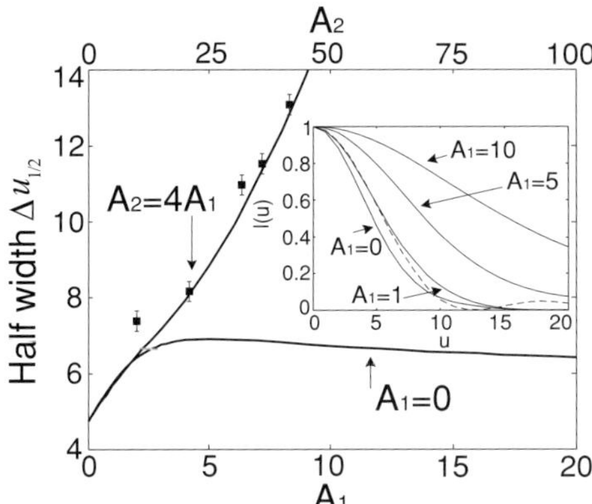

Fig. 4.19 Half width at half maximum of the axial response, $\delta u_{\frac{1}{2}}$ as a function of the normalised fibre spot size parameter when $A_2 = 4A_1$ (bottom axis) and when $A_1 = 0$ (left axis). The squares are experimental results for $A_1 = 2.0$, 4.2, 6.4, 7.3 and 8.4, respectively. Inset: normalised axial response of a perfect SHG reflector in fibre-optical SHG microscopy using a fibre coupler ($A_2 = 4A_1$) for different values of the normalised optical spot size parameter A_1. The dashed curve represents the case for $A_1 = 0$ and $A_2 \to \infty$. Reprinted with permission from Ref. [57], M. Gu and L. Fu, *Opt. Express* **14**, 1175 (2006). © 2006, Optical Society of America.

is enhanced while that for $s > 1/3$ is reduced. Eventually (when $4A_1 = A_2 \to \infty$), the CTF approaches a delta function at $s = 0$, which means that there is no axial imaging ability. To characterise axial resolution, one usually considers imaging of a perfect SHG reflector scanning through the focus of the objective. This axial response is a measure of axial resolution or the optical sectioning property [9] and can be calculated using the modulus squared of the Fourier transform of the axial cross section of the 3D CTF at $l = 0$. After mathematical manipulations, such an axial response can be expressed as

$$I(u) = \left| \int_0^1 \left\{ \int \frac{\exp\{-l^2[(A_1 - iu/2) + (A_2 - iu/2)]\}}{A_1 - iu/2} \right. \right.$$

$$\left. \left. \times \{1 - \exp[-\rho_0^2(A_1 - iu/2)]\}d\theta \right\} l dl \right|^2, \tag{4.23}$$

where $\rho_0 = -l\cos\theta + \sqrt{1 - l^2 \sin^2\theta}$ and $u = (8\pi/\lambda_0)z\sin^2(\alpha/2)$.

The normalised SHG axial response and the half width at half maximum (HWHM, $\Delta u_{1/2}$) as a function of the normalised fibre spot size parameters A_j are shown in Fig. 4.19. It shows that the SHG axial resolution approaches 5.57 for $A_1 = 0$ and $A_2 \to \infty$. In this case, Eq. 4.23 reduces to $I(u) = [\sin(u/4)(u/4)]^2$ and is depicted as a dashed curve in the inset of Fig. 4.19, which confirms that SHG microscopy exhibits an inherent optical section property without necessarily using finite-sized detection. This critical feature also implies that SHG microscopy has an improvement in axial resolution by 35% compared with two-photon fluorescence microscopy without any pinhole.

In the case of SHG microscopy using a fibre coupler [55], the HWHM is approximately 4.72 for $4A_1 = A_2 = 0$. The HWHM as a function of A_1 exhibits a linear dependence when $A_1 > 5$. Under the experimental conditions of $A_1 = 2.0$, 4.2, 6.4, 7.3 and 8.4, the measured values of $\Delta u_{1/2}$ are shown as square spots in Fig. 4.19. $\Delta u_{1/2}$ is derived from the axial response of the fibre-optic SHG microscope to a thin layer of AF-50 dye. Parameter A_1 is varied by using various values of numerical apertures of the objective to couple the SHG signal into the single-mode fibre coupler. It is shown that experimental results further confirm the dependence of the axial resolution on the normalised fibre spot size parameters. The deviations between the theory and the experimental data might be due to the presence of spherical aberration and the finite thickness of the SHG layer. Compared with the HWHM in fibre-optical two-photon fluorescence microscopy (Section 4.2.5), the axial resolution in fibre-optical SHG microscopy is increased by approximately 7%.

4.3.4 Polarisation anisotropy

The coherent process of SHG microscopy enables the polarisation dependence of harmonic light that provides information about molecular organisation and nonlinear susceptibilities not available from fluorescence light with random phase to be exploited [56]. In contrast to conventional polarisation microscopy examining the linear birefringence of samples, SHG microscopy can obtain the absolute orientation of molecules by use of arbitrary combinations of fundamental and harmonic polarisation states. Due to the polarisation anisotropy nature, SHG microscopy has been combined with multi-photon fluorescence microscopy and optical coherent tomography [50, 52] to enhance the imaging contrast for morphology identification in biological tissue. Historically, SHG polarisation anisotropy has been studied with bulk optical systems that permit precise control of polarisation states of light. The difficulty in achieving SHG polarisation anisotropy through fibre-optic microscope systems is maintaining linear polarisation of both ultrashort pulses and SHG signals over the wide wavelength range. To this point, the practical imaging capabilities and theory of SHG microscopy using a single-mode fibre coupler have been discussed at length. Now let us consider the polarisation characteristics of fibre-optic SHG microscopy using a single-mode fibre coupler, particularly with regard to the performance in polarisation anisotropy measurements; exemplified with KTP microcrystals, fish scale and rat tail tendon.

A schematic diagram of a typical experimental setup suitable for measuring the polarisation anisotropy of a fibre-optic SHG microscope is similar to Fig. 4.12 [56]. It is important to ensure that the objective used in the system has a low NA in order to ensure that it has no depolarisation effect on the polarised light. The fibres must be placed in natural state, i.e. arranged to avoid strain and stress because the experimental results may be sensitive with respect to fibre bending and stress.

To investigate the evolution of input polarisation states, an arbitrary linear polarisation direction is set by rotating a $\lambda/4$ plate and a Glan Thompson polariser placed in the excitation arm (i.e. before coupling into port 1) of the instrument [56]. For a given incident polarisation angle θ_i at the input port of fibres, determine the maximum (I_{max})

and minimum (I_{min}) intensity of the output beam through an analyser (a second Glan Thompson polariser placed in the signal arm) and calculate the degree of polarisation as

$$\gamma = \frac{I_{\text{max}} - I_{\text{min}}}{I_{\text{max}} + I_{\text{min}}}. \tag{4.24}$$

The SHG polarisation anisotropy is measured by obtaining images through rotations of the analyser before the PMT while maintaining the excitation polarisation after the fibre. In experiments, the initial rotation angle of the analyser, corresponding to the maximum SHG intensity, is parallel to the excitation polarisation. A 400/9 nm bandpass filter is placed before the PMT to ensure that only the SHG signal is detected and commonly the sample is scanned two-dimensionally by a scanning stage.

To determine the practical ability of a fibre-optic SHG microscope to undertake polarisation anisotropy measurements, it is important to understand the polarisation characteristics of the single-mode fibre coupler under various illumination conditions. Prior results show that the linear polarisation states of pulsed and continuous laser beams over a range from near infrared to visible wavelengths can be maintained in the conventional single-mode fibre coupler due to the birefringence effect [59]. Furthermore, polarisation preservations appear at an angular interval of approximately $90°$ of the incident polarisation angle with respect to the transverse axes of the fibre. This implies that the single-mode fibre coupler enables the delivery of a linearly polarised excitation beam and the propagation of the SHG signal. To apply this knowledge to fibre-optic SHG microscopy imaging, a standard nonlinear optical crystal, KTP microcrystals are used as a sample to give well-polarised SHG emission under the linear excitation polarisation.

It is useful to first quantitatively analyse the polarisation anisotropy of the KTP microcrystals in a commercial nonlinear laser scanning microscope so that the "gold standard" result can then be compared with the results obtained using the SMF coupler-based SHG microscope. The data presented here were obtained using an Olympus, Fluoview 300 in epi-detection. When the laser excitation polarisation is fixed, SHG signals are expected to have parallel polarisation with the laser and therefore should yield a $\cos^2 \theta$ pattern by rotating the analyser before the PMT. Successive SHG images of the microcrystals are recorded from both microscope systems when the laser with linear polarisation is delivered and the analyser is rotated by $180°$ in steps of $10°$. Figures 4.20(a), (b) and 4.20(c), (d) show SHG images of the KTP microcrystals at orthogonal polarisation orientations of the analyser obtained from the standard laser scanning microscope and the SMF coupler-based microscope, respectively. In both cases, the extracted SHG intensity as a function of the analyser rotation angle is well consistent with the prediction based on a $\cos^2 \theta$ pattern, which can be observed from Fig. 4.20(e). This demonstrates that the SHG microscope using an SMF coupler exhibits the same manner of SHG polarisation anisotropy compared with that in the conventional SHG microscope with bulk optics. The deviation of the experimental data from the theoretical expectation may arise from the depolarisation effects of galvanometric mirrors and the imaging objective.

The capability of the system for the polarisation anisotropy measurement is ultimately exemplified by the SHG signals obtained from a tetra fish scale (Figs. 4.21(a), (b)).

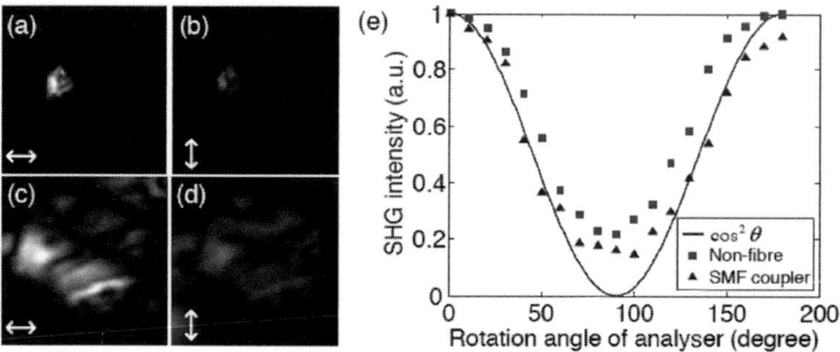

Fig. 4.20 SHG polarisation anisotropy measurement with the KTP microcrystals. SHG images are obtained with orthogonal polarisation orientations of the analyser in a standard laser scanning microscope (a), (b) and an SHG microscope using a single-mode fibre coupler (c), (d), respectively. (e) Dependence of the SHG intensity on the rotation angle of the analyser in a laser scanning (non-fibre) microscope and a single-mode fibre coupler-based microscope, where the results fit a $\cos^2 \theta$ function. Each image has a dimension of 30 μm × 30 μm. Reprinted with permission from Ref. [56], L. Fu and M. Gu, *Opt. Express* **16**, 5000 (2008). © 2008, Optical Society of America.

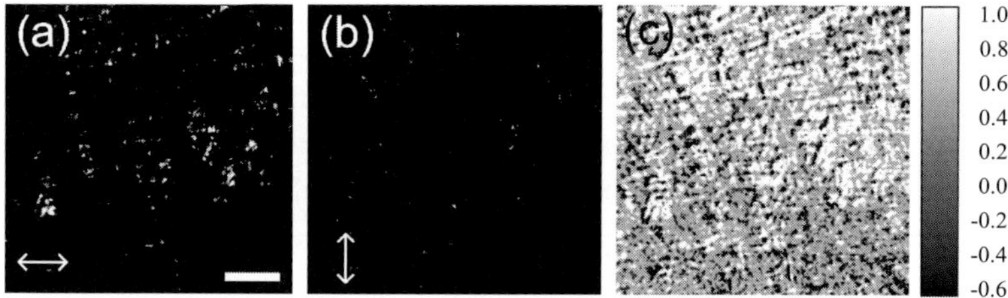

Fig. 4.21 SHG polarisation anisotropy measurement with a fish scale in an SHG microscope using a single-mode fibre coupler. (a), (b) SHG images obtained with orthogonal polarisation orientations of analyser. Scale bar is 20 μm. (c) Image of the anisotropy parameter derived from (a) and (b). Reprinted with permission from Ref. [56], L. Fu and M. Gu, *Opt. Express* **16**, 5000 (2008). © 2008, Optical Society of America.

The fish scale consists of an abundance of well-structured collagen fibrils, which are corresponded to SHG signals. The observation from Figs. 4.21(a) and (b) reveals that the collagen fibrils in fish scale are highly anisotropic. In particular, molecular orientation in fish scale can be quantified by the anisotropy parameter

$$\beta = \frac{I_{\max} - I_{\min}}{I_{\max} + 2I_{\min}},$$
(4.25)

where $I_{\max}$ and $I_{\min}$ are the SHG intensity with the polarisation parallel and perpendicular to the incident polarisation (Fig. 4.21(c)). Measured β values in most areas in Fig. 4.21(c) are greater than 0.7, which is consistent with the prior results obtained from a standard SHG microscope and indicates a good alignment of collagen fibrils relative to the incident polarisation.

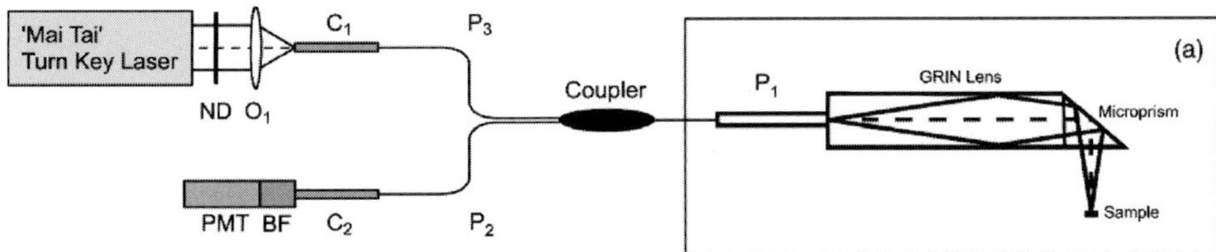

Fig. 4.22 Schematic diagram of an experimental two-photon fluorescence endoscope using a single-mode optical fibre coupler, a microprism and a GRIN rod lens. O_1: 0.25 NA 10× microscope objectives, ND: neutral density filter, BF: band pass filter, C_1, C_2: fibre chucks, P_1, P_2, P_3: coupler ports. Inset (a) shows the detailed component structure of the microoptic scanning head. Reprinted with permission from Ref. [60], D. Bird and M. Gu, *Opt. Lett.* **28**, 1552 (2003). © 2003, Optical Society of America.

4.4 Towards nonlinear endoscopic imaging

While the nonlinear fibre-based systems presented in the preceding sections are a vast improvement over conventional benchtop apparatus, application of the instruments to clinical endoscopy is not entirely feasible, primarily due to the large number of bulk components at the sample site including a conventional imaging objective, optics mounts and a scanning stage. A key technology that is necessary for the application of nonlinear fibre-optic microscopy to imaging of internal organ systems is an endoscope that is capable of delivering, focusing and scanning an ultrashort pulse laser beam for two-photon excitation or SHG. Further to this requirement, the endoscope must be capable of collecting the emitted fluorescence or harmonic signal with suitable efficiency. To meet these requirements one can exploit the feasibility of using gradient index (GRIN) rod lenses and microoptic components in order to miniaturise the imaging optics and realise a practical endoscopic tool.

This section provides a detailed description of the construction of an all-fibre two-photon fluorescence endoscope using a microscanning head and discusses the dependence of imaging resolution on the illumination power.

An experimental arrangement of a two-photon fluorescence microendoscope is shown in Fig. 4.22, and is in every respect identical to that described in Section 4.2.2, Fig. 4.3 except for several modifications at the delivery arm of the coupler (port 1) [60]. These adaptations are necessary in order to facilitate entry of the endoscope into internal channels of the body. The imaging optics are combined to form a single microscanning head, which is similar to that used in optical coherence tomography and is depicted in inset (a) of Fig. 4.22. It consists of a 1.0 mm diameter Plano-Plano, 0.25 pitch, 0.46 NA GRIN lens designed for a wavelength of 830 nm used for focusing incident illumination and a 1.0 mm BK7 right angle microprism that is used as a beam directing element. The pitch of the lens is chosen to yield the required Gaussian beam parameters. The single-mode optical fibre that was port 1 of the fibre coupler is now attached to the microscanning head (comprising the microprism and the GRIN rod lens) with an ultraviolet curing optical adhesive to form a single unit, which has an approximate working distance of 2.0 mm.

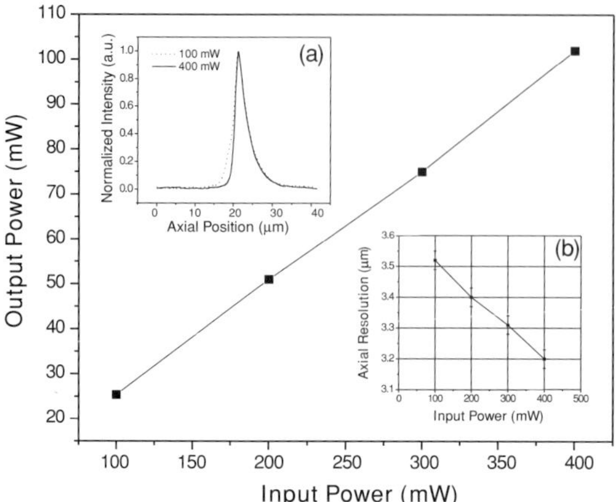

Fig. 4.23 Coupling efficiency of the two-photon fibre-optic microendoscope for the output power measured at the focus of the scan head and the input power measured at port 3 of the fibre coupler. Inset (a) shows the axial responses to a fluorescent polymer block for an input power of 100 mW (dashed curve) and 400 mW (solid curve). Inset (b) shows the dependence of the FWHM of the axial response as a function of the excitation power. Reprinted with permission from Ref. [60], D. Bird and M. Gu, *Opt. Lett.* **28**, 1552 (2003). © 2003, Optical Society of America.

In this embodiment, image acquisition is achieved by circumferential scanning a sample structure into which the microendoscope head can be inserted. The emitted fluorescence from the sample is collected by the micro-optic scanning head and is returned via the same optical path used for pulse delivery. The collected signal is delivered via port 2 of the fibre coupler to a photomultiplier tube masked with a band-pass filter, BF, with an operating wavelength of 550 nm (20 nm).

To determine the magnitude of illumination power that can be delivered to a sample through the fibre coupler and the microendoscope, it is necessary to measure the coupling efficiency for illumination of wavelength 800 nm. Typical results are shown in Fig. 4.23, which depicts the measured output power at the focus as a function of input power to port 3 of the coupler. The coupling efficiency of the endoscope is reduced to approximately 28% compared with 35% obtained with the system using bulk imaging components (Section 4.2.1, Fig. 4.4). This decrease is due to numerous factors including the slight mismatch in the refractive indices of the fibre core, the GRIN rod lens, the microprism and the optical adhesive. In the return path, the signal level of the instrument would be increased through dichromatic coating of the microprism surfaces. Further, it can be seen from Fig. 4.22 that a portion of the incident beam is transmitted through the microprism along the principal axis and lost. This portion was measured to be approximately 8% of the input power to port 3 of the fibre coupler.

One way to determine the ability of an imaging system of this type to discriminate between axial depths is to measure the axial response to a fluorescent polymer sample. Typical measurements of the fluorescence intensity as a function of the axial distance perpendicular to the endoscope scanned through the micro-optic focus are shown in inset (a) of Fig. 4.23. The full width at half maximum of the axial response (solid curve in inset (a) of Fig. 4.23) measured with the new two-photon fluorescence endoscope, z, is approximately 3.2 µm for an input power of 400 mW to port 3, which is increased

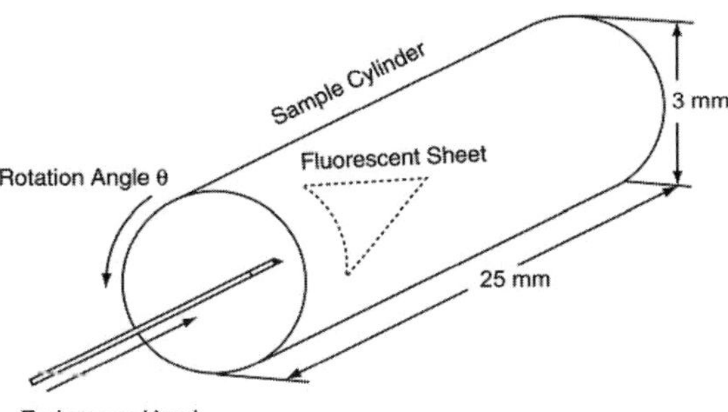

Fig. 4.24 Schematic diagram of the cylindrical sample holder. Reprinted with permission from Ref. [60], D. Bird and M. Gu, *Opt. Lett.* **28**, 1552 (2003). © 2003, Optical Society of America.

by approximately 37% compared with that measured by the fibre-coupler-based bulk optic microscope described in Section 4.2.1. The increase in the FWHM indicates a degradation of axial resolution, which is attributed to two main factors. First, in this case the pinhole effect provided by the fibre tip aperture of port 1 is affected by the GRIN rod lens and microprism assembly. Even a small fabrication defect and mismatch between the refractive indices will affect the alignment of the device. Second, the NA of the GRIN rod lens on the imaging side is effectively reduced. In inset (b) of Fig. 4.23, the dependence of the FWHM z on the incident power to port 3 is depicted, showing that the resolution is improved as the input power is increased. This feature is caused by the fact that the spectrum of the pulsed beam is broadened and blue-shifted due to self-phase modulation and self-steepening, respectively (See Fig. 4.7).

The feasibility of using the microendoscope head for imaging an internal structure can be demonstrated by simulated rotation of the distal optics inside an internal body cavity by rotating a hollowed cylindrical sample holder around a fixed microendoscope head. The sample holder as shown in Fig. 4.24 was machined from a 25 mm length of Perspex to form a hollowed cylindrical rod of 3 mm inner diameter and mounted to the shaft of a variable speed 12V DC rotary motor. The speed of rotation could be accurately controlled by the potential supplied to the motor. A two-dimensional image was constructed by progressively increasing the insertion depth of the microendoscope head into the cylinder with a step size of 0.5 mm and performing a single rotational scan. The sample consisted of a fluorescent paper sheet cut into the shape of a 5.0 mm equilateral triangle.

The sheet can be prepared by uniformly coating a piece of $75g\ m^{-2}$ paper with a mixture of AF-50 fluorescent dye and isopropyl alcohol. Double-sided adhesive tape is then used to secure the sample inside the cylinder oriented such that the base of the triangle is parallel with the circumference of the cylinder. Figure 4.25 shows five representative acquisitions into the cylinder. The first series is obtained by rotating the cylinder with the focus of the microendoscope positioned 1.0 mm along the fluorescent sample. The trigonometric shape of the top constructed by the five curves reveals that the region of the detected fluorescence intensity as a function of the angular displacement of the cylindrical sample holder is consistent with the actual shape of the fluorescent material. It clearly shows that as the microendoscope head is inserted further into the cavity, the rotation angle range over which fluorescence is detected decreases accordingly.

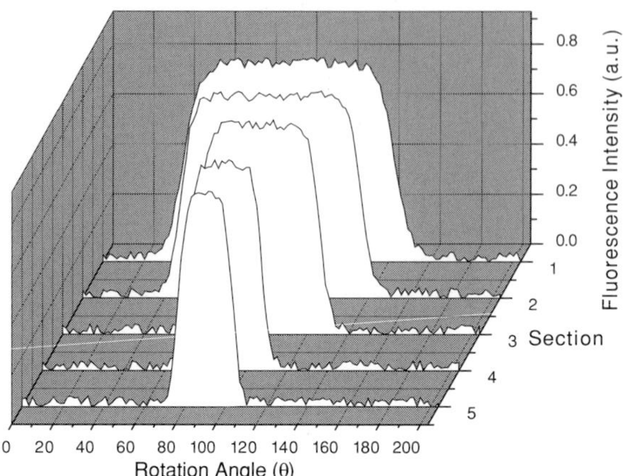

Fig. 4.25 A series of transverse cross-sectional images taken at 0.5 mm steps into a cylindrical channel having a 5 mm equilateral triangle fluorescent sheet attached to the internal surface. The initial series (No. 1) was acquired 1.0 mm into the cylinder relative to the base of the fluorescent triangle. Reprinted with permission from Ref. [60], D. Bird and M. Gu, *Opt. Lett.* **28**, 1552 (2003). © 2003, Optical Society of America.

It should be pointed out that for practical clinical implementation of the device it is clearly not feasible to rotate the sample around the endoscope. One method that may address this issue is close contact coupling of two optical fibres using a fibre ferrule. In this manner it is possible for the scanning head at the distal end of the endoscope to be rotated independently of the fixed fibre coupler. Further, a high coupling efficiency of the incident illumination across the small air gap at the interface of the two fibres can be maintained. For precision placement of the scanning head at internal lesion sites, the fibre may be incorporated into an existing endoscope. Such an instrument usually consists of a white light source for visual aid and other apparatus, including magnification optics for targeting a lesion site within the viewing area.

4.5   Summary

This chapter has introduced the topic of nonlinear fibre-optical microscopy and presents two new novel technologies: fibre based two-photon and SHG microscopy. The feasibility of compact, low-cost systems that use a single-mode optical fibre coupler have been discussed. The coupler is unique in that it acts as a low-pass filter for an infrared ultrashort pulse laser beam, so that both the delivery of the pulsed beam and the collection of two-photon fluorescence or SHG signal can be optimised. The aperture of the fibre acts as an effective confocal pinhole, which leads to self-alignment, reduction of multiple scattering and an enhanced optical sectioning effect for high-resolution three-dimensional imaging. Compared with a conventional two-photon system employing a large area detector without pinhole mask, the resolution improvement is as high as 30% and even higher in an SHG mode. These instruments may make two-photon fluorescence and SHG endoscopy possible for point-of-care *in vivo* medical diagnostic applications.

References

[1] M. Minsky. Memoir on inventing the confocal scanning microscope. *Scanning*, **10**:128–138, 1988.

[2] C. J. R. Sheppard. Scanning optical microscopy. In *Advances in Optical and Electron Microscopy*, volume **10**, pages 1–98. London, Academic Press, 1987.

[3] A. F. Slomba, D. E. Wasserman, G. I. Kaufman and J. F. Nester. A laser flying spot scanner for use in automated fluorescence antibody instrumentation. *J. Assoc. Adv. Med. Instrum.*, **6**:230–234, 1972.

[4] P. M. Delaney, M. R. Harris and R. G. King. Fiber-optic laser scanning confocal microscope suitable for fluorescence imaging. *Appl. Opt.*, **33**:573–577, 1994.

[5] S. Kimura and T. Wilson. Confocal scanning optical microscope using single-mode fiber for signal detection. *Appl. Opt.*, **30**:2143–2150, 1991.

[6] J. Benschop and G. Von Rosmalen. Confocal compact scanning optical microscope based on compact disc technology. *Appl. Opt.*, **30**:1179–1184, 1991.

[7] T. Dabbs and M. Glass. Single-mode fibers used as confocal microscope pinholes. *Appl. Opt.*, **31**:705–706, 1992.

[8] T. Dabbs and M. Glass. Fiber-optic confocal microscope: FOCON. *Appl. Opt.*, **31**:3030–3035, 1992.

[9] Min Gu. *Principles of Three-Dimensional Imaging in Confocal Microscopes*. Singapore, World Scientific, 1996.

[10] M. Gu, C. J. R. Sheppard and X. Gan. Image formation in a fiber-optical confocal scanning microscope. *J. Opt. Soc. Am. A*, **8**:1755–1761, 1991.

[11] M. Gu, X. Gan and C. J. R. Sheppard. Three-dimensional coherent transfer functions in fiber-optical confocal scanning microscopes. *J. Opt. Soc. Am. A*, **8**:1019–1025, 1991.

[12] M. Gu and C. J. R. Sheppard. Experimental investigation of fibre-optical confocal scanning microscopy: including a comparison with pinhole detection. *Micron*, **24**:557–565, 1993.

[13] A. W. Snyder and J. D. Love. *Optical Waveguide Theory*. London, Chapman and Hall, 1983.

[14] M. Gu and C. J. R. Sheppard. Fibre-optical confocal scanning interference microscopy. *Opt. Commun.*, **100**:79–86, 1993.

[15] D. K. Hamilton and C. J. R. Sheppard. A confocal interference microscope. *Opt. Acta*, **29**:1573–1577, 1982.

[16] D. K. Hamilton and T. Wilson. Surface profile measurement using the confocal microscope. *J. App. Phys.*, **53**:5320–5322, 1982.

[17] H. J. Matthews, D. K. Hamilton and C. J. R. Sheppard. Aberration measurement by confocal interferometry. *J. Mod. Opt.*, **36**:233–250, 1989.

[18] Y. Fujii and Y. Yamazaki. A fibre-optic 3-D microscope with high depth sensitivity. *J. Microscopy*, **158**:145–151, 1989.

[19] M. Gu, H. Zhou and C. J. R. Sheppard. Fibre-optical confocal scanning interferometry. In *19th ACOFT Proceedings (IREE Society)*, p. 318–321. Sydney, 1994.

[20] H. Zhou, C. J. R. Sheppard and M. Gu. A compact confocal interference microscope based on a four-port single-mode fibre coupler. *Optik*, **103**:45–48, 1996.

[21] M. Gu and C. J. R. Sheppard. Signal level of the fibre-optical confocal scanning microscope. *J. Mod. Opt.*, **38**:1621–1630, 1991.

[22] C. J. R. Sheppard and T. Wilson. Multiple traversing of the object in the scanning microscope. *Opt. Acta*, **27**:611–624, 1980.

[23] M. Gu and D. K. Bird. Fibre-optical double-pass confocal microscopy. *Optics and Laser Technology*, **30**:91–93, 1998.

[24] M. V. Iravani. Fibre-optic scanning differential interference contrast optical microscope. *Electron. Lett.*, **22**:103–105, 1986.

[25] R. Juskaitis and T. Wilson. Differential confocal scanning microscope with a two-mode optical fiber. *Appl. Opt.*, **31**:898–903, 1992.

[26] D. Bird and M. Gu. Resolution improvement in two-photon fluorescence microscopy using a single-mode fiber. *Appl. Opt.*, **41**:1852–1857, 2002.

[27] R. Wolleschensky, T. Feurer, R. Sauerbrey and U. Simon. Characterisation and optimization of a laser-scanning microscope in the femtosecond regime. *Appl. Phys. B*, **67**:87–94, 1998.

[28] D. Bird. *Fiber-Optic Two-Photon Fluorescence Microscopy*. Ph.D. thesis, Faculty of Engineering and Industrial Research, Swinburne University of Technology, Australia, 2002.

[29] W. Denk, J. H. Strickler and W. W. Webb. Two-photon laser scanning fluorescence microscopy. *Science*, **248**:73–75, 1990.

[30] D. Bird and M. Gu. Compact two-photon fluorescence microscope using a single-mode optical fiber coupler. *Opt. Lett.*, **27**:1031–1033, 2002.

[31] D. Bird and M. Gu. Fibre-optic two-photon scanning fluorescence microscopy. *J. Micros.*, **208**:35–48, 2002.

[32] G. P. Agrawal. *Nonlinear Fiber Optics*. California, Academic Press, 1989.

[33] D. Day and M. Gu. Effects of refractive-index mismatch on three-dimensional optical data storage density in a two-photon bleaching polymer. *Appl. Opt.*, **37**:6299–6304, 1998.

[34] C. J. R. Sheppard. The spatial frequency cut-off in three-dimensional imaging. *Optik*, **72**:131–133, 1986.

[35] C. J. R. Sheppard. The spatial frequency cut-off in three-dimensional imaging ii. *Optik*, **74**:128–129, 1986.

[36] S. Kimura and C. Munakata. Calculation of three-dimensional optical transfer function for a confocal scanning fluorescent microscope. *J. Opt. Soc. Am. A*, **6**:1015–1019, 1989.

[37] S. Kimura and C. Munakata. Three-dimensional optical transfer function for the fluorescent scanning optical microscope with a slit. *Appl. Opt.*, **29**:1000–1007, 1990.

[38] S. Kimura and C. Munakata. Dependence of 3-D optical transfer functions on the pinhole radius in a fluorescent confocal optical microscope. *Appl. Opt.*, **29**:3007–3011, 1990.

[39] O. Nakamura and S. Kawata. Three-dimensional transfer function analysis of the tomographic capability of a confocal fluorescent microscope. *J. Opt. Soc. Am. A*, **7**:522–526, 1990.

[40] S. Kawata, R. Arimoto and O. Nakamura. Three-dimensional optical-transfer function analysis for laser-scan fluorescent microscope with an extended detector. *J. Opt. Soc. Am. A*, **8**:171–175, 1991.

[41] S. Kawata and R. Arimoto. Laser-scan fluorescent microscope with annular excitation. *Optik*, **86**:7–10, 1992.

[42] M. Gu and C. J. R. Sheppard. Confocal fluorescent microscopy with a finite-sized circular detector. *J. Opt. Soc. Am. A*, **9**:151–153, 1992.

[43] M. Gu and C. J. R. Sheppard. Three-dimensional imaging in confocal fluorescent microscopy with annular lenses. *J. Mod. Opt.*, **38**:2247–2763, 1991.

[44] M. Gu and C. J. R. Sheppard. Three-dimensional optical transfer function in a fiber-optical confocal fluorescent microscope using annular lenses. *J. Opt. Soc. Am. A*, **9**:1991–1999, 1992.

[45] B. R. Frieden. Optical transfer of a three-dimensional object. *J. Opt. Soc. Am.*, **57**:56–66, 1967.

[46] M. Gu and D. Bird. Three-dimensional optical-transfer-function analysis of fibre-optical two-photon fluorescence microscopy. *J. Opt. Soc. Am. A*, **20**:941–947, 2003.

[47] I. S. Gradstein and I. M. Ryshik. *Tables of Series, Products and Integrals*. Frankfurt, Verlag Harri Deutsch, 1981.

[48] D. Huang, E. A. Swanson, C. P. Lin *et al.* Optical coherence tomography. *Science*, **254**:1178–1181, 1991.

[49] G. J. Tearney, M. E. Brezinski, B. E. Bouma *et al. In vivo* endoscopic optical biopsy with optical coherence tomography. *Science*, **276**:2037–2039, 1997.

[50] E. Beaurepaire, L. Moreaux, F.Amblard and J. Mertz. Combined scanning optical coherence and two-photon-excited fluorescence microscopy. *Opt. Lett.*, **24**:969–971, 1999.

[51] Y. Guo, P. P. Ho, H. Savage *et al.* Second-harmonic tomography of tissue. *Opt. Lett.*, **22**:1323–1325, 1997.

[52] W. E. Zipfel, R. M. Williams, R. Christie, A. Y. Nikitin, B. T. Hyman and W. W. Webb. Live tissue intrinsic emission microscopy using multiphoton-excited native fluorescence and second harmonic generation. *Proc. Natl. Acad. Sci. USA*, **100**:7075–7080, 2003.

[53] P. J. Campagnola and L. M. Loew. Second harmonic imaging microscopy for visualizing biomolecular arrays in cells, tissues and organisms. *Nat. Biotech.*, **21**:1356–1360, 2003.

[54] L. Fu. *Fibre-Optical Nonlinear Endoscopy*. Ph.D. thesis, Faculty of Engineering and Industrial Research, Swinburne University of Technology, Australia, 2006.

[55] L. Fu, X. Gan and M. Gu. Use of a single-mode fiber coupler for second-harmonic-generation microscopy. *Opt. Lett.*, **30**:385–387, 2005.

[56] L. Fu and M. Gu. Polarization anisotropy in fiber-optic second harmonic generation microscopy. *Opt. Express*, **16**:5000–5006, 2008.

[57] M. Gu and L. Fu. Three-dimensional image formation in fiber-optical second harmonic generation microscopy. *Opt. Express*, **14**:1175–1181, 2006.

[58] P. J. Campagnola, M. Wei, A. Lewis and L. M. Loew. High-resolution nonlinear optical imaging of live cells by second harmonic generation. *Biophys. J.*, **77**:3341–3349, 1999.

[59] L. Fu, X. Gan, D. Bird and M. Gu. Polarisation characteristics of a 1 × 2 fiber coupler under femtosecond-pulsed and continuous wave illumination. *Opt. Laser Tech.*, **37**:494–497, 2005.

[60] D. Bird and M. Gu. Two-photon fluorescence endoscopy with a micro-optic scanning head. *Opt. Lett.*, **28**:1552–1554, 2003.

5   Nonlinear optical endoscopy

Ever since researchers realised that microscopy based on nonlinear optical effects can provide information that is blind to conventional linear techniques, applying nonlinear optical imaging to *in vivo* medical diagnosis in humans has been the ultimate goal [1,2]. The development of nonlinear optical endoscopy that permits imaging under conditions in which a conventional nonlinear optical microscope cannot be used is the primary method to extend applications of nonlinear optical microscopy toward this goal. Fibre-optic approaches that allow for remote delivery and collection in a minimally invasive manner are normally used in nonlinear optical endoscopy. In Chapter 4, a compact nonlinear optical microscope based on a single-mode fibre (SMF) coupler to replace complicated bulk optics was described.

There are several key challenges involved in the pursuit of *in vivo* nonlinear optical endoscopy. First, an excitation laser beam with an ultrashort pulse width should be delivered efficiently to a remote place where efficient collection of faint nonlinear optical signals from biological samples is required. Second, laser-scanning mechanisms adopted in such a miniaturised instrumentation should permit size reduction to a millimetre scale and enable fast scanning rates for monitoring biological processes. Finally, the design of a nonlinear optical endoscope based on micro-optics must maintain great flexibility and compact size to be incorporated into endoscopes to image internal organs.

This chapter summarises major technologies in fibre-optic nonlinear optical microscopy (Section 5.1), followed by the design and characterisation of a nonlinear optical endoscope that implements a double-clad photonic crystal fibre (PCF) in Section 5.2, a microelectromechanical system (MEMS) mirror and a gradient index (GRIN) lens is given in Section 5.3. The three-dimensional (3D) high-resolution imaging from *in vitro* internal organ and cancer tissues demonstrates the feasibility of nonlinear optical endoscopy that aims for *in vivo* 3D imaging in deep tissue. In Section 5.4 a double-clad fibre coupler is applied into a nonlinear optical microscope.

5.1   An introduction to nonlinear optical endoscopy

5.1.1   Optical fibres and ultrashort pulse delivery

Optical fibres have been extensively used in imaging and sensing systems due to their mechanical flexibility and compact size [3,4]. Single-mode fibres are the most common

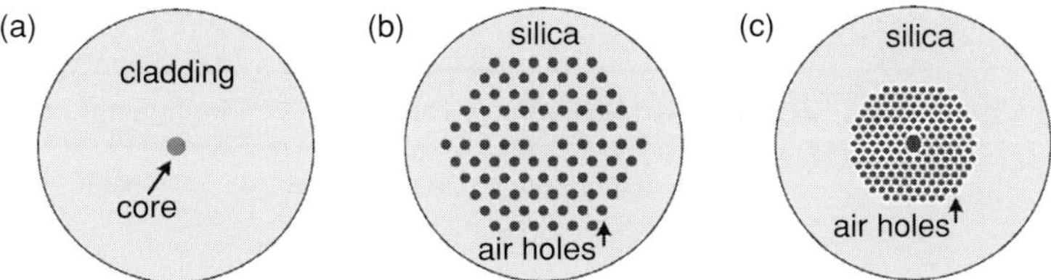

Fig. 5.1 Schematic diagram of conventional fibres and PCFs. (a) A standard fibre consisting of two bulk materials. (b) A high-index guiding PCF with a solid silica core. (c) A hollow-core PBF guiding light with photonic bandgaps. Reprinted with permission from Ref. [12], L. Fu and M. Gu, *J. Micros.* **226**, 195 (2007). © 2007, John Wiley & Sons.

type of fibres in fibre-optic imaging systems. However, a key problem with using an SMF is its limited ability for efficient excitation delivery and signal collection. On the one hand, ultrashort pulses ($\sim$80 – 200 fs in duration) through an SMF experience severe temporal and spectral broadenings due to group-velocity dispersion (GVD), self-phase modulation (SPM), and self-steepening [5], leading to significant reductions in the efficiency of nonlinear excitation and penetration depth [6, 7]. On the other hand, a low numerical aperture (NA $\sim$ 0.1) and a small core size ($\sim$5 µm) of an SMF make the fibre sensitive to optical aberrations in imaging systems and therefore limit the collection efficiency of the nonlinear emission at sample sites. Furthermore, the small fibre core acting as a confocal pinhole blocks multiply scattered photons outside the focal volume in biological tissue, resulting in fewer photons contributing to the image formation. An alternative fibre is a multi-mode fibre that is superior for signal collection as a result of its relatively greater NA and large core size. Unfortunately, multiple spatial modes from multi-mode fibres cannot be focused to a near diffraction-limited spot to produce efficient nonlinear excitation for high optical resolution. For these reasons, the use of appropriate fibres is the heart of the endoscope design. Photonic crystal fibres (PCFs), which have revolutionised fibre optics over the last decade, can overcome the limitations of a conventional fibre [8, 9].

Unlike the conventional fibre consisting of two bulk materials (Fig. 5.1(a)), PCFs using photonic crystals as a new physical mechanism to guide light have broken limitations of the total internal reflection (TIR) principle in conventional fibres. In 1996, Russell *et al.* realised the first microstructured silica fibre with a periodic array of several hundred air holes running down their length, and founded the field of PCFs [9]. Photonic crystal fibres may be divided into two classes, high-index core fibres (Fig. 5.1(b)) and photonic bandgap fibres (PBFs) (Fig. 5.1(c)). In a high-index guiding fibre, a two-dimensional (2D) photonic crystal having a lower effective refractive index than a core material can be used as a fibre cladding. As a result, modes can be guided in the solid silica core surrounded by silica–air photonic crystal cladding in a form of modified TIR (Fig. 5.1(b)). Alternatively, a hollow-core PBF enables the mode propagation in vacuum with the photonic bandgap material as a cladding layer (Fig. 5.1(c)), because the photonic bandgaps of the photonic crystal support no propagation modes. Due to the 2D photonic crystals in the fibre, PCFs can control the light in the ways that are not possible or even imaginable for standard optical fibres, such as endless single-mode operation, dispersion

Table 5.1 PCFs related to nonlinear optical microscopy.

Features	LMA PCFs	Hollow-core PBFs	Double-clad PCFs	Highly nonlinear PCFs
Operation wavelength	Wide range	Near zero-dispersion wavelength	Wide range	Near zero-dispersion
Advantages	Reduced SPM effect for ultrashort pulse delivery; endless single mode over wide wavelength range	Low loss; high power threshold for nonlinear effects; no prechirping for high energy pulse delivery	Reduced SPM effect in the core; high NA in the inner cladding; dual function for pulse delivery and collection	Extremely high non-linear coefficient; ideal media to generate supercontinuum for multi-spectral imaging
Limitations	Low NA for signal collection; dispersion compensation is required	Narrow operating wavelength window; low NA for signal collection	Dispersion compensation is required	Broadened pulse durations; inefficient collection with small core

engineering, and supercontinuum generation [10, 11]. To date, several types of PCF have been applied to or have shown great potential for nonlinear optical microscopy (Table 5.1).

Since the efficiency of self-phase modulation in fibres is strongly dependent on the peak power of pulses and the nonlinearity coefficient that is inversely proportional to the effective core area, one way to reduce nonlinearity is to use a large-core fibre to minimise peak intensities over a large area [5, 13]. In contrast with SMFs, the accurate control of hole size and distribution in PCFs can tailor the exact effective refractive index of the cladding and result in the design with a larger mode area or with endless single-mode operation [9]. A large-mode-area (LMA) PCF has a large core size (up to 35 μm in diameter) with the single-mode guidance for any wavelength at which silica is transparent, and therefore significantly reduces nonlinear effects for ultrashort pulse delivery (Table 5.1). An investigation into femtosecond pulse propagation through LMA PCFs has shown that pulse durations after the LMA PCFs (15 μm and 25 μm in diameter) are less broadened compared with conventional optical fibres [14]. The delivery of 3 nJ pulses as short as 140 fs at wavelength 800 nm is achievable over 1.3 m of the LMA PCF. Although LMA PCFs have not been applied to any imaging modality, the near-diffraction-limited output beam with a femtosecond timescale from the fibre makes it very suitable for the excitation of nonlinear optical processes. However, the low NA of these types of fibre will give rise to a limited collection efficiency of nonlinear optical signals.

An ideal alternative fibre for ultrashort pulse delivery is a hollow-core PBF. The air core in PBFs is created by locally breaking the periodicity of a photonic crystal (Table 5.1), supporting modes with frequencies falling inside the photonic crystal bandgap of the surrounding photonic crystals. Thereby, hollow-core PBFs can circumvent material nonlinearity associated with SPM since the light is guided in the air core. Furthermore, by carefully designing the size and distribution of air holes, the dispersion profile of hollow-core PBFs could change from normal to anomalous dispersion in the transmission window [10]. Thus the zero-dispersion wavelength of the hollow-core fibre can be shifted from the conventional 1310 nm (zero-dispersion wavelength for bulk silica) to a shorter wavelength, such as 800 nm or down to a visible wavelength. It has been demonstrated that the hollow-core PBF having a zero-dispersion wavelength near 810 nm enables nearly distortion-free propagation for the pulse energy of $\sim$4.6 nJ without the use of a prechirping unit [15]. To date, a number of two-photon excited fluorescence (TPEF) microscopes and endoscopes using hollow-core PBFs to deliver femtosecond pulses have been reported [16–18]. The operation wavelength range of the fibre, however, is typically several tens of nanometres around the central wavelength, making the propagation of visible nonlinear optical signals impossible. Therefore a hollow-core PBF is ideally suitable for high energy ultrashort pulse delivery, but it could not be used in a single-fibre-based endoscope system to facilitate simultaneous pulse delivery and backward signal collection.

Double-clad PCFs can play a dual role of ultrashort pulse delivery and efficient collection of nonlinear optical signals [19–22]. Double-clad PCFs are actually the manifestation of the concept of double-clad fibres in PCFs, enabling the double-clad structure with pure silica [9]. The LMA core of the fibre is placed in a microstructured inner cladding, offering single-mode guidance for near infrared beam and reducing nonlinearity significantly for excitation. The inner cladding has a high NA up to 0.6 and a diameter of hundreds of micrometres to propagate light in visible and near infrared wavelength ranges with a high efficiency. The high NA of the inner cladding is achieved by separating the inner and outer claddings with a web of silica bridges that are substantially narrower than the wavelength of the guided light (Table 5.1). Therefore, the hybrid of single-mode operation of near infrared light and the multimode guidance in the visible wavelength range in a single fibre so far can only be offered by double-clad PCFs, leading to the possibility of nonlinear optical imaging for internal organs with flexible fibre-optic endoscopes, which will be presented in the following subsections. More importantly, it has been demonstrated that comparing the signal achieved by a standard SMF, the use of double-clad PCFs in a nonlinear optical microscope improves the detection efficiency significantly by two orders of magnitude [20,22]. However, the application of a prechirp unit might be required for dispersion compensation.

As opposed to the PCFs discussed above, highly nonlinear PCFs can exploit the nonlinearity in fibres to alter the pulse spectrum and width for nonlinear optical imaging applications rather than avoid the nonlinear pulse distortion. Highly nonlinear PCFs are today the most commonly used type in various PCFs. They guide light in a very small core (diameters down to 1 μm) surrounded by the cobweb-like microstructure or the air-hole cladding (Table 5.1), exhibiting a high nonlinearity. For example, a typical value of the effective nonlinearity for a standard fibre is 1 $W^{-1}Km^{-1}$, whereas nonlinear PCFs can be designed to have a value at 215 $W^{-1}Km^{-1}$ [9]. In addition to its high nonlinearity, zero-dispersion wavelengths of the fibre can be designed over a wide range in the visible

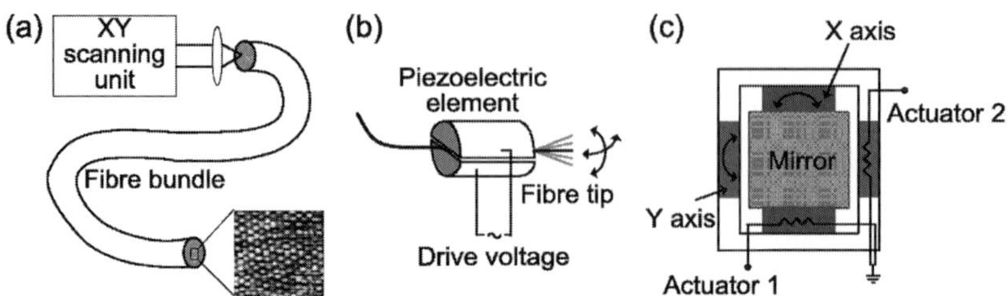

Fig. 5.2 Scanning mechanisms in miniaturised microscopy. (a) Proximal scanning of a fibre bundle. (b) Mechanical resonance of fibre tip with piezoelectric actuators. (c) A MEMS mirror. Reprinted with permission from Ref. [12], L. Fu and M. Gu, *J. Micros.* **226**, 195 (2007). © 2007, John Wiley & Sons.

and near infrared spectra to efficiently generate supercontinuum using low-peak-power pulses and a short length (several centimetres to several tens of centimetres) of the fibre [11]. It has been demonstrated that the supercontinuum generated in PCFs has the bandwidth of sunlight but is 104 times brighter [8]. The broad bandwidth and high spectral brightness of the supercontinuum generation (SCG) based on highly nonlinear PCFs have been applied to a number of imaging modalities, including optical coherence tomography [23], coherent anti-Stokes Raman scattering (CARS) [24], confocal [25] and TPEF microscopy [26–29]. The combination of a highly nonlinear PCF and a grating pair has been used for pulse compression to enhance the two-photon excitation efficiency [26]. The doubled frequency of the broadband spectra after the nonlinear PCFs can also be used as a complementary light source to excite dyes whose two-photon absorption spectra are below the tunable range of a Ti:sapphire laser [27, 28]. In particular, the SCG enables a multi-spectral excitation beam to acquire multicolour TPEF imaging for multi-functional visualisation [29].

5.1.2   Scanning mechanisms

Although optical fibres allow for remote delivery of light to a given spot, they have to be combined with a scanning mechanism to form a 2D image. Scanning mechanisms in miniaturised microscopy can be divided into two categories as proximal scanning and distal scanning depending on the position of the unit relative to laser sources (Fig. 5.2). Proximal scanners do not have to be inserted into a narrow working channel of an endoscope and enable a compact probe geometry. Distal scanners close to the tip of an endoscope probe are usually combined with a single fibre, offering versatile optical designs and high resolution for optical imaging.

Proximal scanning typically involves a pair of galvanometer mirrors and an imaging fibre bundle (Fig. 5.2(a)) [4, 30]. A fibre bundle consists of up to 100,000 individual fibres closely packaged within an overall diameter of less than 3 millimetres and with an NA of approximately 0.3. The laser beam is raster scanned on the proximal end of the bundle and transferred to the distal end before being focused on a specimen. Each fibre within the bundle serves as a point source as well as a detection pinhole for imaging. However, pixilation of the fibre bundle gives rise to a limited lateral resolution dependent on the spacing of adjacent fibres. Furthermore, due to the thin cladding layer between the fibres, the leakages of excitation laser beam or collected signal to adjacent fibres can result in a reduced contrast of imaging. Nonetheless, this effect can be reduced by use of a spatial light modulator to provide sequential illumination on each fibre [31].

For illumination delivery through a single fibre, either the fibre tip or the light coupled from the fibre is scanned by a proximal scanner inside the endoscope probe [17,32,33]. A piezoelectric or an electromagnetic actuator can be excited near the mechanical resonance frequency of the fibre end to generate a 2D scanning pattern, as shown in Fig. 5.2(b). A piezoelectric bending element has been used to drive the motion of the fibre tip in a form of a Lissajous pattern [32]. Alternatively, a spiral scan pattern can be produced by a tubular piezoelectric actuator consisting of two pairs of drive electrodes at a scanning frequency of approximately 1,300 Hz [33]. Furthermore, a concept of the rotational scanning method has been experimentally demonstrated to achieve circumferential information inside internal organs [34].

Microfabricated mirrors based on MEMS technology have rapidly emerged recently. A MEMS mirror can facilitate endoscopic beam scanning because of its small size, low power consumption and excellent microbeam manipulating capability [35–39]. Two-dimensional scanning can be achieved with a single MEMS mirror having a mirror plate size of approximately 0.5–2 mm (Fig. 5.2(c)). MEMS mirrors are typically based on electrostatic actuation or electrothermal actuation, providing angular rotations of up to 30 degrees with low driving voltages [21,40]. MEMS scanners are created through sequential material etching and deposition processes, which are similar to the technology used for integrated circuit fabrication. Although the fabrication of MEMS devices requires complicated processes and expensive facilities, MEMS technology can integrate microcomponents and give enormous benefit for construction of a compact microscope, such as microscopes-on-a-chip. Therefore, a mirror based on MEMS technology is an ideal scanning mechanism not only for nonlinear optical endoscopic imaging but also for the advancement of compact microscope systems in the future. The feasibility of a MEMS mirror in nonlinear optical endoscopy has been demonstrated in this chapter.

In addition to MEMS scanning mirrors, microfabricated components such as chevron beam thermal actuators have been proposed to facilitate scanning of a GRIN lens and a microprism [41]. Furthermore, some novel scanning methods such as multiple laser foci produced by a microlens array enable high-speed multi-focal imaging. However, the signal detection requires a multi-anode photomultiplier tube and the cross talk between pixels usually reduces the imaging contrast [17].

5.1.3   Geometries of fibre-optical nonlinear optical microscopy

Although the technologies summarised above have enormous advantages on their own, the difficulty in creating a fibre-optical nonlinear optical endoscope is the design and integration of these technologies for an optimised nonlinear optical imaging performance. To date, fibre-optical TPEF microscopy is the primary embodiment of the development

Table 5.2 Various applications of fibre-optical nonlinear optical microscopy.

Geometries	Fibres	Scanning	Advantages	Limitations
Portable TPEF microscopy	Mechanical flexibility. Reduced size and weight for freely moving animals			
Head mounted TPEF microscopy	SMF	Distal, Lissajous	First demonstration of miniaturised 2P microscopy	Unstable imaging during sudden movements
Portable TPEF microendoscopy	Hollow-core PBF	Proximal, Lissajous	Very light weight, no pulse distortion and reduced SPM	Design of two fibres hinders miniaturisation
Handheld TPEF microscopy	Hollow-core PBF	Proximal, multi-focal	High-speed multi-focal imaging, no pulse distortion and reduced SPM	Relatively large and heavy among the miniaturised TPEF microscopes
Rigid TPEF endoscopy	Combination of GRIN lenses enables minimally invasive and deep imaging in tissues			
TPEF micro-endoscopy	None	Proximal, raster	Minimal pulse distortion, deep tissue imaging	Rigid GRIN probes, abberations in long GRIN probes
Flexible TPEF endoscopy	Potential applications in imaging of internal organs with great mechanical flexibility			
SMF coupler TPEF and SHG endoscopy	SMF coupler	Distal, rotational	Mechanical flexibility, all fibre design	Dispersion compensation needed, limited signal
Fibre-bundle TPEF endoscopy	Fibre bundle	Proximal, raster	Mechanical flexibility stationary and compact probe	Dispersion compensation needed, limited resolution
Double-clad fibre TPEF endoscopy	Double-clad fibre	Distal, spiral	Mechanical flexibility	Dispersion compensation needed, limited signal
Double-clad fibre PCF TPEF and SHG endoscopy	Double clad PCF	Distal, MEMS	Mechanical flexibility, reduced SPM, improved signal level	Dispersion compensation needed
CARS endoscopy	SMF	None	First demonstration of fibre optic CARS microscopy	Scanning mechanism needed, limited signal level

of fibre-optical nonlinear optical endoscopy. The present research in fibre-optical TPEF microscopy may be classified by applications into three categories: portable TPEF microscopy, rigid TPEF endoscopy and flexible TPEF endoscopy (Table 5.2).

Portable TPEF microscopy has been designed as a compact and light-weight imaging device, which can be carried by animals or used for epithelial tissue imaging (Table 5.2). A miniature head-mounted TPEF microscope has been first demonstrated for studies of dendritic morphology and calcium transients in brains of anaesthetised or awake animals [32]. In their design, an SMF fibre is used for both delivery of laser pulses and illumination scanning. Prior studies have shown that a spectral blue shift in an SMF leads to an improvement in resolution under two-photon excitation although an SMF in TPEF microscopy experiences temporal and spectral broadenings [7,42]. This miniature microscope has sufficient resolution and flexibility to measure the neural activity on a cellular scale. Recently, a portable TPEF microendoscope having a size of 3.5 cm × 1.2 cm × 1.5 cm and a mass of only 3.9 gram has been reported [18]. It is based on a compound GRIN lens probe, a hollow-core PBF for near distortion-free delivery of femtosecond pulses and another high NA multi-mode fibre for fluorescence collection, exhibiting micrometre-scale resolution for brain imaging.

Like the traditional rigid endoscopy, TPEF microendoscopy uses the combination of several GRIN lenses as a probe to provide enough length for imaging in deep tissues (Table 5.2). The compound GRIN lens probe typically has a relay lens and an objective lens with a diameter of sub-millimetre size, enabling the insertion into solid tissue with minimal invasion as well as translating an image plane from proximal scanning mirrors to the focal plane in tissue [43–45]. Since there is no fibre employed in the system and the GRIN endoscopes do not suffer from SPM with an illumination power less than 200 mW, the endoscope has a minimal pulse distortion compared with other geometries of fibre-optical TPEF microscopy. However, it must be combined with a conventional TPEF microscope to facilitate the pulse delivery and laser scanning.

Flexible TPEF endoscopy has an optical probe with a few millimetres diameter that can be incorporated into a traditional endoscope (Table 5.2). It is well suited to image either epithelial tissue such as skin or internal organs such as gastrointestinal tracts and cervix. Flexible TPEF endoscopy was first demonstrated by use of an SMF coupler [34, 46], which has been shown in Chapter 4. The application of this geometry has been extended to second harmonic generation (SHG) imaging to analyse the molecular orientations of structural proteins [47]. More importantly, an introduction of a double-clad PCF to the endoscopy has resulted in significant enhancement of signal level [20]. The integration of such a double-clad PCF and a MEMS mirror offers a great flexibility and a superior system performance [21]. The design and characterisation of double-clad PCF-based endoscopy will be presented in the following subsections. It also has been demonstrated that a fibre bundle combined with a compound GRIN lens enables two-photon imaging through a 1 mm diameter probe [30]. Its proximal scanning method also avoids vibrations of the fibre or the scanning unit during rapid accelerations.

It should be noticed that the concept of CARS endoscopy has been experimentally demonstrated recently [48]. A step-index SMF has been used to simultaneously deliver pump and Stokes beams having a pulse width of a few picoseconds and to collect the CARS signal in the backward direction. More recently, supercontinuum generated by a highly nonlinear PCF has enabled simultaneous CARS and TPEF images [24]. Organelles in a living yeast cell such as mitochondria and nucleus can be visualised through the combination of CARS and two-photon processes. These studies have demonstrated a great potential for the multicolour and multi-function endoscopy by combining all these nonlinear optical mechanisms through fibre-optics.

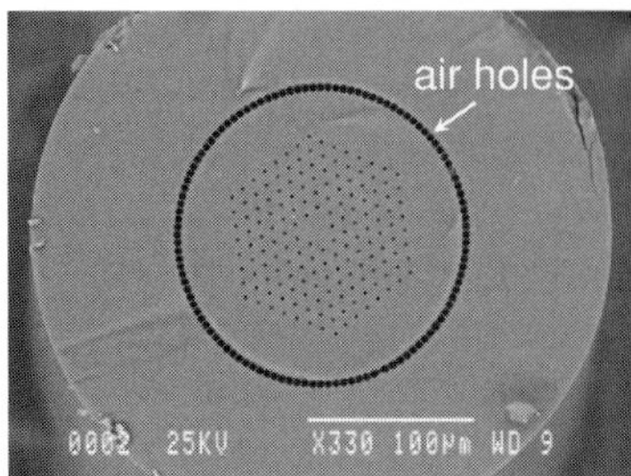

Fig. 5.3 Scanning electron microscopy image of the double-clad PCF used for nonlinear optical microscopy.

5.2 Nonlinear optical microscopy using double-clad PCFs

5.2.1 Characterisation of double-clad PCFs

The double-clad PCF is shown in Fig. 5.3. This fibre was originally designed for multi-mode pumping in the inner cladding and single-mode emission through the core doped with Yb [49, 50]. For endoscopy, the doped elements in the core have been removed to produce a passive double-clad PCF for the nonlinear optical imaging. In this way, the fibre is designed to confine near infrared light in the central core as a single transverse mode and confine visible light in the multi-mode inner cladding. Furthermore, the NA of the inner cladding produced by the PCF technology is higher than that provided by any other fibre fabrication technology.

This new double-clad PCF was successfully fabricated by Crystal Fibre A/S. The double-clad PCF has a core diameter of 20 μm (i.e. a 17 μm mode field diameter at wavelength 780 nm), an inner cladding with a diameter of 165 μm and an NA of 0.6 at wavelength 800 nm. The fibre core is surrounded by air holes with a hole to hole pitch ratio of 0.26. Within the outer cladding region of 340 μm in diameter, a ring of air holes is used to efficiently guide and collect light in the pure silica multi-mode inner cladding. The background propagation losses are as low as 10 dB/Km at wavelength 800 nm. It should be pointed out that compared with the double-clad fibre used for spectrally encoded endoscopy [51] and the double-clad PCF for TPEF biosensing [19, 22], the size of both the core and the inner cladding of our fibre is three times larger. Therefore, it gives rise to a further reduced SPM effect and the more efficient collection of nonlinear signals.

As the double-clad PCF in a nonlinear optical microscope is used to deliver a near infrared excitation laser beam and collect nonlinear optical signals in the visible range, it is important to understand the properties of the fibre under various operating conditions. Figure 5.4(a) shows the coupling efficiency of the fibre for the three given values of the NA of the coupling objectives over the wavelength range between 410 and 870 nm. Moreover, the output patterns of the fibre at different wavelengths are depicted in Fig. 5.4(b)–(g). It shows that the double-clad PCF offers the robust single-mode guidance of near infrared light in the central core and the efficient propagation of visible light within the multi-mode inner cladding. Figure 5.4(h) illustrates an intensity profile across

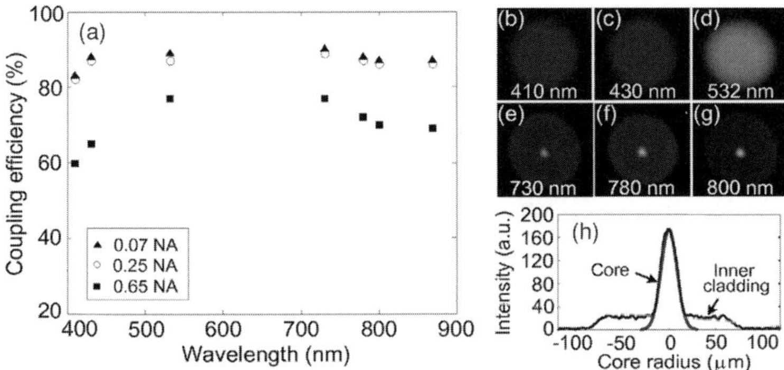

Fig. 5.4 (a) Coupling efficiency of the double-clad PCF in the wavelength range of 410–780 nm for three values of the NA (0.07, 0.25 and 0.65) of coupling objectives. (b)–(g) Digital camera photographs of the output pattern from the double-clad PCF between wavelengths 410 and 800 nm. (h) Gaussian fit of an intensity profile at the output of the fibre. Reprinted with permission from Ref. [20], L. Fu, X. Gan and M. Gu, *Opt. Express* **13**, 5528 (2005). © 2005, Optical Society of America.

the output pattern of the fibre at wavelength 800 nm (Fig. 5.4(g)). In spite of the leakage from the core to the inner cladding and a triangular shape of the fibre core, the beam profile of the central core has a nearly Gaussian intensity distribution, indicating the single-mode operation in the fibre core at a near infrared wavelength.

A coupling efficiency of over 80% with a maximum of approximately 90% in the wavelength range of 410–870 nm is achievable, if the NA of coupling objectives is 0.07 or 0.25 to match the low NA of the central core. However, there is a 20% degradation in the coupling efficiency by using the coupling objective with an NA of 0.65 due to the mode leakage in the inner cladding. It is found that approximately 28% of the output power from the double-clad PCF is guided in the central core at wavelength 800 nm when a coupling objective of NA 0.07 is used, whereas only 10% and 8% are in the core for a coupling objective of NA 0.25 and NA 0.65, respectively. Consequently, the use of a coupling objective with an NA of 0.07 can optimise the coupling efficiency in both the near infrared and the visible wavelength ranges. In particular, the coupling efficiency in the visible wavelength range is approximately twice that obtained with the SMF coupler [47], confirming that the double-clad PCF can lead to collection improvement for both TPEF and SHG imaging.

5.2.2   Experimental arrangement

To investigate the imaging performance of the double-clad PCF, let us consider a nonlinear optical imaging system as shown in Fig. 5.5 [20]. A laser beam generated from a Ti:sapphire laser with a repetition rate of 80 MHz and a pulse width of approximately 80 fs is coupled through an iris diaphragm and a microscope objective O_1 (0.65 NA, 40×) into the double-clad PCF with a length of approximately 1 m. The size of the iris diaphragm is adjusted to achieve the maximum laser power guided in the central core. The output beam from the fibre is collimated by the objective O_2 of NA 0.07 before

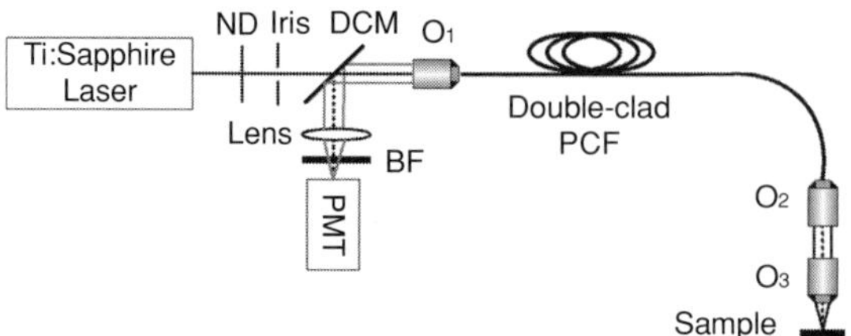

Fig. 5.5 Schematic diagram of the nonlinear optical microscope based on a double-clad PCF. O_1: 0.65 NA 40× microscope objective, O_2: 0.07 NA 4× microscope objective, O_3: 0.85 NA 40× imaging objective, ND: neutral density filter, BF: bandpass filter, DCM: dichroic mirror. Reprinted with permission from Ref. [20], L. Fu, X. Gan and M. Gu, *Opt. Express* **13**, 5528 (2005). © 2005, Optical Society of America.

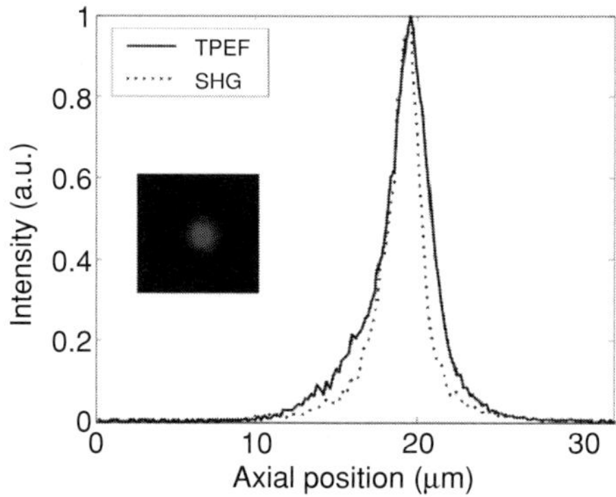

Fig. 5.6 Axial responses of the TPEF and SHG signals from a thin layer of AF-50 dye at an excitation wavelength of 800 nm in a microscope using a double-clad PCF. The inset shows the photograph of the beam profile on the back aperture of the imaging objective. The power on the sample is approximately 1.5 mW. Reprinted with permission from Ref. [20], L. Fu, X. Gan and M. Gu, *Opt. Express* **13**, 5528 (2005). © 2005, Optical Society of America.

being launched into the imaging objective O_3 (0.85 NA, 40×). The backward nonlinear signal via the PCF is collected by objective O_1 to match the high NA of the PCF inner cladding. The choice of a low NA objective O_2 and a high NA objective O_1 maximises the collection efficiency of the nonlinear signals. A dichroic mirror (DCM) reflects the TPEF and SHG signals, which are further filtered by a bandpass filter (BF) and focused onto a photomultiplier tube (PMT).

5.2.3   Axial resolution

An experimental investigation into the axial response of the system for characterising the three-dimensional imaging performance of the nonlinear optical microscope consists of scanning a thin layer of AF-50 fluorescence dye in the z direction. The result is shown in Fig. 5.6, where the full width at half maximum (FWHM) of the axial responses of TPEF and SHG at an excitation wavelength of 800 nm is 2.8 µm and 2.5 µm, respectively, obtained by placing a 510/20 nm bandpass filter or a 400/9 nm bandpass filter before the PMT. It reveals a degradation of axial resolution of approximately 33% in the double-clad PCF-based microscope, compared with that in a microscope using an SMF coupler [47]. This may result from the centrally localised light distribution before the imaging objective (see the inset of Fig. 5.6), which effectively decreases the NA of the imaging objective, and the large area of the inner cladding, which effectively increases the pinhole size. Furthermore, in both systems, an SMF and a double-clad PCF, a slight enhancement of SHG axial resolution has been shown due to its shorter wavelength [52].

It should be pointed out that the laser beam from the inner cladding experiences a stronger effect of waveguide dispersion than that from the central core. As a result, the ultrashort pulses delivered outside the central core should be broadened to a larger timescale. This feature is confirmed by the experimental investigation of the TPEF axial response, demonstrating that the peak intensity of the TPEF axial response when the excitation beam is well coupled in the central core is approximately 39 times as high as that when the excitation beam is decoupled transversely in the inner cladding. This result indicates that the laser beam delivered outside the central core contributes little to the nonlinear excitation. Furthermore, the fibre core should have a similar dispersion property to the silica material, enabling the delivery of ultrashort pulses for nonlinear optical processes.

5.2.4 Improvement of signal level

As described in the above subsections, the double-clad PCF having large size and NA can lead to a significant improvement in the signal level of the imaging system. To investigate the signal level of the double-clad PCF-based nonlinear microscope, we compare the strength of the axial responses of TPEF and SHG from the double-clad PCF and a standard SMF. The fused-silica SMF has an operation wavelength of 820 nm, a core/cladding diameter ratio of approximately 4/125 and NA 0.16. The coupling efficiency of the SMF of a 1 m length is approximately 30% at wavelength 800 nm. The FWHM of the axial response with the two types of fibre is kept the same for a given excitation power. Although an alternative comparison could be performed based on the given pulsewidth on the sample, our treatment is of importance in practical imaging when the excitation power is given from a laser.

The TPEF and SHG axial responses of the system using the double-clad PCF and a standard SMF are measured, in which case the excitation power before the imaging objective O_3 is varied but the gain of the PMT is identical. The intensity of nonlinear signals detected by the PMT is not normalised. Based on these results, the peak intensity of the axial responses from the two fibres as a function of the power before the imaging objective is shown in Fig. 5.7 on a log–log scale. The slope of two demonstrates the quadratic dependence of the TPEF and SHG intensity on the excitation power. It is clearly observed that the detected intensity of the nonlinear signals from the double-clad PCF is approximately 6.8 times stronger than that from the SMF in the case of the same excitation power delivered to the sample. As a result, if one considers that the excitation beams in the central core of the two types of fibre actually result in the

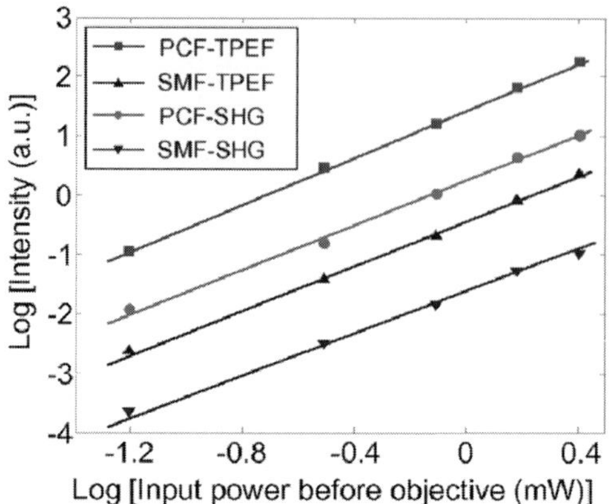

Fig. 5.7 Detected intensity of TPEF and SHG from the double-clad PCF-based microscope and a standard SMF-based microscope as a function of the power before the imaging objective. Reprinted with permission from Ref. [20], L. Fu, X. Gan and M. Gu, *Opt. Express* **13**, 5528 (2005). © 2005, Optical Society of America.

nonlinear process, an enhancement of approximately 40 times in the nonlinear signal intensity detected through the double-clad PCF is achieved. This result is obtained with an objective for 3D nonlinear optical imaging, and is thus applicable for imaging a thick sample when an optical sectioning property is critical. It also reveals that double-clad PCFs can support efficient propagation for the incoherent TPEF signal as well the coherent SHG signal.

5.2.5   Nonlinear optical imaging

The 3D imaging capability of the nonlinear optical microscope using a double-clad PCF is demonstrated through a series of imaging experiments. First, a scale of black tetra fish is imaged in order to make a direct comparison with those images obtained with an SMF coupler system [47]. The result is shown in Fig. 5.8(a), where a series of SHG sections taken at a 2 μm depth step into the fish scale are depicted. By considering the performance of the microscope systems using the PCF and the SMF coupler, the signal level of the PCF-based microscope system is increased by a factor of approximately 65. Second, the feasibility of the system for biomedical study is shown in Fig. 5.8(b), which displays a series of SHG images of rat tail tendon with a 2 μm depth step. The tendon is obtained from an 8-week old Sprague-Dawley rat tail, attached to the coverslip directly, and imaged within 2 hours after extraction. The image sections in Fig. 5.8(b) clearly resolve the morphology of mature, well-organised collagen fibrils even at an imaging depth of 20 μm, showing the pronounced optical sectioning property of the system. The result implies that the efficient PCF-based nonlinear microscope can be a potential tool for direct visualisation of collagen-related diseases. Furthermore, both image sets exhibit high contrast and significantly improved signal level of the system as a result of the efficient signal collection through the double-clad PCF.

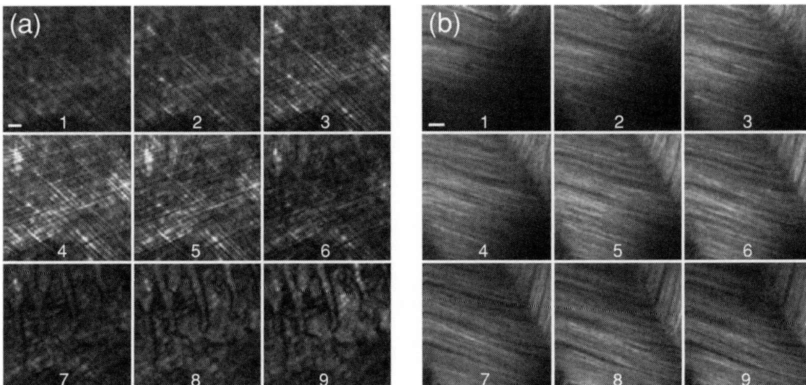

Fig. 5.8 Series of SHG imaging sections from (a) a fish scale and (b) a rat tail tendon in the nonlinear optical microscope using a double-clad PCF. The image section spacing is 2 μm. The scale bars represent 10 μm. Reprinted with permission from Ref. [20], L. Fu, X. Gan and M. Gu, *Opt. Express* **13**, 5528 (2005). © 2005, Optical Society of America.

5.2.6 SHG polarisation anisotropy measurement

The coherent emission feature in SHG microscopy enables the polarisation dependence of harmonic light that provides information about molecular organisation and nonlinear susceptibilities not available from fluorescence light with random phase. Generally, polarisation anisotropy measurements are made by obtaining images through rotations of the analyser before the detector while maintaining the excitation polarisation. Alternatively, absolute molecular orientations can be determined by rotating the excitation polarisation.

The investigation into polarisation characteristics of the double-clad PCF demonstrates that a degree of polarisation of 0.84 can be preserved in the fibre core at given incident polarisation angles under the illumination of 800 nm [53]. Under the experimental condition where the linearly ploarised light for excitation is delivered by the fibre core, SHG polarisation anisotropy measurements in microscopy using a double-clad PCF are shown in Fig. 5.9. For both fish scale and rat tail tendon imaging, SHG images are obtained in the case of no analyser before PMT and orthogonal polarisation orientations of the analyser. Molecular orientation in the sample can be quantified by the anisotropy parameter [54]

$$\beta = \frac{I_{\max} - I_{\min}}{I_{\max} + 2I_{\min}}, \tag{5.1}$$

where $I_{\max}$ and $I_{\min}$ are the SHG intensity with the polarisation parallel and perpendicular to the incident polarisation. It is found that the average β value is approximately 0.15 for the fish scale and 0.2 for the rat tail tendon, demonstrating that SHG signals from the two well-ordered samples experience depolarisation through the double-clad PCF. This result implies that photonic crystal structures in the inner cladding of the fibre, which enable the enhancement of the SHG collection efficiency, however result in a significant depolarisation effect over the near infrared and the visible wavelength ranges. Therefore,

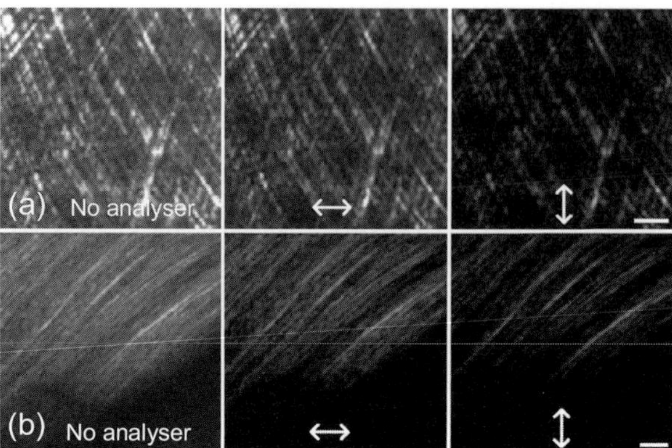

Fig. 5.9 SHG polarisation anisotropy measurements with (a) a fish scale and (b) rat tail tendon in a nonlinear optical microscope using a double-clad PCF. Each set of SHG images is obtained without an analyser or with orthogonal polarisation orientations of the analyser. Scale bars are 10 μm. Reprinted with permission from Ref. [53], L. Fu and M. Gu, *Opt. Express* **16**, 5000 (2008). © 2008, Optical Society of America.

the development of a polarisation-maintained double-clad PCF would benefit the SHG polarisation anisotropy measurement in nonlinear optical microscopy.

5.3   A nonlinear optical endoscope based on a double-clad PCF and a MEMS mirror

5.3.1   Endoscope design

In Section 5.2, the enhancement of the collection efficiency and the 3D high-resolution imaging ability of a double-clad PCF have been demonstrated. To develop a nonlinear optical endoscope, a compact scanning mechanism and a miniaturised objective should be integrated into the imaging system. As a result, an ultrasmall probe head for nonlinear optical imaging can be designed, as shown in Fig. 5.10 [21].

A laser beam generated from a Ti:sapphire laser with a repetition rate of 80 MHz and a pulsewidth of approximately 80 fs is negatively prechirped by a pair of gratings, which act as a prechirp unit (1,200 grooves/mm, 28.7° blaze angle), before being coupled through an iris diaphragm, a dichroic mirror (DCM) and a microscope objective CO (0.65 NA, 40×) into the double-clad PCF. This PCF plays a dual role to offer robust single-mode guidance of the near infrared light in the central core and efficient propagation of the visible light within the multi-mode inner cladding. The excitation laser beam coupled from the fibre is reflected and scanned by a MEMS mirror and then focused onto a sample through a GRIN lens. The backscattering nonlinear signals propagated through the fibre are collected by a photomultiplier tube (PMT) via the DCM and a bandpass filter (BF). It should be pointed out that a single GRIN lens is chosen in the design in order to achieve the greatest flexibility and minimise the aberration and the SPM effect occurring in the probe. As a consequence, the endoscope head of the nonlinear optical

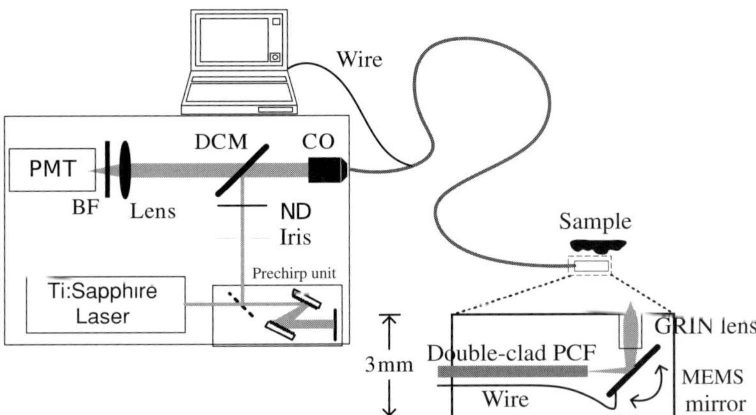

Fig. 5.10 Schematic diagram of the nonlinear optical endoscope based on a double-clad PCF, a MEMS mirror and a GRIN lens. CO: 0.65 NA 40× microscope objective. Reprinted with permission from Ref. [21], L. Fu, A. Jain, H. Xie, C. Cranfield and M. Gu, *Opt. Express* **14**, 1027 (2006). © 2006, Optical Society of America.

imaging system is approximately 3 mm in diameter, equipped with the MEMS mirror and the GRIN lens.

5.3.2 Axial resolution and signal level

In a nonlinear optical microscope system based on a GRIN lens and a fibre, the NA of a GRIN lens is effectively changed as the gap between the fibre and the GRIN lens varies [55]. Such an effective change in NA of a GRIN lens can affect the performance of fibre-optic nonlinear optical endoscopy in three aspects. The first aspect is imaging resolution that depends on the square of the effective NA [52], the second one is the illumination and collection efficiency that shows a complicated dependence on the effective NA, and the third one is the SHG and TPEF strength that depends on the fourth power of the effective NA [56]. To investigate these effects in the system based on a double-clad PCF and a GRIN lens, a 0.2 pitch GRIN lens (810 nm, 0.5 NA) with a diameter of 0.5 mm is used to characterise the performance of TPEF axial resolution and signal level, which is shown in Fig. 5.11(a). When the gap length is approximately 5 mm to fill the back surface of the GRIN lens, the optimised axial resolution of TPEF and SHG at an excitation wavelength of 800 nm for the system is approximately 6 μm and 5.4 μm, respectively, depicted in Fig. 5.11(b). In this case, lateral resolution for nonlinear optical imaging is approximately 1 μm and the working distance is approximately 150 μm.

The dependence of axial resolution on the gap length exhibits a similar manner to that which has been shown in a system using an SMF coupler and a GRIN lens [55]. However, the collected signals through the GRIN lens and the double-clad PCF are more efficient, and therefore do not exhibit significant decrease as the gap length increases. As a result, the system does not show the trade-off feature and simultaneous optimisations

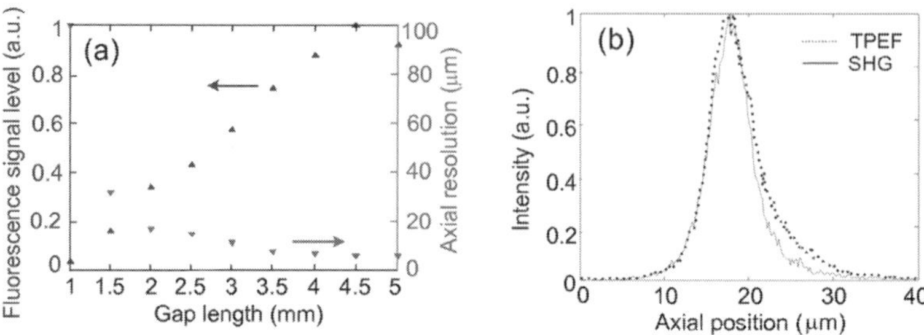

Fig. 5.11 (a) Detected intensity and axial resolution of TPEF as a function of the gap length between the double-clad PCF and a GRIN lens. (b) Axial responses of TPEF and SHG in a nonlinear optical microscope using a double-clad PCF and a GRIN lens. The excitation wavelength is 800 nm. Reprinted with permission from Ref. [21], L. Fu, A. Jain, H. Xie, C. Cranfield and M. Gu, *Opt. Express* **14**, 1027 (2006). © 2006, Optical Society of America.

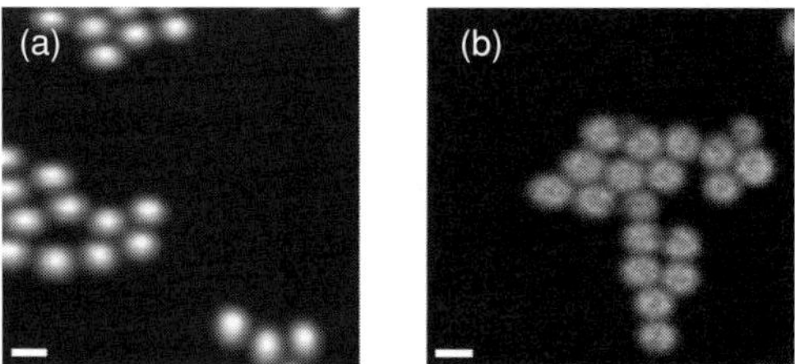

Fig. 5.12 TPEF images obtained from two geometries of fibre-optic nonlinear microscopes: (a) using a double-clad PCF and a GRIN lens (1 mm diameter, 0.2 pitch, 0.5 NA) and (b) using an SMF coupler and a GRIN lens (1 mm diameter, 0.25 pitch, 0.46 NA). Scale bars represent 10 μm.

of axial resolution and signal level that can be obtained by use of double-clad PCFs. In contrast, the optimisations of the axial resolution and the signal level have to be achieved at different gap lengths in the system based on an SMF coupler and a GRIN lens. This is due to the fact that the double-clad PCF has a high NA and a large core diameter, which are not sensitive to the coupling alignment and aberrations for a given GRIN lens.

To further confirm the high spatial resolution and the enhancement of signal level of the system, a direct comparison is made based on the TPEF images of 10 μm microspheres obtained from the PCF-GRIN lens-based system and the SMF coupler-GRIN lens-based system (Fig. 5.12). In both geometries, the gap length between the fibre and the GRIN lens is set to obtain the maximum detected signal. After the normalisations of the excitation power and the gain of the PMT, it is found that optimised signal level of the endoscope using a double-clad PCF is approximately 160 times higher than that of the SMF coupler-based endoscope. Additionally, when the grating pair in the prechirp unit has a diffraction angle of approximately 15° and a distance of approximately 1.2 cm for a 1 metre double-clad PCF, the signal level in the PCF-based system can be further increased by one order of magnitude as a result of the group-velocity dispersion compensation through the fibre.

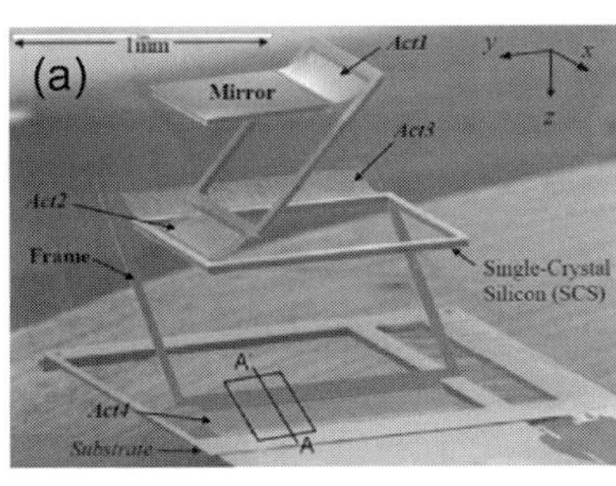
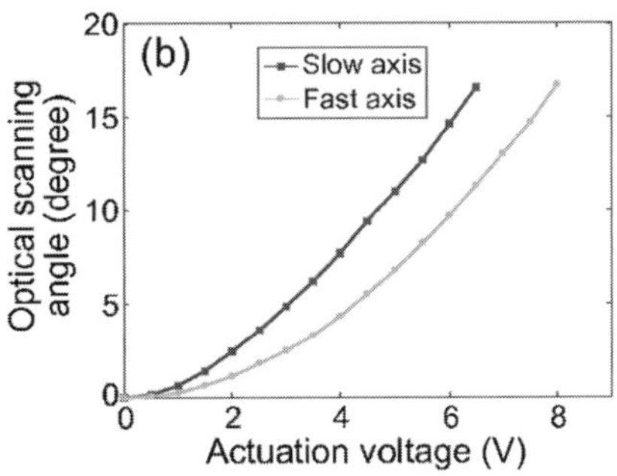

Fig. 5.13 (a) Scanning electron microscopy image of a 2D MEMS mirror in a nonlinear optical endoscope. (b) Optical scanning angle of slow and fast axes as a function of the applied voltage.

5.3.3 Endoscopic imaging

A two-axis scanning mirror based on MEMS technology can two-dimensionally scan the laser beam from the fibre for 2D image formation in conjunction with a GRIN lens. Various single-crystal silicon-based 2D micromirror devices have been reported for biomedical applications [35, 39]. Most of these optical scanners require high actuation voltages for large rotation angles due to their use of electrostatic actuation. However, optical scanners needed by endoscopic biomedical imaging applications are required to scan large optical angles with a high scanning speed, but at low driving voltages [57]. The 2D MEMS mirror based on electrothermal actuation can perform large bi-directional 2D optical scans over $\pm 30°$ at less than 12 V DC [57]. Figure 5.13(a) is the scanning electron microscopy image of the 2D MEMS mirror and used for our nonlinear optical endoscope. The aluminium-coated mirror plate is 0.5 mm by 0.5 mm in size and has a reflection efficiency of approximately 80% at wavelength 800 nm [57]. Furthermore, the small initial tilt angle of the mirror simplifies the endoscope design and packaging.

Line scans can be obtained by applying DC voltages on each actuator individually. We choose two actuators as slow and fast scanning axes for 2D image formation. The corresponding optical scanning angle versus actuation-voltage characteristics for the two actuators are shown in Fig. 5.13(b). It can be observed that the optical scanning angles have a linear dependence on the actuation voltage in the range between 2.5 and 7.5 V DC.

However, the difficulty in driving a MEMS mirror for smooth scanning is that any sudden alteration of the applied voltage on actuators will cause damage or irregular scanning patterns. The problem is solved by the design of a special ramp waveform to increase or decrease the actuation voltage gradually. A 2D raster scanning pattern can be generated by this MEMS mirror. Two actuators are synchronised and excited to create their linear angular displacement by applying driving voltages based on the designed ramp waveform. The scanning rate of the mirror is 7 lines/second for our experiments, though its resonance frequency is approximately 480 Hz that could be used to increase the speed for imaging acquisition. It should be pointed out that the low driving voltage, the simple structure and ruggedness to operation make the 2D MEMS mirror suitable for the endoscopic system and safe for clinical applications.

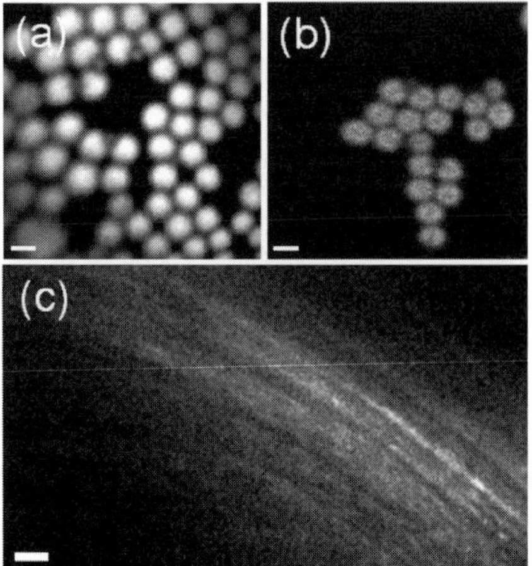

Fig. 5.14 TPEF images obtained (a) through a double-clad PCF, a 2D MEMS mirror and a GRIN lens, and (b) through a single-mode fibre coupler, a GRIN lens and a bulk scanning stage. The sample is fluorescent microspheres with a diameter of 10 μm. The power in (a) for TPEF is approximately 1.8 mW. (c) Z projection of 11 SHG sections obtained from the rat tail tendon with an imaging spacing of 10 μm. Scale bars represent 10 μm. Reprinted with permission from Ref. [59], L. Fu, A. Jain, C. Cranfield, H. Xie and M. Gu, *J. Biomed. Opt.* **12**, 040501 (2007). © 2007, International Society for Optical Engineering.

The capability of a 2D MEMS mirror for nonlinear optical endoscopic imaging is demonstrated by the TPEF image (Fig. 5.14(a)) with 10 μm diameter fluorescent microspheres. It shows that the 2D MEMS mirror enables efficient delivery of the light beam over the broadband wavelength range and smooth response for image acquisition. In this case, a GRIN lens with a large diameter (1.8 mm diameter, 0.6 NA, 0.23 pitch) is used to produce a field of view up to 150 μm. The performance of the system is also confirmed by the comparison between this image and the TPEF image (Fig. 5.14(b)) obtained from a system based on an SMF coupler, a GRIN lens and a bulk scanning stage. It is shown that the signal level of the nonlinear optical endoscope is approximately 120 times higher than the microscope system using an SMF coupler and a bulk scanning mechanism due to the large collection area and the high NA of the inner cladding of the PCF. Furthermore, combined with the previous characterisation experiment in which case a MEMS mirror is not used and the focal spot is the same [21], the result in Fig. 5.14(a) also implies that 80 fs pulses are broadened by approximately 25% after the MEMS mirror as the TPEF efficiency is inversely proportional to the pulse width [58]. The pulse broadening might be due to the GVD caused by the aluminium coating of the mirror plate.

The ability to obtain SHG images through the system using a 2D MEMS mirror is shown by SHG optical sections from rat tail tendon (Fig. 5.14(c)). The tendon is dissected from a Sprague-Dawley rat tail and imaged directly. It can be seen that in terms of the contrast pattern and axial resolution, the nonlinear optical images produced by the MEMS mirror is consistent with the images obtained with a double-clad PCF, a GRIN lens, and a scanning stage [20].

5.3.4   3D tissue imaging

There are a number of potential applications for the nonlinear optical endoscope described in this section [12, 59, 60]. Since the majority of cancers are epithelial tissues in origin, we have focused our efforts on imaging of epithelial tissues. All experiments were approved by the University Animal Experimentation Ethics Committee. The excitation power is approximately 40 mW at 800 nm. Based on the GRIN lens (1.8 mm diameter, 0.6 NA, 0.23 pitch) and the MEMS mirror (actuation voltages of 2.5–7 V on the fast axis and 3–6.9 V on the slow axis, the scanning rate of 7 lines/second), the imaging field of view is approximately 140 μm $\times$ 100 μm.

Epithelial pre-cancers and cancers are associated with morphological and functional alterations of cells, normally assessed by invasive biopsy. Due to the unique features of nonlinear optical imaging, the use of nonlinear optical endoscopy in the normal endoscopic procedure has the following advantages. (1) Combined with the macroscopic imaging capability of an endoscope, nonlinear optical microscopy provides microscopic imaging to identify metaplasia or flat lesions that are not easily resolved by traditional endoscopy. (2) The visualisation of the size and spatial arrangement of nuclei and vascular structures in three dimensions within tissue may reduce or remove the need for tissue excision for histopathologic interpretation. (3) The 3D high-resolution imaging ability enables a more accurate differentiation between normal and malignant tissues to give better diagnosis. Therefore, we explore the potential of our nonlinear optical endoscope in clinical applications by imaging through gastrointestinal tract and cancer tissues.

Figure 5.15(a) is a series of *in vitro* TPEF images of rat large intestine tissue. The large intestine is extracted from a Sprague-Dawley rat. To enhance the contrast in endoscopic imaging, the luminal epithelial tissue of the rat large intestine is stained with 1% Acridine Orange in Ringer's solution. TPEF images are obtained through the thick tissue with a penetration depth of 100 μm. The ability of the endoscope system to discriminate the microscopic anatomy is also investigated by comparing these measurements with the TPEF image (Fig. 5.15(b)) which is taken from the same large intestine tissue in a standard laser scanning two-photon microscope with a 0.85 NA objective. It is shown that the image pattern obtained from both the endoscope system and the standard microscope are similar, and surface epithelial cells surrounding intestinal crypts (see arrows in frame 3 and Fig. 5.15(b)) can clearly be observed in both cases. The structural details in the rat large intestine tissue can be further visualised in the 3D reconstruction of the image stack (Fig. 5.15(c)), where the intestinal crypts are displayed (see arrows) by cropping the 3D volume. In addition, high resolution of the nonlinear optical endoscope also enables the visualisation of the openings to the gastric pits of the rat stomach columnar mucosal tissue (data not shown). These results demonstrate that the nonlinear optical endoscope that offers the micrometre-scale resolution in deep tissue may differentiate various tissue types and identify early mucosal lesions in the gastrointestinal tract.

To demonstrate the imaging ability of the nonlinear optical endoscope for cancer tissue, human u-87 MG glioblastoma cells (a kind of human breast cancer cell) are

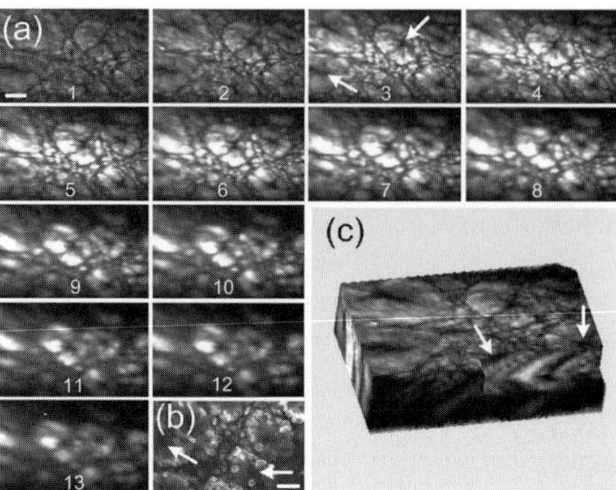

Fig. 5.15 (a) A series of *in vitro* images of rat large intestine tissue. Each section is recorded at an axial step of 7.5 μm into the sample. The excitation power is approximately 40 mW at wavelength 800 nm. (b) A TPEF image of the rat large intestine tissue taken from a standard laser scanning microscope with an objective (0.85 NA). (c) 3D visualisation of the rat large intestine tissue based on the image stack in (a). Image reconstruction is performed using AMIRA. Arrows are pointing to the intestinal crypts. Scale bars are 20 μm. Reprinted with permission from Ref. [59], L. Fu, A. Jain, C. Cranfield, H. Xie and M. Gu, *J. Biomed. Opt.* **12**, 040501 (2007). © 2007, International Society for Optical Engineering.

xenografted into the hind leg of a nude Balb/c mouse and cultured for one week before the dissection from the nude mouse for imaging. The breast cancer tissue is stained with 1% Acridine Orange to visualise the cell nuclei. TPEF images of human u-87 MG glioblastoma tissue are taken from the endoscope system (Fig. 5.16(a)) with a penetration depth of approximately 75 μm and the standard microscope (Fig. 5.16(b)), respectively. Figure 5.16(c) is the 3D reconstruction of TPEF images in Fig. 5.16(a). It is found that the endoscope system and the standard microscope produce a consistent pattern with the same sample. Furthermore, the contrast pattern of the cancer tissue is fundamentally different from that achieved in gastrointestinal tract tissue, showing the extremely dense distribution of cell nuclei in the cancer tissue. Beyond the morphological identification, specific labels and methods such as fluorescence proteins and fluorescence lifetime measurement could be combined with the nonlinear optical endoscope for multi-dimensional cancer tissue imaging in future studies.

5.4   Nonlinear optical microscopy using a double-clad PCF coupler

We have demonstrated that the double-clad photonic crystal fibre is ideally suitable for nonlinear optical endoscopy due to its dual function of single-mode and multi-mode delivery through the central core and the inner cladding region with a high NA, respectively, within a single piece of fibre. A further compact imaging system could be achieved by use of a multi-port fibre coupler to gain self-alignment and replace bulk

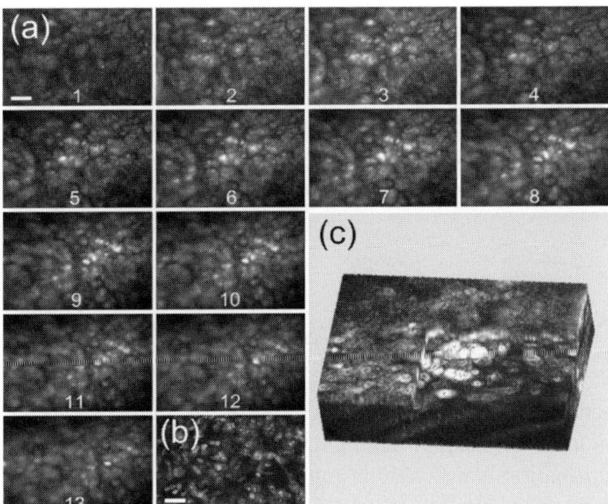

Fig. 5.16 (a) A series of *in vitro* images of the human u-87 MG glioblastoma tissue. Each section is recorded at an axial step of 5 μm into the sample. The excitation power is approximately 40 mW at wavelength 800 nm. (b) A TPEF image of the breast cancer tissue from the standard microscope used in Fig. 5.15(b). (c) 3D visualisation of the breast cancer tissue. Scale bars are 20 μm. Reprinted with permission from Ref. [59], L. Fu, A. Jain, C. Cranfield, H. Xie and M. Gu, *J. Biomed. Opt.* **12**, 040501 (2007). © 2007, International Society for Optical Engineering.

optics for an all-fibre microscopy system. However, there has been no report on the fabrication of a double-clad PCF coupler or any applications.

Recent theoretical study and fabrication of PCF couplers have been reported for a number of PCF structures [61–64]. The side-polished method has been used to produce a single-mode PCF coupler with a tunable splitting ratio [63]. By removing a part of the cladding region of each PCF, two PCFs were mated close enough to achieve evanescent field coupling through the sides of the PCFs. In addition, the fused biconical taper method has been successful in generating single-mode and multi-mode air–silica PCF couplers according to the principle of mode expansion [62, 64]. In particular, the multi-port multi-mode air-clad holey fibre coupler can deliver optical power of up to 35 W at wavelength 850 nm [64]. Compared with the polishing method [63], the fused biconical taper method can protect the complicated structures of PCFs with a more manoeuvrable fabrication process [62, 64]. The difficulty in fabricating a double-clad PCF coupler is to achieve mode coupling through two cladding regions and preserve the single-mode and multi-mode guidance in the core and inner cladding regions, respectively. In particular, for applications in nonlinear optical microscopy, one arm of the double-clad PCF coupler should permit the single-mode propagation of a near infrared laser beam in the core, while the other arm can facilitate the multi-mode collection of the visible nonlinear optical signals. This section presents a three-port double-clad PCF coupler by the fused tapered method, which shows the single-mode and multi-mode separation at different wavelengths [65]. A compact nonlinear optical microscope is implemented by this PCF coupler and a GRIN lens, demonstrating high-resolution 3D TPEF and SHG images.

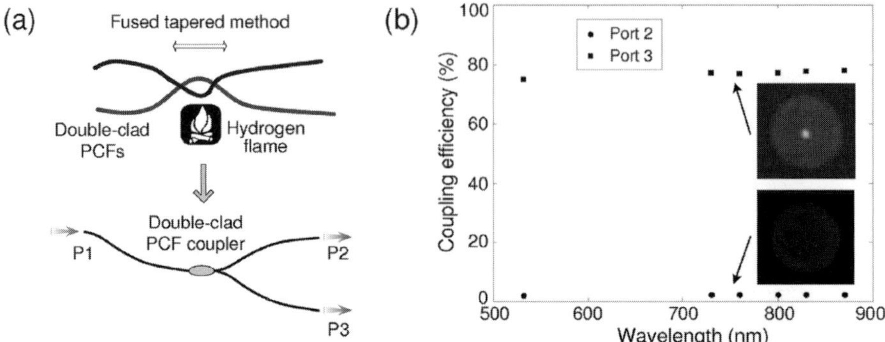

Fig. 5.17 (a) Fabrication process of the double-clad PCF coupler. (b) Coupling efficiency at output ports 2 and 3 of the double-clad PCF coupler having a coupling-starting pulling length of 6.1 mm as a function of the illumination wavelength at input port 1. Insets: digital camera photographs of output patterns of the double-clad PCF coupler at wavelength 800 nm.

5.4.1   A double-clad PCF coupler

Since the fused taper method facilitates the coupling of single-mode and multi-mode PCFs, it is adopted to make a double-clad PCF coupler. As shown in Fig. 5.17(a), two lengths of the double-clad PCFs are first twisted, heated by a hydrogen flame with a flame size of approximately 10 mm, and then drawn gradually in the fused region. The temperature and the size of the flame can be controlled by adjustment of the gas-flow rate. The fabrication of the PCF coupler was conducted in Fovice in Korea. Acrylate coating is removed only in the fusion region. As the fibres are elongated and the fibre diameter is reduced, mode coupling occurs if the confined mode field in a PCF is extended to its neighbouring double-clad PCF. Further, it is important to control the pulling length in the fused region, as it determines the splitting ratio of the coupler as well as the mode coupling condition in the core and cladding regions. In the case of the short pulling length, where the rings of air holes in two fibres are very close or partially collapsed, the multi-mode in the inner cladding is coupled to the neighbouring fibre. Light propagated in the outer cladding leaks out quickly beyond the fusion region where the acrylate coating is still remaining. The length that yields 1% coupling is defined as the coupling-starting pulling length. In the fabrication system various coupling-starting pulling lengths for the coupler are obtained, at which one arm of the coupler is terminated to form a 1×2 PCF coupler.

For a double-clad PCF coupler of a coupling-starting pulling length of approximately 6.1 mm, its coupling efficiency is measured from ports 2 and 3 (see Fig. 5.17(a)), while a laser beam in the wavelength range 532–870 nm is coupled to port 1. As shown in Fig. 5.17(b), the coupling efficiencies at ports 2 and 3 are approximately 2.2% and 77.5%, respectively, with a splitting ratio of 97/3 between ports 3 and 2 over the visible and near infrared wavelength range. Under this condition, the coupler reveals a low insertion loss of 1.1 dB between ports 1 and 3, while the insertion loss is as high as 16.6 dB between ports 1 and 2. The coupling efficiency of this coupler is different from that

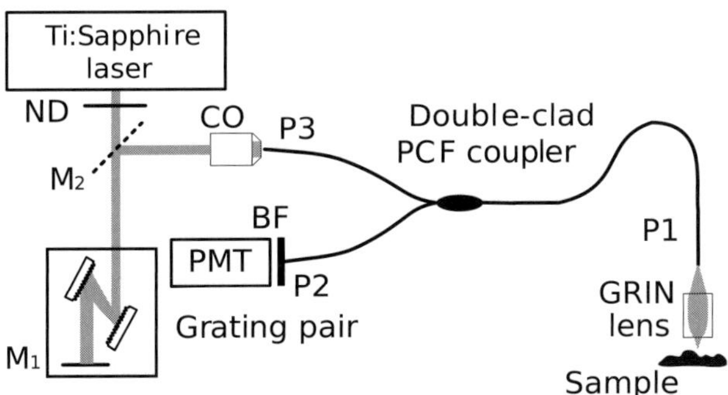

Fig. 5.18 Schematic diagram of the nonlinear optical microscope with a three-port double-clad PCF coupler and a GRIN lens. Mirror M_2 (shown dashed) is located on a different plane from mirror M_1. ND: neutral density filter, BF: bandpass filter, CO: coupling objective.

of a single-mode fibre coupler [46, 47, 66], which exhibits a lower coupling efficiency at port 3 in the visible wavelength region.

It has been shown in our previous study that a single-mode laser beam in the central core of the double-clad PCF makes a significant contribution to the excitation of nonlinear optical signals [20]. Therefore it is essential to confirm the single-mode delivery feature of the double-clad PCF coupler. To this end, output patterns at ports 2 and 3 of the same coupler are recorded by a camera and are shown in the insets of Fig. 5.17(b) when port 1 is illuminated at wavelength 800 nm. It can be seen that the arm (port 3) guiding the most power of the laser beam still maintains the single-mode propagation in the central core, whereas only the multi-mode propagation is observed from the other arm (port 2) in the near infrared wavelength range. This result, together with the splitting ratio shown in Fig. 5.17(b), implies that using port 3 for the delivery of the pulsed excitation beam and port 2 for the visible signal collection facilitates the feature of the efficient single-mode propagation in the core and the multi-mode collection through the inner cladding. This unique feature, which cannot be obtained from a single-mode fibre coupler, may prove advantageous for compact all-fibre nonlinear optical microscopy.

It should be pointed out that a double-clad PCF coupler having a longer coupling-starting pulling length shows a higher splitting ratio of 74/26 between ports 3 and 2 in the near infrared wavelength range. Compared with the coupler having a shorter coupling-starting pulling length (Fig. 5.17(b)), this coupler exhibits a higher excess loss of 5.8 dB. Furthermore, the measurement of the output patterns shows that the multi-mode guidance of the laser beam at a near infrared wavelength is observed at both ports. This outcome might be due to the complete collapse of air holes surrounding the fibre core in the fused region.

5.4.2   Experimental arrangement

We construct an all-fibre nonlinear optical microscope based on the double-clad PCF coupler (shown in Fig. 5.17) and a GRIN lens, which is depicted in Fig. 5.18. A turnkey Ti:sapphire of the wavelength range between 730 nm and 870 nm provides short pulses with a pulse width of 80 fs and a repetition rate of 80 MHz. The laser beam is prechirped by double passing a pair of gratings (1,200 grooves/mm, 28.7° blaze angle) before they are coupled to port 3 of the fibre coupler with a 4×/0.12 NA objective, CO. For a double-clad PCF fibre with a length of 1 m, negative prechirping of approximately $-32,200$ fs^2 is required to compensate for the group velocity dispersion in the fused silica material. Such a dispersion compensation arrangement leads to an improvement of the nonlinear optical signal level by a factor of approximately 11 for a given average power (<10 mW).

The output laser beam at port 1 is focused through a 1 mm diameter, 0.2 pitch, 0.5 NA GRIN lens onto a sample. The excited nonlinear optical signals from the sample are collected via the detection arm (port 2) of the coupler attached to a PMT. Therefore, the double-clad PCF coupler acts as a beam splitter to separate the nonlinear optical signals from the excitation laser beam delivered by the fibre core. Furthermore, a coupling objective with a low NA (0.07 or 0.12) can be used to maximise the excitation power delivered in the fibre core without the combination of an iris diaphragm and a high NA coupling objective (see Fig. 5.5), which collects the backward signals through a single piece of double-clad PCF. In this case, the delivered power in the fibre core can be 50 mW for nonlinear excitation.

5.4.3   System performance

To characterise the depth discrimination of the fibre-optic nonlinear optical microscope using a double-clad PCF coupler, we measure the TPEF axial response to a thin layer of AF-50 fluorescence dye. As axial resolution and signal level of the system are dependent on the gap length between the fibre and the back surface of the GRIN lens, a gap length of approximately 8 mm is chosen to optimise the signal level of the system. A typical TPEF axial response to the thin layer reveals that the axial resolution of the nonlinear optical microscope is approximately 10 μm. This value is consistent with the result we have obtained with a double-clad PCF and the same GRIN lens [20]. A slight improvement of axial resolution can be achieved when the back aperture of the GRIN lens is over-filled. However, it results in an approximately 25% degradation of the signal level of the system.

The optical sectioning ability of the new system is further demonstrated by sets of TPEF images of 10 μm diameter microspheres and SHG images of KTP crystal powder, which are displayed in Figs. 5.19(a) and (b), respectively. It can be seen that the system can collect high-resolution 3D images so that defects in the microspheres are well visualised. The results also imply that the miniature microscope could be used as a probe for the simultaneous detection of TPEF and SHG signals in tumour tissue. Furthermore, compared with the images obtained with an SMF coupler and a GRIN lens [55], the signal level of the images obtained with the new system is improved by a factor of approximately 5 due to the simultaneous optimisations of signal level and axial resolution in a nonlinear optical microscope using a double-clad PCF and a GRIN lens.

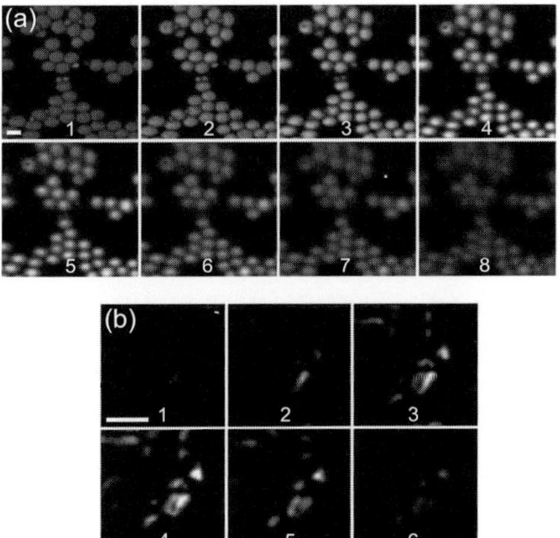

Fig. 5.19 (a) TPEF image sections of 10 µm diameter fluorescent microspheres and (b) SHG image sections of KTP crystal powder. Each set of images has an axial depth of 5 µm into the sample. The excitation wavelength is 800 nm and a 400/9 nm bandpass filter is placed before the PMT when SHG images are acquired. The excitation power is approximately 10 mW on the sample. Scale bars represent 10 µm.

5.5   Summary

The work presented in this chapter has demonstrated a new horizon for nonlinear optical endoscopy. Imaging modalities of both TPEF and SHG have been demonstrated in fibre-optic nonlinear optical microscopy and endoscopy based on a double-clad PCF, a MEMS mirror and a GRIN lens. The signal level of fibre-optic nonlinear optical endoscopy has been significantly enhanced by two orders of magnitude. As a result, 3D nonlinear optical endoscopic imaging through tissue has become possible. Taking advantages of compact size and high resolution, fibre-optic nonlinear optical endoscopy can produce an imaging tool with new functionalities, high resolution, multiple imaging mechanisms, which will enable *in vivo* visualisations of functional and morphological changes of tissue at the microscopic level rather than direct observations with a traditional instrument at the macroscopic level.

References

[1] F. Helmchen and W. Denk. Deep tissue two-photon microscopy. *Nat. Methods*, **2**:932–940, 2005.

[2] B. A. Flusberg, E. D. Cocker, W. Piyawattanametha, J. C. Jung, E. L. M. Cheung, and M. J. Schnitzer. Fiber-optic fluorescence imaging. *Nat. Methods*, **2**:941–950, 2005.

[3] S. Kimura and T. Wilson. Confocal scanning optical microscope using single-mode fiber for signal detection. *Appl. Opt.*, **30**:2143–2150, 1991.

[4] T. Dabbs and M. Glass. Fibre-optic confocal microscope: FOCON. *Appl. Opt.*, **31**:3030–3035, 1992.

[5] G. P. Agrawal. *Nonlinear Fiber Optics*. San Diego, Academic Press, 1989.

[6] R. Wolleschensky, T. Feurer, R. Sauerbrey and U. Simon. Characterization and optimization of a laser-scanning microscope in the femtosecond regime. *Appl. Phys. B*, **67**:87–94, 1998.

[7] D. Bird and M. Gu. Fibre-optic two-photon scanning fluorescence microscopy. *J. Microscopy*, **208**:35–48, 2002.

[8] J. C. Knight and P. S. J. Russell. New ways to guide light. *Science*, **296**:276–277, 2002.

[9] A. Bjarklev, J. Broeng and A. S. Bjarklev. *Photonic Crystal Fibers*. Norwell, Kluwer Academic Publishers, 2003.

[10] W. H. Reeves, D. V. Skryabin, F. Biancalana, J. C. Knight, P. St. J. Russell, F. G. Omenetto, A. Efimov and A. J. Taylor. Transformation and control of ultrashort pulses in dispersion-engineered photonic crystal fibres. *Nature*, **424**:511–515, 2003.

[11] J. M. Dudley, L. Provino, N. Grossard, J. Maillotte, R. S. Windeler, B. J. Eggleton and S. Coen. Supercontinuum generation in air-silica microstuctured fibers with nanosecond and femtosecond pulse pumping. *J. Opt. Soc. Am. B*, **19**:765–771, 2002.

[12] L. Fu and M. Gu. Fibre-optical nonlinear optical microscopy and endoscopy. *J. Micros.*, **226**:195–206, 2007.

[13] F. Helmchen, D. W. Tank and W. Denk. Enhanced two-photon excitation through optical fiber by single-mode propagation in a large core. *Appl. Opt.*, **41**:2930–2934, 2002.

[14] D. G. Ouzounov, K. D. Moll, M. A. Foster, W. R. Zipfel, W. W. Webb and A. L. Gaeta. Delivery of nanojoule femtosecond pulses through large-core microstructured fibers. *Opt. Lett.*, **27**:1513–1515, 2002.

[15] W. Göbel, A. Nimmerjahn and F. Helmchen. Distortion-free delivery of nanojoule femtosecond pulses from a Ti:sapphire laser through a hollow-core photonic crystal fiber. *Opt. Lett.*, **29**:1285–1287, 2004.

[16] S. P. Tai, M. C. Chan, T. H. Tsai, S. H. Guol, L. J. Chen and C. K. Sun. Two-photon fluorescence microscope with a hollow-core photonic crystal fiber. *Opt. Express.*, **12**:6122–6128, 2004.

[17] D. Kim, K. H. Kim, S. Yazdanfar and P. T. C. So. Optical biopsy in high-speed handheld miniaturized multifocal multiphoton microscopy. In *Multiphoton Microscopy in the Biomedical Sciences V*, volume **5700**, pages 14–22. Bellingham, SPIE, 2005.

[18] B. A. Flusberg, J. C. Jung, E. D. Cocker, E. P. Anderson and M. J. Schnitzer. *In vivo* brain imaging using a portable 3.9 gram two-photon fluorescence microendoscope. *Opt. Lett.*, **30**:2272–2274, 2005.

[19] M. T. Myaing, J. Y. Ye, T. B. Norris, T. Thomas, J. R. Baker, W. J. Wadsworth, G. Bouwmans, J. C. Knight and P. St. J. Russell. Enhanced two-photon biosensing with double-clad photonic crystal fibers. *Opt. Lett.*, **28**:1224–1226, 2003.

[20] L. Fu, X. Gan and M. Gu. Nonlinear optical microscopy based on double-clad photonic crystal fibers. *Opt. Express.*, **13**:5528–5534, 2005.

[21] L. Fu, A. Jain, H. Xie, C. Cranfield and M. Gu. Nonlinear optical endoscopy based on a double-clad photonic crystal fiber and a MEMS mirror. *Opt. Express.*, **14**:1027–1032, 2006.

[22] J. Y. Ye, M. T. Myaing , T. P. Thomas, I. Majoros, A. Koltyar, J. R. Baker, W. J. Wadsworth, G. Bouwmans, J. C. Knight, P. St. J. Russell and T. B. Norris. Development of a double-clad photonic crystal fiber based scanning microscope. In *Multiphoton Microscopy in the Biomedical Sciences V*, volume **5700**, pages 23–27. Bellingham, SPIE, 2005.

[23] N. Nishizawa, Y. Chen, P. Hsiung, E. P. Ippen and J. G. Fujimoto. Real-time, ultrahigh-resolution, optical coherence tomography with an all-fiber, femtosecond fiber laser continuum at 1.5 μm. *Opt. Lett.*, **29**:2846–2848, 2004.

[24] H. Kano and H. Hamaguchi. *In vivo* multi-nonlinear optical imaging of a living cell using a supercontinuum light source generated from a photonic crystal fibre. *Opt. Express.*, **14**:2798–2804, 2006.

[25] K. Shi, P. Li, S. Yin and Z. Liu. Chromatic confocal microscopy using supercontinuum light. *Opt. Express.*, **12**:2096–2101, 2004.

[26] G. McConnell and E. Riis. Two-photon laser scanning fluorescence microscopy using photonic crystal fibre. *J. Biomed. Opt.*, **9**:922–927, 2004.

[27] G. McConnell and E. Riis. Photonic crystal fibre enables short-wavelength two-photon laser scanning fluorescence microscopy with fura-2. *Phys. Med. Biol.*, **49**:4757–4763, 2004.

[28] J. A. Palero, V. O. Boer, J. C. Vijvergerg, H. C. Gerritsen and H. J. C. M. Sterenborg. Short-wavelength two-photon excitation fluorescence microscopy of tryptophan with a photonic crystal fiber based light source. *Opt. Express.*, **13**:5363–5368, 2005.

[29] K. Isobe, W. Watanabe, S. Matsunaga, T. Higashi, K. Fukui and K. Itoh. Multi-spectral two-photon excited fluorescence microscopy using supercontinuum light source. *Jpn. J. Appl. Phys.*, **44**:L167–L169, 2005.

[30] W. Göbel, J. N. D. Kerr, A. Nimmerjahn and F. Helmchen. Miniaturized two-photon microscope based on a flexible coherent fiber bundle and a gradient-index lens objective. *Opt. Lett.*, **29**:2521–2523, 2004.

[31] P. M. Lane, A. L. P. Dlugan, R. Richards-Kortum and C. E. MacAulay. Fibre-optic confocal microscopy using a spatial light modulator. *Opt. Lett.*, **25**:1780–1782, 2000.

[32] F. Helmchen, M. S. Fee, D. W. Tank and W. Denk. A miniature head-mounted two-photon microscope: High-resolution brain imaging in freely moving animals. *Neuron*, **31**:903–912, 2001.

[33] M. T. Myaing, D. J. MacDonald and X. Li. Fiber-optic scanning two-photon fluorescence endoscope. *Opt. Lett.*, **31**:1076–1078, 2006.

[34] D. Bird and M. Gu. Two-photon fluorescence endoscopy with a micro-optic scanning head. *Opt. Lett.*, **28**:1552–1554, 2003.

[35] A. Jain, A. Kopa, Y. Pan, G. K. Fedder and H. Xie. A two-axis electrothermal micromirror for endoscopic optical coherence tomography. *IEEE J. Quant. Elect.*, **10**:636–642, 2004.

[36] H. Xie, Y. Pan and G. K. Fedder. Endoscopic optical coherence tomographic imaging with a CMOS-MEMS micromirror. *Sensors and Actuators*, **103**:237–241, 2003.

[37] P. Tran, D. Mukai, M. Brenner and Z. Chen. *In vivo* endoscopic optical coherence tomography by use of a rotational microelectromechanical system probe. *Opt. Lett.*, **29**:1236–1238, 2004.

[38] P. Herz, Y. Chen, A. Aguirre, K. Schneider, P. Hsiung, J. Fujimoto, K. Madden, J. Schmitt, J. Goodnow and C. Petersen. Micromotor endoscope catheter for *in vivo*, ultrahigh-resolution optical coherence tomography. *Opt. Lett.*, **29**:2261–2263, 2004.

[39] D. Aguirre, P. R. Herz, Y. Chen, J. G. Fujimoto, W. Piyawattanametha, L. Fan, S. Hsu, M. Fujino, M. C. Wu and D. Kopf. Ultrahigh resolution OCT imaging with a two-dimensional mems scanning endoscope. In *Advanced Biomedical and Clinical Diagnostic Systems III*, volume **5692**, pages 277–282. Bellingham, SPIE, 2005.

[40] W. Piyawattanametha, R. P. J. Barretto, T. Ko, B. A. Flusberg, E. D. Cocker, H. Ra, D. Lee, O. Solgaard and M. J. Schnitzer. Fast-scanning two-photon fluorescence imaging based on a microelectromechanical systems (mems) two-dimensional scanning mirror. *Opt. Lett.*, **31**:2018–2020, 2006.

[41] H. Choi, S. Chen, D. Kim, P. T. C. So and M. L. Culpepper. Design of a non-linear endomicroscope biopsy probe. In *Biomedical Optics Topical Meeting*. Fort Lauderdale, 2006.

[42] D. Bird and M. Gu. Resolution improvement in two-photon fluorescence microscopy with a single-mode fiber. *Appl. Opt.*, **41**:1852–1857, 2002.

[43] J. C. Jung, A. D. Mehta, E. Aksay, R. Stepnoski and M. J. Schinitzer. *In vivo* mammalian brain imaging using one- and two-photon fluorescence microendoscopy. *J. Neurophysiol.*, **92**:3121–3133, 2004.

[44] J. C. Jung and M. J. Schnitzer. Multiphoton endoscopy. *Opt. Lett.*, **28**:902–904, 2003.

[45] M. J. Levene, D. A. Dombeck, K. A. Kasischke, R. P. Molloy and W. W. Webb. *In vivo* multiphoton microscopy of deep brain tissue. *J. Neurophysiol.*, **91**:1908–1912, 2004.

[46] D. Bird and M. Gu. Compact two-photon fluorescence microscope based on a single-mode fiber coupler. *Opt. Lett.*, **27**:1031–1033, 2002.

[47] L. Fu, X. Gan and M. Gu. Use of a single-mode fiber coupler for second-harmonic-generation microscopy. *Opt. Lett.*, **30**:385–387, 2005.

[48] F. Legare, C. L. Evans, F. Ganikhanov and X. S. Xie. Towards CARS endoscopy. *Opt. Express.*, **14**:4427–4432, 2006.

[49] J. Limpert, T. Schreiber, S. Nolte, H. Zellmer, A. Tunnermann, R. Iliew, F. Lederer, J. Broeng, G. Vienne, A. Petersson and C. Jakobsen. High-power air-clad large-mode-area photonic crystal fiber laser. *Opt. Express.*, **11**:818–823, 2003.

[50] J. Limpert, A. Liem, M. Reich, T. Schreiber, S. Nolte, H. Zellmer, A. Tunnermann, J. Broeng, A. Petersson and C. Jakobsen. Low-nonlinearity single-transverse-mode ytterbium-doped photonic crystal fiber amplifier. *Opt. Express.*, **12**:1313–1319, 2004.

[51] D. Yelin, B. E. Bouma, S. H. Yun and G. J. Tearney. Double-clad fiber for endoscopy. *Opt. Lett.*, **29**:2408–2410, 2004.

[52] M. Gu. *Principles of Three-Dimensional Imaging in Confocal Microscopes*. Singapore, World Scientific, 1996.

[53] L. Fu and M. Gu. Polarization anisotropy in fiber-optic second harmonic generation microscopy. *Opt. Express.*, **16**:5000–5006, 2008.

[54] P. J. Campagnola, A. C. Millard, M. Terasaki, P. E. Hoppe, C. J. Malone and W. A. Mohler. Three-dimensional high-resolution second-harmonic generation imaging of endogenous structural proteins in biological tissues. *Biophys. J.*, **81**:493–508, 2002.

[55] L. Fu, X. Gan and M. Gu. Characterization of the GRIN lens-fiber spacing toward applications in two-photon fluorescence endoscopy. *Appl. Opt.*, **44**:7270–7274, 2005.

[56] C. Xu and W. W. Webb. Measurement of two-photon excitation of cross sections of molecular fluorophores with data from 690 to 1050 nm. *J. Opt. Soc. Am. B*, **13**:481–491, 1996.

[57] A. Jain and H. Xie. An electrothermal SCS micromirror for large bi-directional 2D scanning. In *13th International Conference on Solid-state Sensor, Actuators and Microsystems*, pages 988–991. Korea, 2005.

[58] W. Denk, J. H. Strickler and W. W. Webb. Two-photon laser scanning fluorescence microscopy. *Science*, **248**:73–75, 1990.

[59] L. Fu, A. Jain, C. Cranfield, H. Xie and M. Gu. Three-dimensional nonlinear optical endoscopy. *J. Biomed. Opt.*, **12**:040501, 2007.

[60] L. Fu. *Fibre-Optical Nonlinear Endoscopy*. Ph.D. thesis, Faculty of Engineering and Industrial Research, Swinburne University of Technology, Australia, 2006.

[61] J. R. Salgueiro and Y. S. Kivshar. Nonlinear dual-core photonic crystal fiber couplers. *Opt. Lett.*, **30**:1858–1860, 2005.

[62] B. H. Lee, J. B. Eom, J. Kim, D. S. Moon, U. Paek and G. Yang. Photonic crystal fiber coupler. *Opt. Lett.*, **27**:812–814, 2002.

[63] H. Kim, J. Kim, U. Paek, B. Lee and K. T. Kim. Tunable photonic crystal fiber coupler based on a side-polishing technique. *Opt. Lett.*, **29**:1194–1196, 2004.

[64] Y. Kim, Y. Jeong, K. Oh, J. Kobelke, K. Schuster and J. Kirchhof. Multiport n×n multimode air-clad holey fiber coupler for high-power combiner and splitter. *Opt. Lett.*, **30**:2697–2699, 2005.

[65] L. Fu and M. Gu. A double-clad photonic crystal fiber coupler for compact nonlinear optical microscopy imaging. *Opt. Lett.*, **31**:1471–1473, 2006.

[66] D. Bird. *Fiber-Optic Two-Photon Fluorescence Microscopy*. Ph.D. thesis, Faculty of Engineering and Industrial Research, Swinburne University of Technology, Australia, 2002.

6   Trapped-particle near-field scanning optical microscopy

The aim of this chapter is to provide a comprehensive understanding of trapped-particle near-field scanning optical microscopy (NSOM). The principle of optical trapping and laser tweezers is briefly explained in Section 6.1. Section 6.2 summarises the motivation of using a laser-trapped microsphere as a probe in NSOM. The basic principle of trapped-particle NSOM is described in Section 6.3. Two major aspects of this technique, laser trapping performance and near-field Mie scattering of dielectric and metallic particles, are discussed in Sections 6.4 and 6.5, respectively. Experimental results on image formation in trapped-particle NSOM are described in Section 6.6. In Section 6.7, some prospects for the future development of this technique are put forward.

6.1   Optical trapping and laser tweezers

Photons carry momentum. When the change in momentum occurs upon reflection, refraction, transmission and absorption of a light beam, the rate of change of momentum results in a force being exerted on an object. The origin of this force can be understood from Newton's laws. A light ray that is refracted through a dielectric particle changes its direction due to the refraction process. Since light carries momentum, a change in light direction implies that there must exist a force associated with that change. The resulting force, manifested as a recoil action due to the momentum redirection, draws mesoscopic particles toward the highest photon flux in the focal region. This recoil is unnoticeable for refraction by macroobjects such as lenses, but it has a substantial and measurable influence on mesoscopic refractive objects such as small dielectric particles.

An illustration of the optical trapping principle is shown in Fig. 6.1(a). The original demonstration of laser trapping was reported in the 1970s. When Ashkin started an experiment with optical traps [1], he realised that an unfocused laser beam would draw objects of high refractive index towards the centre of the beam and propel them in the direction of propagation. An arrangement of two counter-propagating beams allowed objects to be trapped in three dimensions.

In 1986, Ashkin discovered that a single, tightly focused laser beam could be used to capture small dielectric particles and confine them three-dimensionally [2]. An optical tweezers system consists of a laser beam tightly focused into a very small region, generating an extremely large electric field gradient, using a microscope objective. When such a tightly focused laser beam interacts with a mesoscopic particle, piconewton forces

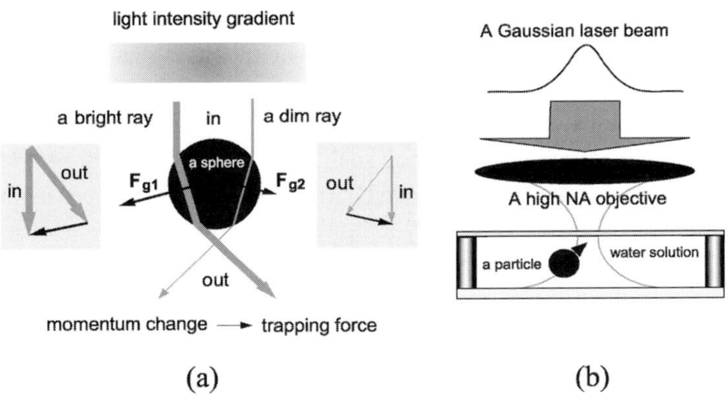

Fig. 6.1 Principle of (a) optical trapping and (b) laser tweezers.

are exerted on the particle and it is attracted toward the highest intensity region by the so-called gradient force, while the radiation force, also known as the scattering force, acts in the direction of the light propagation. Under the conditions when the gradient force dominates, the particle with a refractive index larger than that of the surrounding medium is trapped in three dimensions. For this reason, optical tweezers are usually constructed with a microscope objective, as shown in Fig. 6.1(b).

Optical tweezers provide a tool for non-invasive manipulation of microscopic matter [3–6] and are finding an increasing use in various scientific disciplines including the rapidly expanding fields of single-molecule biophysics, cell biology, studying of molecular motors and manipulation of atoms and particles. In particular, optical tweezers are useful purely as manipulators and positioning devices, owing to their ability to confine, organise and assemble microobjects. The focus in the rest of this chapter is the utilisation of a trapped microobject for optical near-field imaging [5, 6].

6.2   Laser trapped microsphere as a near-field probe

Near-field scanning optical microscopy provides a new opportunity for extracting fine information from a sample under inspection with resolution beyond the diffraction limit [7, 8]. It is fundamentally different from conventional optical microscopy in the sense that it relies on the probing of evanescent photons localised in a region close to a sample under study, while the latter is based on the diffraction property of light in the far-field region.

To detect evanescent photons, one needs a near-field probe which can convert evanescent photons into a far-field region for detection. The characteristics of a near-field probe, in the form of an open aperture or a scatterer, determine primarily image quality such as image resolution and contrast in NSOM [9]. Recently, a Mie particle ($\phi \gg \lambda$, where ϕ and λ are the diameter of a particle and the wavelength of illumination, respectively) or a Rayleigh particle ($\phi \ll \lambda$), which is trapped by a laser beam focused by a high numerical aperture objective, has been utilised as a near-field probe [7]. Imaging of a sample is then achieved by collecting the evanescent photons scattered by a laser-trapped particle. The capability of this type of probe in near-field imaging has been experimentally demonstrated [10–13].

This imaging scheme, termed as trapped-particle NSOM in this chapter, has inherited the merits of the laser trapping technique such as its non-invasive access to a sample. In the case of a dielectric particle [11], it can be trapped in two dimensions against a sample under the radiation pressure caused by a highly focused laser beam. Consequently, the distance between the particle and the sample can be maintained to be zero during a scanning process. Thus, no distance regulation is needed. The advantages of trapped-particle NSOM over conventional NSOM consisting of a tapered tip can be summarised as follows.

Trapped-particle NSOM possesses a high signal-collection efficiency because a high numerical aperture objective is used for trapping. It also exhibits high resolution. Hypothetically, one would think the scattered signal might come from the entire region of a trapped particle. However, the signal from the contacting point of a trapped particle is much stronger than that from the other regions. Resolution is therefore not determined by the size of a trapped particle but by the contacting part between a trapped particle and a sample. An instrument based on this principle can operate at a constant distance (zero) between a probe and a sample because a trapped particle can be pushed toward the surface of a sample. Finally, this imaging system also gives rise to a force sensitivity. This property can be used as a feedback mechanism in the trapped-particle NSOM.

Although a trapped dielectric particle is an appealing probe for NSOM, there are a few problems associated with this technique. First, the signal scattered by a trapped dielectric particle is not strong enough for image construction since a dielectric particle is refractive. Second, although imaging of trapped-particle NSOM uses a high numerical aperture (NA) objective, the transverse trapping efficiency for a dielectric particle decreases with the numerical aperture of a trapping objective [7]. In other words, a compromise must be made between signal strength and scanning speed.

However, these problems may be solved with the employment of a trapped metallic particle. Compared with dielectric particles, metallic particles, especially noble metallic particles, have small skin depths and are less transmissive [14]. In other words, metallic particles may scatter more efficiently than dielectric particles. Further, the transverse trapping force on a metallic particle increases with the numerical aperture of a trapping objective [15]. With the use of a trapped metallic particle, surface plasmon resonance may be excited [16], which could enhance the strength of the scattered signal. As a result, fast imaging and a high signal-to-noise ratio can be obtained simultaneously for near-field imaging with a trapped metallic particle.

6.3   Near-field scanning optical microscopy using a trapped-particle probe

As is well known, conventional optical microscopy offers a useful non-invasive tool for viewing the microscopic world with resolution limited by the diffraction property of light [14]. The growing demand for high-resolution imaging has inspired the emergence

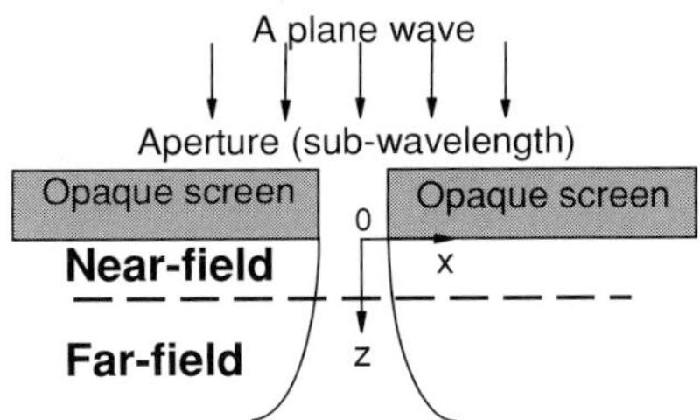

Fig. 6.2 Demonstration of near-field and far-field regions.

of such advanced imaging techniques as confocal microscopy [17, 18] and NSOM [7, 8, 19]. In particular, NSOM has significantly pushed imaging resolution down to the level of tens of nanometres, which is far beyond the diffraction limit [8, 14].

6.3.1 Principle of near-field scanning optical microscopy

To understand the principle of NSOM, we start with the concept of the optical near-field and far-field regions. As an example shown in Fig. 6.2, let us consider that an opaque screen with an aperture of sub-wavelength is illuminated with a plane wave [20]. The light field transmitted through the aperture comprises both longitudinal evanescent (non-propagating) and transverse (propagating) components. The intensity of the evanescent wave along the longitudinal direction is given by

$$I = I_0 \exp(-2\beta z), \tag{6.1}$$

where I_0 is the light intensity at the aperture and β the attenuating factor of the evanescent wave related to the physical properties of the incident plane wave (e.g. wavelength and polarisation) and the medium (e.g. absorption and scattering) where an evanescent wave arises. Usually, the magnitude of β is tens of nanometres [14].

An evanescent wave originates from fine structures or high spatial frequency components of a medium, which are smaller than the illumination wavelength. Because of its exponential decay nature, an evanescent wave can barely travel beyond a particular distance originating from the aperture. The regions inside and outside this critical range, which is usually less than half the illumination wavelength, are termed as the near-field and far-field regions of the aperture, respectively. The propagating components, carrying the information of the structure variation larger than the scale of the illuminating wavelength, are collected by an objective to form a far-field image of the object [14]. It is clear that if the evanescent (non-propagating) components are imaged, the resulting image can have high resolution which is not limited by the diffraction effect. To obtain a signal of high spatial frequencies, it is essential that a small probe can be physically placed in the near-field region of a sample. The idea of NSOM is to utilise these evanescent photons by introducing a near-field probe of sub-wavelength in close proximity to a sample under study. In the presence of this type of probe, the evanescent photons from the sample are converted to propagating ones by scattering or waveguiding. Imaging of NSOM is then realised by scanning a probe across a sample.

Although there are many forms of near-field probe reported so far, one can group near-field probes into three divisions: aperture probes, apertureless probes and trapped-particle probes. For aperture-type probes, the commonly used probes are tapered fibre tips and sub-wavelength apertures. The typical probes for apertureless-type NSOM are semiconductor tips, metallic needles, and semiconductor or metallic cantilevers. A trapped-particle probe can be trapped dielectric or metallic Mie particles.

6.3.2 Principle of near-field scanning optical microscopy using a trapped-particle probe

The use of a trapped particle as a near-field probe for imaging was reported by Ghislain and Webb [21]. Since a trapped particle can be remotely controlled in two or three dimensions with a high accuracy and can act as an efficient scatterer, laser trapping has provided a novel approach for near-field signal acquisition in both force microscopy and optical microscopy.

For force microscopy, a trapped particle is used as a probe to detect the interaction between a sample and the probe [21]. Compared with a mechanical cantilever that converts force into displacement, the spring constant (stiffness) of a trapped probe is reduced by three to four orders of magnitude while a high resonance frequency is retained. This feature is desirable because a small spring constant improves force sensitivity as well as isolating vibration noise. In addition, laser trapping has an advantage that the spring constant of a trap can be instantly changed during the measurement by varying the laser power. By employing a three-dimensional trapped dielectric bead as a position sensor, trapped-particle scanning force microscopy has become possible [22, 23].

The idea of utilising laser trapping for optical imaging was proposed by Malmqvist and Hertz [10]. In this method, an axially trapped dielectric or metallic particle is used as an illumination source for a sample under study [10]. By placing the trapping beam at the bottom of a dielectric particle, the particle can be trapped only in two dimensions [11]. In other words, a transversely trapped dielectric particle is pushed towards the direction of the trapping beam. With this form of particle-trapped NSOM, spatial resolution of less than 100 nm has been demonstrated. This type of NSOM can also be achieved using a metallic particle [13].

A schematic diagram of trapped-particle NSOM is shown in Fig. 6.3. A particle is trapped with a laser beam focused by a high numerical aperture microscope objective. The particle is laterally pulled toward the optical axis of the focused laser beam, while it is vertically pushed down toward the sample surface by the radiation force of the laser beam. As a result, the distance between the sample and the particle is maintained to be zero.

In addition to the trapping laser beam, another laser beam of a different wavelength is incident at the interface between the sample and its substrate (normally a prism) under the condition of total internal reflection. An evanescent wave from the sample is generated with this illumination laser beam and is scattered by the trapped particle to a photomultiplier tube (PMT) placed in the far-field region. To remove stray light scattered from locations other than the trapped particle and therefore improve the signal-to-noise ratio of imaging, a pinhole is mounted at the conjugate point of the particle with respect

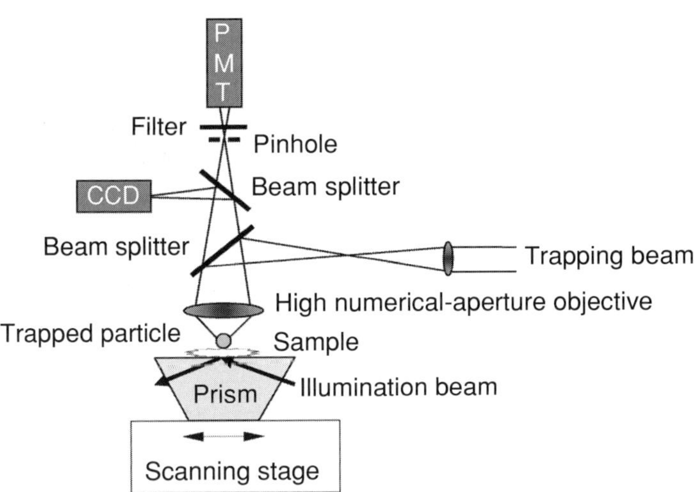

Fig. 6.3 Schematic diagram of trapped-particle NSOM.

to the trapping objective. The reflection of the trapping beam on the particle surface is blocked with a bandpass filter. The image of the sample is constructed by scanning the particle across the sample.

6.3.3 Physical parameters affecting trapped-particle near-field scanning optical microscopy

In comparison with the other forms of NSOM, trapped-particle NSOM is unique since it relies on collecting evanescent waves scattered by a particle for image formation. This process, i.e. the scattering of evanescent waves by a small particle, is termed near-field Mie scattering in analogy to the concept of Mie scattering [14].

Two major aspects related to image formation of trapped-particle NSOM are trapping performance and scattering of evanescent waves. Regarding laser trapping, apodisation of a trapping objective [24], refractive-index mismatching [25] and the numerical aperture of a trapping objective [15, 25] affect the light distribution in the focal region of a trapping objective and consequently the trapping force on dielectric and metallic particles.

It is known from Mie theory that scattering of a propagating wave introduces depolarisation [14, 16]. Depolarisation also occurs to the evanescent wave scattered by small particles [26–28]. It has been shown that both depolarisation and signal strength of scattered evanescent waves affect image formation of particle-trapped NSOM. Such an effect is dependent on the size of trapped dielectric and metallic particles [29, 30].

6.4 Trapping performance of dielectric and metallic particles

The principle of laser trapping has been described in Section 6.1. For a Mie particle, a ray optics model can be used to calculate trapping force on dielectric [31, 32] and

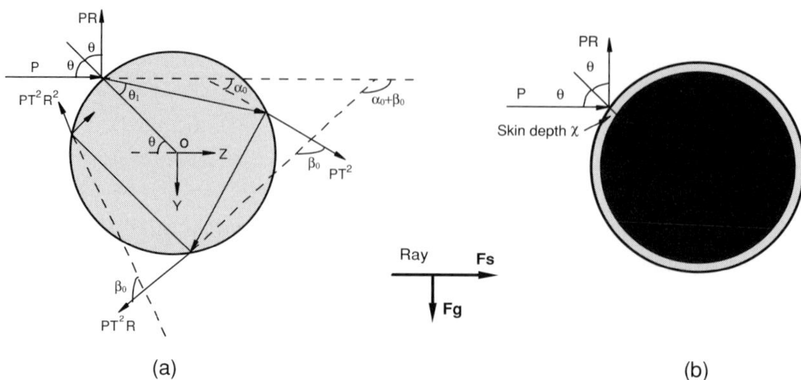

Fig. 6.4 Geometry for the generation of force on dielectric (a) and metallic (b) particles.

metallic [15] particles. In trapped-particle NSOM, the trapping performance is determined by three effects related to a high numerical aperture objective used for trapping, as they determine the scanning speed and stability of trapped-particle NSOM. These effects include the apodisation function of a trapping objective, the numerical aperture of a trapping objective, and the mismatching of refractive indices between the cover glass and the medium where a trapped particle is suspended. Although these effects can be dealt with using an electromagnetic wave model for the trapping force, which is based on the vectorial diffraction theory [33], in the case of Mie particles they can be sufficiently described using the ray optics model with some physical modifications.

6.4.1 Dielectric and metallic particles

According to the ray optics model for a trapping force [31], a trapping laser beam may be simply decomposed into individual rays, and the ray density on an objective lens is assumed to be the same as that of the power density. For a dielectric particle, as shown in Fig. 6.4(a), a single ray of power P gives rise to a series of reflected and refracted rays. Due to the change in momentum per second, the particle experiences a force. The force components resolved into two directions parallel and perpendicular to the incident ray are called the scattering and gradient forces F_s and F_g, respectively, and are given by [31]

$$\begin{cases} F_s = \dfrac{nP}{c}\left\{1 + R\cos 2\Theta - \dfrac{T^2[\cos(2\Theta - 2\Theta_1) + R\cos 2\Theta]}{1 + R^2 + 2R\cos 2\Theta_1}\right\}, \\[4mm] F_g = \dfrac{nP}{c}\left\{R\sin 2\Theta - \dfrac{T^2[\sin(2\Theta - 2\Theta_1) + R\sin 2\Theta]}{1 + R^2 + 2R\cos 2\Theta_1}\right\}. \end{cases} \tag{6.2}$$

Here Θ and Θ_1 are the incident and refractive angles of a single ray with respect to the surface of a particle, R and T are the Fresnel reflectance and transmittance of a particle [14], n is the refractive index of the particle relative to the surrounding medium, and c is the speed of light in vacuum.

When light impinges on a metallic particle, most of the light is reflected due to the high reflectance of the metallic surface, while the rest of the light penetrates through the particle. The energy density of the transmitted light falls to $1/e$ of its original value after the light travels through a skin depth that is usually of several or tens of nanometres [14]. The momentum change of the incident light, or the optical force experienced by a metallic Mie particle, is mainly determined by the reflection at the surface of a particle. The optical force, caused by the multiple reflections on the inner surface of a particle, can be neglected for a metallic Mie particle. The expressions for the scattering force F_s and the gradient force F_g caused by a single ray incident on a metallic particle can be expressed as [15]

$$
\begin{cases}
F_s = \dfrac{nP}{c}\{1 + R\cos 2\Theta\}, \\[2mm]
F_g = \dfrac{nP}{c}\{R\sin 2\Theta\}.
\end{cases}
\tag{6.3}
$$

Equation 6.3 is actually the first term of Eq. 6.2 for the gradient and scattering forces on a dielectric particle.

Assuming that the light intensity distribution over the aperture of a trapping objective is $I(\rho)$, we can express the total trapping force on a particle as

$$
\mathbf{F_t} = \mathbf{F_x} + \mathbf{F_y} + \mathbf{F_z} = \frac{\int_0^{2\pi}\int_0^{\rho_{\max}} \mathbf{F_x} I(\rho)\rho\,d\rho\,d\varphi}{\int_0^{2\pi}\int_0^{\rho_{\max}} I(\rho)\rho\,d\rho\,d\varphi} + \frac{\int_0^{2\pi}\int_0^{\rho_{\max}} \mathbf{F_y} I(\rho)\rho\,d\rho\,d\varphi}{\int_0^{2\pi}\int_0^{\rho_{\max}} I(\rho)\rho\,d\rho\,d\varphi}
$$
$$
+ \frac{\int_0^{2\pi}\int_0^{\rho_{\max}} \mathbf{F_z} I(\rho)\rho\,d\rho\,d\varphi}{\int_0^{2\pi}\int_0^{\rho_{\max}} I(\rho)\rho\,d\rho\,d\varphi},
\tag{6.4}
$$

where ρ and φ are the radial position and the azimuthal angle of a ray over the objective aperture, respectively, and $\rho_{\max}$ is the maximum radius of the objective aperture. $\mathbf{F_x}$, $\mathbf{F_y}$ and $\mathbf{F_z}$ are the sum of the projection of F_s and F_g of a single ray in the x, y, and z directions in a given Cartesian coordinate system. $\mathbf{F_X}$, $\mathbf{F_Y}$ and $\mathbf{F_Z}$ are the total forces in the three directions. In Eq. 6.4, $\mathbf{F_x}$, $\mathbf{F_y}$ and $\mathbf{F_z}$ of a single ray are weighted by the ray density (proportional to the light intensity) before integration and normalised by the total power entering the back aperture of an objective.

The trapping efficiency Q_j, a parameter independent of trapping power for the evaluation of trapping force F_j, is defined as

$$
Q_j = \frac{F_j c}{nP}, \quad j = X, Y, Z.
\tag{6.5}
$$

The projections of the total trapping force (efficiency), $F_t\,(Q_t)$, along the transverse (x or y) and axial (z) directions are defined as the transverse and axial trapping forces (efficiencies) and represented by $F_{tr}\,(Q_{tr})$ and $F_a\,(Q_a)$, respectively, where $Q_t = \sqrt{Q_{tr}^2 + Q_a^2}$.

Using Eqs. 6.2–6.5, we can calculate the total trapping efficiency Q_t for dielectric and metallic particles. As an example, a polystyrene particle of refractive index $n = 1.59$ is considered for the demonstration of dielectric particles. For metallic particles, three types of particle (gold, nickel and silver) are chosen to exhibit high, medium and low absorption properties at a wavelength of 488 nm. Their refractive indices are $n_g = 0.82 + 1.59\mathrm{i}$, $n_n = 1.67 + 2.93\mathrm{i}$ and $n_s = 0.24 + 3.09\mathrm{i}$, respectively [34]. All particles are suspended in water of $n_2 = 1.33$.

The intensity distribution $I(\rho)$ over the aperture of a trapping objective will be discussed in detail in Section 6.4.2. As a demonstration, we assume that $I(\rho)$ satisfies the sine condition [14,35,36] and that a particle is trapped with an oil immersion objective of NA $= 1.25$. The trapping laser beam of wavelength 488 nm is linearly polarised parallel to the x-axis. The trapping laser beam propagates in the direction of the z-axis.

The distribution of the total trapping efficiency Q_t for polystyrene, gold, nickel and silver particles is shown Fig. 6.5. The distribution of Q_t on the right half of the plane is omitted due to its symmetry with respect to the plane at $x = 0$. The particle radius is normalised to be unity. Each individual arrow originates from the trapping position of the laser beam and points towards the direction of the total trapping force. The lengths of the arrows are proportional to the strength of Q_t. The following features can be derived from Fig. 6.5.

It is seen from Fig. 6.5(a) that in the axial direction, the trapping force on a dielectric particle can be either lifting $(-z)$ or pushing $(+z)$ depending on the axial trapping position S above or below the central equatorial plane of the particle. Along the transverse direction, the force on a dielectric particle tends to attract the particle back to the beam axis. A maximum transverse trapping force occurs when the transverse trapping position $|S'| \rightarrow 1$ in the equatorial plane of the particle. This feature implies that a dielectric particle can be trapped in three dimensions in trapped-particle NSOM. However, if the laser beam is focused near the bottom of the particle, only two-dimensional trapping occurs. In other words, a dielectric particle can be pushed against a sample under inspection.

Unlike a dielectric particle where the axial trapping force can be either a lifting or a pushing force depending on the focal position of the trapping laser beam, the axial trapping force on a metallic particle (Figs. 6.5 (b), (c) and (d)) always pushes the particle along the optical axis regardless of the focal position of the trapping laser beam. On the other hand, when the focal position of a laser beam is shifted downward and away from the centre of the particle, the transverse force changes its direction with the transverse position. Transverse trapping of a metallic particle only occurs when the projection of the total force along the transverse direction has a negative value (attractive force). Repulsive force dominates otherwise. The area where transverse trapping exists is located below the central equatorial plane of the particle and away from the beam axis. This result is consistent with the experimental observation that a metallic Mie particle can be trapped only in two dimensions near the bottom of the particle [37, 38].

When a particle is trapped, the maximum transverse trapping efficiency Q_{tr} for metallic particles is larger than that for dielectric particles. In other words, a metallic particle in trapped-particle NSOM provides a stronger signal as well as a faster speed than a dielectric particle, as will be seen in Section 6.6.

Because of high reflection and small skin depth, the scattering force on a metallic particle is dominant over the gradient force. It can be seen from simple geometry (Fig. 6.4) that when the angle θ of a ray of convergence is increased, the net transverse trapping force on a metallic particle is increased. In other words, rays at larger angles of convergence contribute more to the transverse trapping force on a metallic particle. As a result, the use of an obstructed beam leads to the enhancement of the transverse trapping force [39]. For the same reason, the net axial trapping force on a metallic particle may

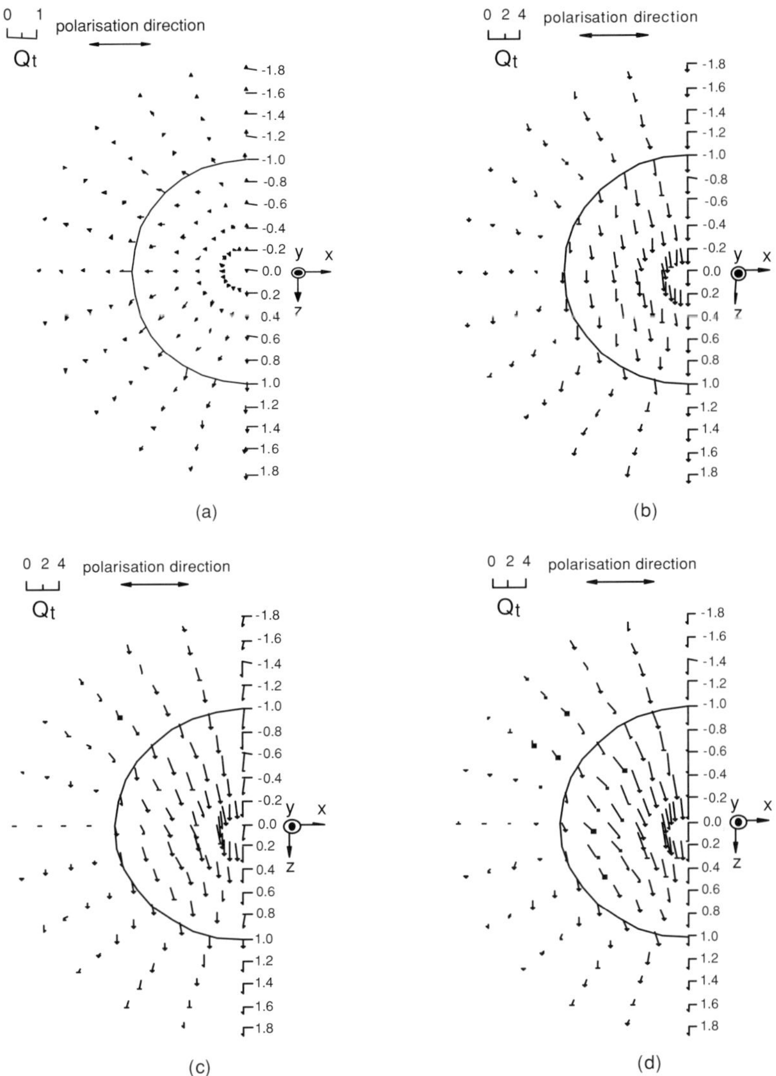

Fig. 6.5 Distributions of the total trapping efficiency Q_t in the x–z plane for polystyrene (a), gold (b), nickel (c) and silver (d) particles, respectively. The polarisation direction of the trapping laser beam is parallel to the x-axis. The laser beam is uniform over the aperture of an objective (oil-immersion, NA $= 1.25$ and linearly-polarised, $\lambda = 488$ nm).

become negative when an obstructed beam is focused near the bottom of the particle, resulting three-dimensional trapping of a metallic particle [40].

6.4.2 Effect of apodisation

In Fig. 6.6, we have assumed the so-called sine condition [14, 35] to determine the intensity distribution $I(\rho)$ over the aperture of a trapping objective. This condition means that when the ray density over the objective aperture is projected into the ray

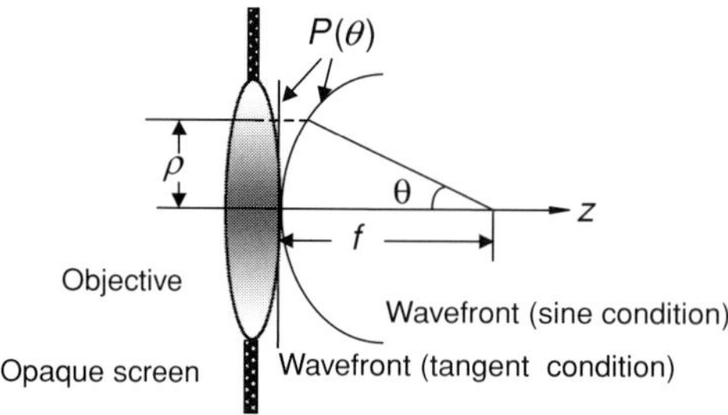

Fig. 6.6 Illustration of the refraction of an incident ray at position ρ by an objective with an apodisation function $P(\theta)$.

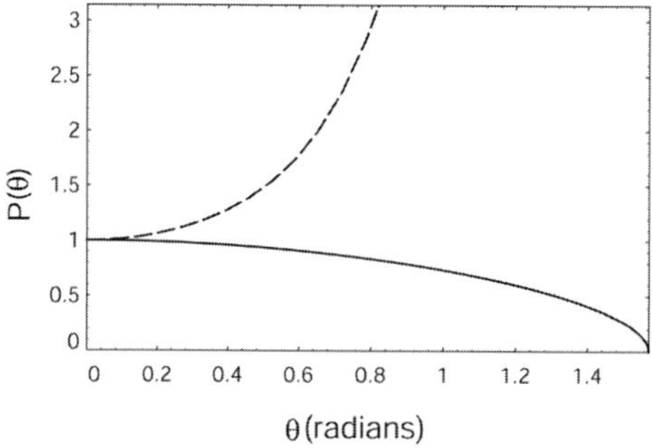

Fig. 6.7 Apodisation function $P(\theta)$ of an objective obeying the sine (solid) and tangent (dash) conditions.

density over the angular aperture of ray convergence, the radial coordinate ρ is projected to the angle of convergence of a ray, θ, via

$$\rho = f \sin\theta, \tag{6.6}$$

where f is the focal length of an objective. An objective obeying the sine condition produces a spherical wavefront (see Fig. 6.6) after the incident rays are refracted by the objective, so that all the incident rays can be converged into a single point and therefore the two-dimensional space-invariant condition is satisfied [36].

In the original ray optics model for the calculation of trapping force [31], the ray projection relation is given by

$$\rho = f \tan\theta, \tag{6.7}$$

which is called the tangent condition [14, 35, 36]. The physical meaning of the tangent condition is that the Helmholtz invariant [14] is obeyed upon the refraction by an objective, that an objective has a constant magnification factor and that the wavefront after the objective is a plane (see Fig. 6.7). In practice, it is extremely difficult to design and manufacture an objective satisfying the tangent condition. Instead, a commercial high numerical aperture objective is designed to satisfy the sine condition.

The amplitude distribution of light on the wavefront after an objective, $P(\theta)$, is called the apodisation function. The sine and tangent conditions correspond to the apodisation functions $P(\theta) = \sqrt{\cos\theta}$ and $P(\theta) = (1/\sqrt{\cos\theta})^3$, respectively [35, 36]. These two

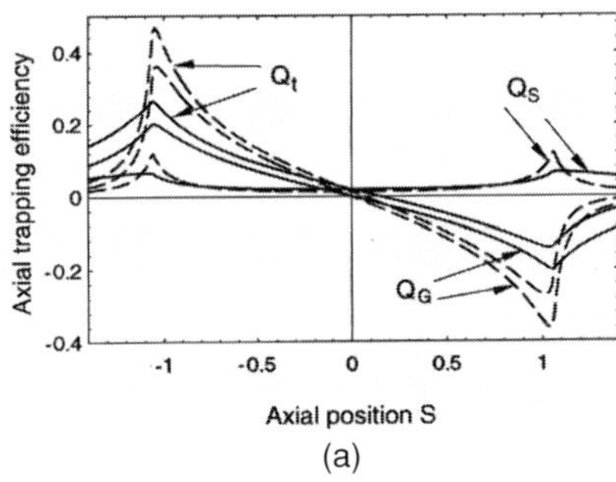

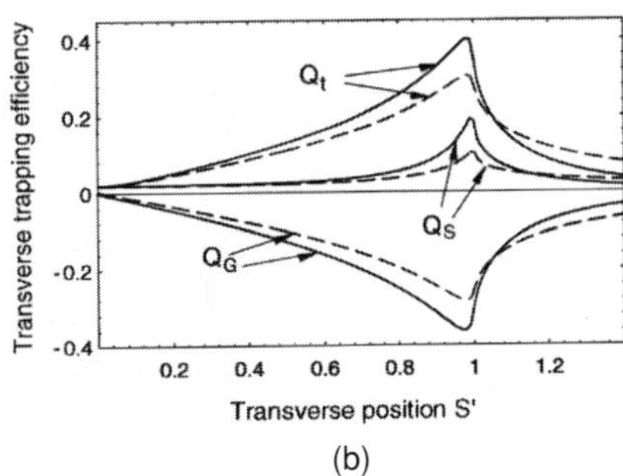

Fig. 6.8 Axial (a) and transverse (b) trapping efficiencies as a function of the particle position under the illumination of a circularly-polarised uniform beam. The solid curves represent the result under the sine condition, while the dashed curves represent that under the tangent condition. For the axial trapping efficiency, $S' = 0$ and Q_t, Q_G, and Q_S correspond to the total, gradient and scattering trapping efficiencies, respectively. For the transverse trapping efficiency, $S = 0$, $Q_{tr} = Q_G$ and $Q_a = Q_S$. The maximum convergence angle of the objective is $70°$ and the relative refractive index is 1.18. Reprinted with permission from Ref. [24], M. Gu, P. Ke and X. Gan, *Rev. Sci. Instrum.* **66**, 3666 (1997). © 1997, American Institute of Physics.

functions give the different light distributions (or ray densities) over the angular aperture in particular when θ is large, as shown in Fig. 6.7. For an objective obeying the sine condition, the apodisation function decreases monotonically with the converging angle θ, while it approaches infinity when the convergence angle θ is close to $90°$ for an objective satisfying the tangent condition.

It is clear that there is a significant difference of trapping force between the sine and tangent conditions when the maximum convergence angle of an objective increases. Figure 6.8 shows an example for a polystyrene dielectric particle [24]. In this case, the magnitude of the axial trapping efficiency under the sine condition is smaller than that under the tangent condition when the axial trapping position $|S|$ is approximately less than 1.02. The difference of the maximum axial trapping efficiency between the two conditions can be up to 50%. However, the magnitude of the transverse trapping efficiency under the sine condition is larger than that under the tangent condition if the transverse trapping position S' is approximately less than 1.02.

6.4.3 Effect of numerical aperture

The ray density under the tangent condition approaches infinity while the ray density under the sine condition approaches zero. As a result, the trapping force in these two cases behaves differently if the numerical aperture of an objective is large. This situation is demonstrated in Fig. 6.9 for a polystyrene dielectric particle.

When the numerical aperture of an objective is low, the difference of the maximum trapping efficiency between the sine and tangent conditions is not significant, but it becomes quite pronounced for a high numerical aperture objective. In comparison with the measured values of the maximum axial trapping efficiency [41], the experimental results fit well to the theoretical curves predicted by the sine condition rather than the

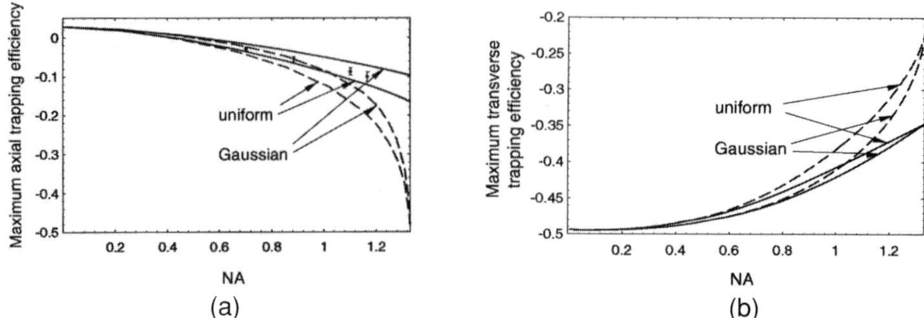

Fig. 6.9 Maximum axial (a) and transverse (b) trapping efficiencies Q_s and Q_{tr} as a function of the numerical aperture of a trapping objective under circularly-polarised uniform and Gaussian beam illumination. For Gaussian beam illumination, the ratio of the beam spot size to the radius of the objective aperture, a, is assumed to be unity. The solid curves correspond to the sine condition, while the dashed curves satisfy the tangent condition. The relative refractive index is 1.18. The black spots in (a) represent the measured values of the maximum axial trapping efficiency. Reprinted with permission from Ref. [24], M. Gu, P. Ke and X. Gan, *Rev. Sci. Instrum.* **66**, 3666 (1997). © 1997, American Institute of Physics.

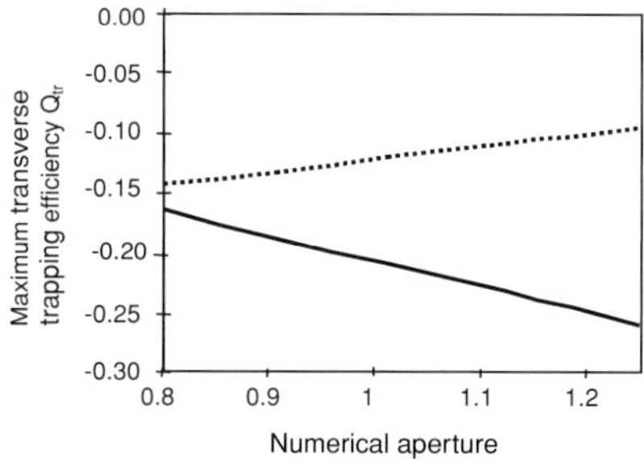

Fig. 6.10 Maximum transverse trapping efficiency Q_{tr} as a function of the numerical aperture of a trapping objective obeying the sine condition for polystyrene (dotted) and gold (solid) particles suspended in water. The trapping beam is s-polarised at wavelength 488 nm.

curves based on the tangent condition, in particular, when the numerical aperture is high. The difference of the maximum transverse trapping efficiency between the sine and tangent conditions becomes pronounced for a high numerical aperture objective. For example, when NA $= 1.2$, the absolute value of the maximum trapping efficiency for the sine condition is approximately 23% larger than that for the tangent condition.

The fact that the value of the maximum transverse trapping efficiency for a dielectric particle decreases with the numerical aperture of a trapping objective is a disadvantage of trapped-particle NSOM. Consequently, there is a trade-off between the signal level determined by the numerical aperture and the scanning speed determined by the trapping force. This problem can be overcome by the use of a metallic particle. As pointed out in Section 6.4.2, the scattering force on a metallic particle is dominant over the gradient force. Therefore, when the angle of a ray of convergence is increased, the net transverse trapping force on a metallic particle is increased. Figure 6.10 shows that the maximum

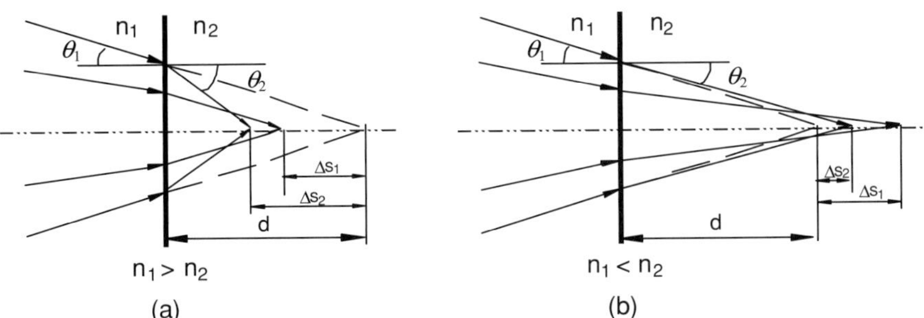

Fig. 6.11 Spherical aberration caused by the refractive-index mismatch at an interface between the first medium (n_1) and the second medium (n_2). The dashed lines illustrate the geometric focus as if there were no interface. The negative and positive axial shifts Δs_i ($i = 1, 2$) are generated under the conditions of $n_1 > n_2$ (a) and $n_1 < n_2$ (b), respectively.

transverse trapping efficiency Q_{tr} for a gold particle increases with the numerical aperture of a trapping objective. This feature was experimentally confirmed [15]. A similar feature for silver and nickel particles was also observed. The increase in the transverse trapping efficiency with the numerical aperture for metallic particles is of importance in trapped-particle NSOM because a trapped metallic particle allows high signal level and high scanning speed to be possible.

6.4.4   Effect of spherical aberration

In trapped-particle NSOM, a particle is usually suspended in a solution such as water. Its refractive index is different from the cover slip (glass) needed for a high numerical aperture objective. Such mismatching of the refractive indices leads to a decrease of the laser power in the focal region of a trapping objective and therefore a decrease in trapping force particularly when the numerical aperture of an objective is large.

To understand this issue, let us consider that a beam is focused from the incident (first) medium of refractive index n_1 into a second medium of refractive index n_2, as illustrated in Fig. 6.11. Because of the refraction at the interface, the position of the geometric focus is axially shifted by a negative and positive value under the conditions of $n_1 > n_2$ (a) and $n_1 < n_2$ (b), respectively. Further, the amount of the shift is dependent on the incident angle of a convergence ray within the aperture of an objective, leading to a distortion of the spherical wavefront produced by an aberration-free objective that obeys the sine condition [35]. Such a wavefront distortion is spherically symmetric and can be quantitatively described, for a high numerical aperture objective, by

$$\Psi(\theta_1, \theta_2, -d) = -d(n_1 \cos \theta_1 - n_2 \cos \theta_2), \tag{6.8}$$

which is called the spherical aberration [35]. Here θ_1 and θ_2 are incident and refractive angles, respectively, and linked by Snell's law [14]. The effect of this function on trapping force cannot be considered in the ray optics model. However, it can be taken into consideration in the vectorial trapping force model introduced in Chapter 8 [33]. Physically, in the presence of spherical aberration, the intensity distribution in the focal

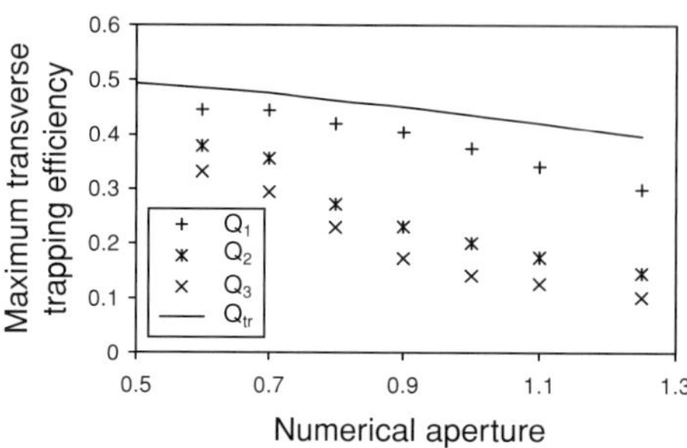

Fig. 6.12 Absolute value of the maximum transverse trapping efficiency as a function of the numerical aperture of a trapping objective for different values of the sample cell thickness d. Q_1, Q_2 and Q_3 correspond to the maximum transverse trapping efficiencies at $d = 34$ μm, 60 μm and 94 μm, respectively. Q_{tr} is the theoretical prediction by the ray optics model under the sine condition [24]. Reprinted with permission from Ref. [25], P. Ke and M. Gu, *J. Mod. Opt.* **45**, 2159 (1998). © 1998, Taylor & Francis, www.informaworld.com.

region of a microscope objective becomes broad in both the axial and transverse directions [25, 42, 43]. As a result, the laser power exerted on a particle is attenuated and the trapping performance becomes poor.

This effect can be experimentally characterised. The measured transverse trapping efficiency as a function of the numerical aperture for different values of the sample cell thickness d is shown in Fig. 6.12. When the thickness of the sample cell increases from 34 μm to 94 μm, the trapping efficiency drops by 25.5% and 66% for NA = 0.6 and NA = 1.25, respectively. This reduction of the trapping efficiency is caused by the spherical aberration resulting from the refractive-index mismatch between the cover slip and the water solution in which a polystyrene particle is trapped. To increase the transverse trapping force of a dielectric particle, one can use an objective of small numerical aperture. However, this method will restrict signal level in particle-trapped NSOM.

6.5　Near-field Mie scattering

In trapped-particle NSOM, scattering of an evanescent wave by a Mie particle is an important process. The characteristics (strength and polarisation) of the scattered evanescent wave resulting from this process determine signal strength, contrast and resolution of trapped-particle NSOM.

6.5.1　Mie scattering and near-field Mie scattering

Mie scattering is referred to as a scattering process of a uniform plane wave by a spherical particle (see Fig. 6.13(a)). When the particle diameter ϕ is small compared with the wavelength λ of the incident light, i.e. when the dimensionless size parameter $q = 2\pi n'\phi/\lambda \ll 1$, Mie scattering reduces to Rayleigh scattering. Here n' is the refractive

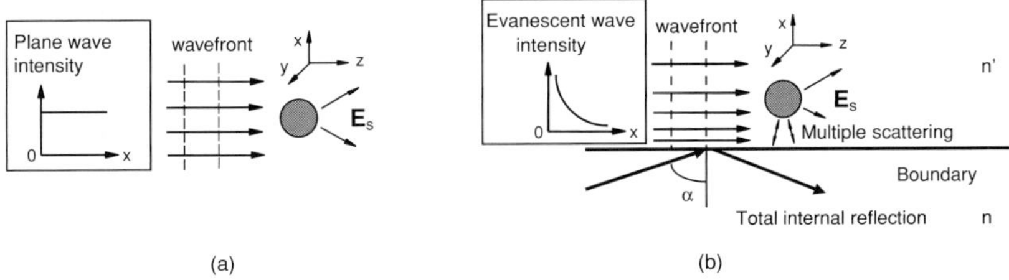

Fig. 6.13 Schematic diagram for Mie scattering (a) and near-field Mie scattering (b). The wavefronts of the plane and evanescent waves are illustrated using dashed lines. The distribution of the light intensity along the x direction is shown. In case (b), a beam of light is incident on a boundary separating two different media under the condition of total internal reflection.

index of the medium surrounding the particle [14]. Near-field Mie scattering is caused by the interaction of a Mie particle with an evanescent wave rather than a plane wave (see Fig. 6.13(b)).

Figure 6.13 illustrates the schemes of Mie scattering and near-field Mie scattering. The spatial distribution of the scattered wave E_s differs between a plane wave and an evanescent wave. For an evanescent wave, the light intensity decays exponentially along the normal to a boundary, while for a plane wave it remains constant at its wavefront. The exponential decay of the evanescent wave above the boundary introduces an asymmetry to the incident field, which results in the presence of certain cross-polarisation components in the scattered field [28]. The damping feature of an evanescent wave causes a strong dependence of the strength of scattered evanescent waves on the distance D between a particle and a boundary where an evanescent wave originates. There is a multiple interaction between a particle and the boundary when the distance D is small [44–46].

6.5.2 Scattered evanescent field in the far-field region

Near-field Mie scattering was treated for the first time by Chew *et al.* [27] in order to attain guidance to the applications of evanescent wave excitation in Raman spectroscopy. This method assumes that a scattering particle is located sufficiently far from the interface at which total internal reflection of illumination occurs, which implies that the boundary effect of the surface is neglected. In Fig. 6.13(b), the incident angle of an electromagnetic wave for total internal reflection is represented by α. The refractive index of the scattering particle is denoted by n_1, while it is immersed into a medium of refractive index n'. The refractive index of the substrate is n and the origin of the x–y–z coordinate system at the centre of the particle. Under these conditions, we can use Chew's method [27] to illustrate the main feature of the scattered evanescent wave.

First, the scattered field of the evanescent wave is depolarised for both dielectric and metallic particles [27]. Second, the scattered field exists in a far-field region, as shown in Fig. 6.14 where the intensity distribution of the scattered evanescent wave along a circle of radius 20 μm in the x–z plane is depicted for dielectric and gold particles. The x–z plane is the plane of incidence of the beam for total internal refraction. It can be noticed

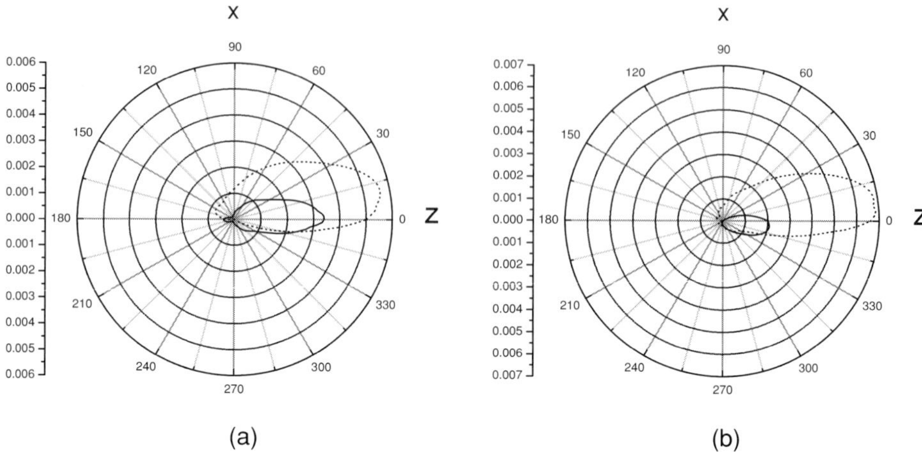

Fig. 6.14 Scattered intensity distribution around dielectric (a) and gold (b) particles of radius 0.5 μm in the x–z plane. The solid and dotted curves correspond to the s- and p-polarisation states of the illumination wave, respectively. $n' = 1.0$, $n = 1.51$, $\lambda = 632.8$ nm and $\alpha = 45°$. (a) $n_1 = 1.6$; (b) $n_1 = 0.13 + i3.16$.

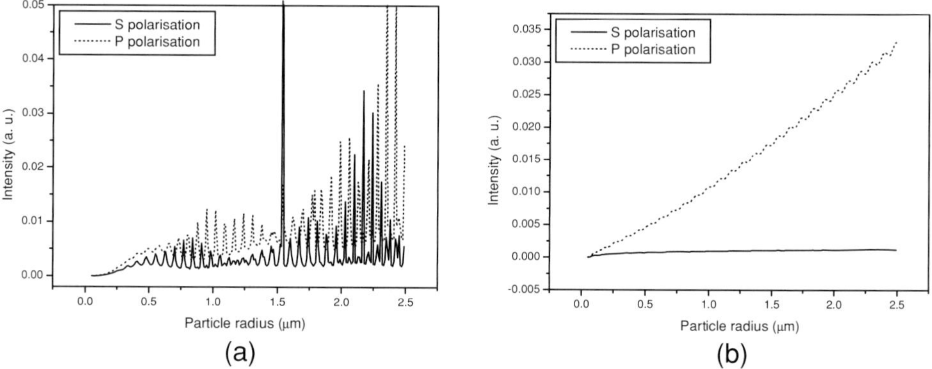

Fig. 6.15 Half-space scattered intensity as a function of particle radius 0.5 μm for s- and p-polarisation illumination. $n' = 1.0$, $n = 1.51$, $\lambda = 632.8$ nm and $\alpha = 45°$: (a) dielectric particle, $n_1 = 1.6$; (b) gold particle, $n_1 = 0.13 + i3.16$.

that the evanescent wave is intensely scattered into certain regions around the particle. Third, in the case of a gold particle, the strength of the scattered field for p-polarised illumination is much stronger than that for s-polarised illumination in particular for a large particle (see Fig. 6.15).

Figure 6.15 shows the intensity integrated over the upper half-space of the particle as a function of the particle size. It is seen that a series of peaks occur as the radius of a dielectric particle increases (see Fig. 6.15(a)). This phenomenon is called morphology-dependent resonance caused by the interference of a light beam propagating inside a dielectric particle confined by total internal reflection [47]. Due to this process, a large energy density can be created inside a particle. Because of the skin depth of a metallic particle, an evanescent wave cannot be effectively refracted and reflected within a metallic particle. As a result, there is no noticeable morphology-dependent resonance for a gold particle, as shown in Fig. 6.15(b).

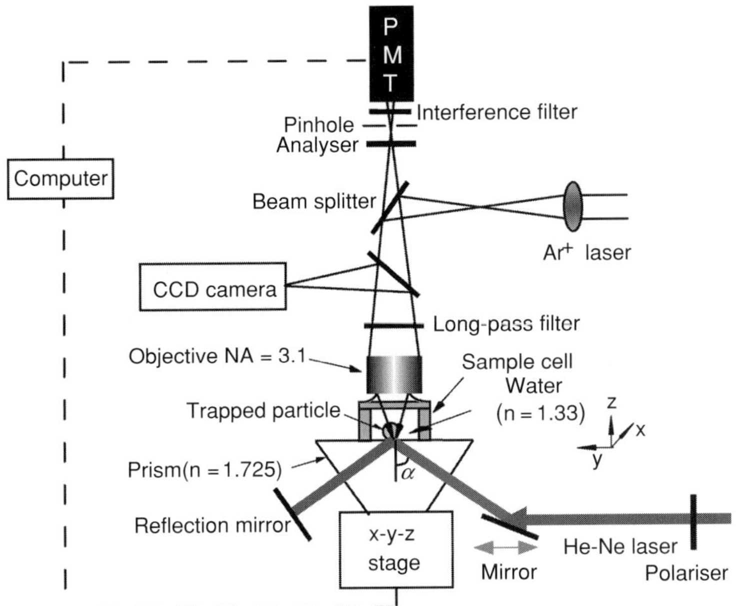

Fig. 6.16 Experimental setup of trapped-particle NSOM. Reprinted with permission from Ref. [28], M. Gu and P. Ke, *App. Phys. Lett.* **75**, 175 (1999). © 1999, American Institute of Physics.

6.6 Image formation in trapped-particle near-field scanning optical microscopy

We have shown that the transverse trapping force on a metallic particle increases with the numerical aperture of a trapping objective. By contrast, the transverse trapping force on a dielectric particle decreases as the numerical aperture increases. As a result, utilising a trapped metallic particle for NSOM may increase the scanning speed for image acquisition while maintaining high signal-to-noise ratio [13]. In addition, the use of a trapped metallic particle as a probe may lead to the enhancement of image quality because of its enhanced scattering effect [13]. If the depolarised scattered photons are removed in detection, image quality in trapped-particle NSOM can be enhanced further [28, 30].

Figure 6.16 shows an experimental implementation of trapped-particle NSOM [13]. Particles for trapping were suspended in water solution ($n' = 1.33$) in a sample cell that was formed of a double-sided tape sandwiched between a cover glass and the prism. The thickness of the sample cell is approximately 120 μm.

An evanescent wave was generated at the surface of an equilateral prism (SF$_{10}$, $n = 1.725$) under the condition of total internal reflection of a He-Ne laser of 17 mW output. The prism was mounted on a three-dimensional scanning stage driven by a piezoelectric controller. The polarisation direction of the He-Ne laser beam was controlled by the polariser. When the reflection mirror (reflectance $R = 85\%$) was in place, an interference evanescent wave pattern was formed on the surface of the prism.

An Ar^+ laser at wavelength 488 nm (5 W) was focused by an oil-immersion objective (NA $= 1.3$) for trapping. A stable two-dimensional trap was achieved for a metallic or dielectric particle by positioning the focal spot of the trapping laser beam near the bottom of the particle. Under this condition, the trapped particle was pushed toward the surface of the prism. The evanescent wave scattered by a trapped particle was then collected by a PMT mounted at the conjugate imaging point of a trapped particle. A CCD camera was used to monitor the trapping process. A narrow bandpass interference filter with a central wavelength of 633 nm was inserted in front of the PMT to block the trapping laser beam. To improve the signal-to-noise ratio of the imaging system, a pinhole of diameter 500 μm was inserted in front of the interference filter to reject the stray light scattered from elsewhere other than the trapped particle. The analyser was used to select the polarisation state of the scattered signal, if necessary.

6.6.1   Effect of scattering strength

The strength of the scattered evanescent wave is determined by the size and material of a particle. To demonstrate these effects on image quality in trapped-particle NSOM, we used dielectric (polystyrene: diameter $\phi = 0.1, 0.2, 0.48, 1$ and 2 μm) and metallic (gold: $\phi = 0.1, 0.5$ and 2 μm; silver: $\phi = 2$ μm) particles [13].

For both dielectric and metallic particles the strength of the scattered signal increases with particle size as predicted by the theory shown in Fig. 6.15. However, no morphology-dependent resonance is observed for dielectric particles because the increment of the particle size in the experiment is larger than the period of morphology-dependent resonance peaks. The monotonic dependence of the strength of the scattered evanescent wave on the size of a particle is due to the increasing interaction (reflection and multiple scattering) between the particle and an evanescent wave when the particle size is increased.

In the case of polystyrene and gold particles, the scattered signal is increased by a factor of approximately 24 and 96, respectively, when the particle size is increased from 0.1 μm to 2 μm. For a particle of diameter 2 μm, the scattered signal from gold and silver particles is approximately 8 and 12 times stronger than that from a polystyrene particle, respectively. The difference of the scattered signal strength between metallic and dielectric particles becomes small as the particle size reduces.

To characterise the image contrast of trapped-particle NSOM, we utilised an evanescent wave interference pattern as a test object. The incident angle of the He-Ne laser (s-polarised) is $\alpha = 60°$. The prism was scanned in the x–y plane to obtain an image. The measured interference evanescent wave patterns shown in Figs. 6.17–6.19 demonstrate the following features.

The interference evanescent wave pattern can be expressed as [48]

$$I(y) = I_0 \left[1 + RT^2 + 2\sqrt{RT} \cos(2ky \sin\theta + \Phi) \right] \exp(-2\beta z), \qquad (6.9)$$

where I_0 is the evanescent wave intensity on the surface of the prism, R and T are the reflectance of the reflection mirror and the transmittance of the prism, respectively, k is

Table 6.1 Strength of the scattered evanescent waves and image contrast of evanescent interference patterns for different types of trapped particles at the surface of an SF$_{10}$ prism. The He-Ne laser was s-polarised and its incident angle inside the prism was set at 65°.

Particle			Dielectric Polystyrene			Metallic			
						Gold	Gold	Gold	Silver
Diameter (µm)	0.1	0.2	0.48	1	2	0.1	0.5	2	2
Signal (a.u.)	0.005	0.01	0.04	0.15	0.25	0.02	0.35	1.95	2.7
Image contrast	1.3%	2.3%	6.2%	10.6%	4.9%	3.9%	6%	12%	14.3%

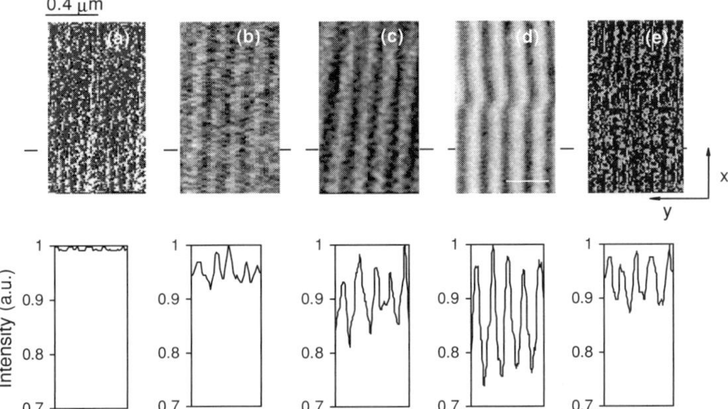

Fig. 6.17 Images (top) and corresponding cross sections (bottom) of interference evanescent wave patterns obtained with a trapped dielectric particle of (a) $\phi = 0.1$ µm; (b) $\phi = 0.2$ µm; (c) $\phi = 0.5$ µm; (d) $\phi = 1$ µm and (e) $\phi = 2$ µm. The acquisition time for each image is approximately 2 min at a speed of 1 µm/s. Reprinted with permission from Ref. [29], P. Ke and M. Gu, *Opt. Comm.* **171**, 205 (1999). © 1999, Elsevier.

the wave number of the incident light inside the prism, and Φ is the phase difference between two interfered waves. The measured fringe spacing measure is approximately 0.21 µm which is consistent with that ($\Delta y = 0.212$ µm) calculated from Eq. 6.9 under the experimental condition.

For dielectric particles (Fig. 6.17), the measured image contrast shown in Table 6.1 exhibits a maximum value for $\phi = 1$ µm. The occurrence of a maximum image contrast is due to the competition of two adverse factors, the signal strength and the background level. Although a larger particle can lead to stronger signal strength that may result in better image contrast, it produces simultaneously a larger background that may degrade image contrast.

A comparison of dielectric and gold particles of given size (Fig. 6.18) shows that the use of a metallic particle leads to a higher scanning speed. Further, a trapped gold particle exhibits an improvement in image contrast from 1.3% to 3.9% due to its higher scattered signal strength.

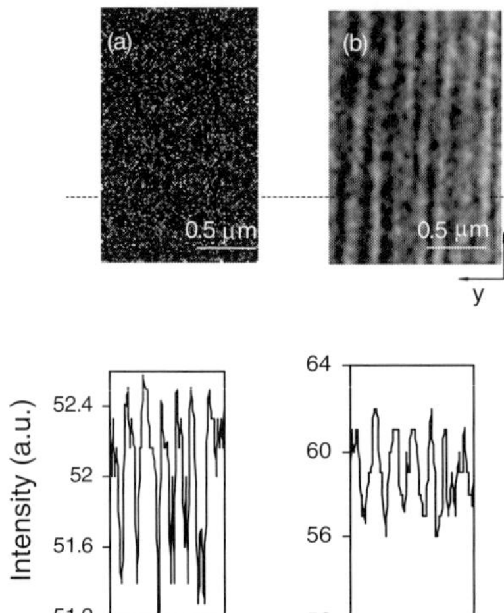

Fig. 6.18 Images (top) and corresponding cross sections (bottom) of interference evanescent wave patterns recorded with a trapped particle of diameter 0.1 μm: (a) polystyrene; (b) gold. The scanning speed of trapped polystyrene and gold particles is 1 μm/s and 1.5 μm/s, respectively, and the corresponding image acquisition time is 2.2 min and 1.6 min.

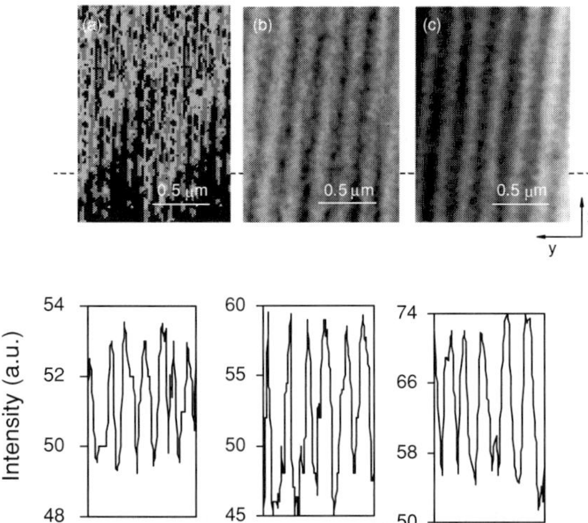

Fig. 6.19 Images (top) and corresponding cross sections (bottom) of interference evanescent wave patterns recorded with a trapped particle of diameter 2 μm: (a) polystyrene; (b) gold; (c) silver. The images are acquired at scanning speeds of 1 μm/s, 1.5 μm/s and 1.5 μm/s for polystyrene, gold and silver particles, respectively.

A comparison between Fig. 6.18 and Fig. 6.19 confirms that for a given material image contrast is improved by the use of a large particle, mainly due to the enhanced signal strength. For example, image contrast is increased by factors of 2.8 and 2.1 for larger polystyrene and gold particles, respectively, and the image contrast for gold and silver particles is improved by factors of 1.5 and 1.9, respectively, as compared with that for a polystyrene particle.

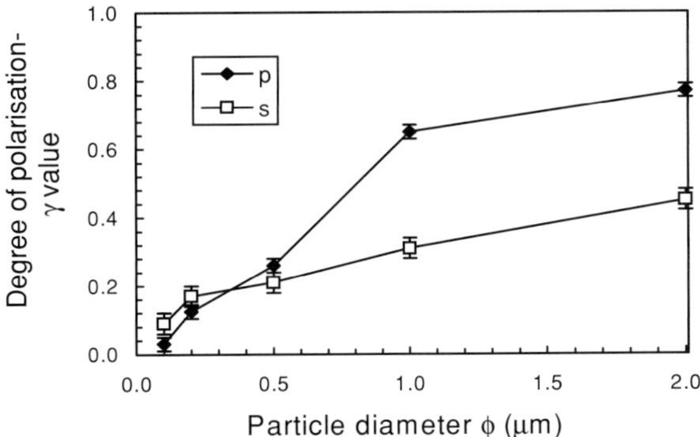

Fig. 6.20 Degree of polarisation as a function of the size of polystyrene particles under s- and p-polarised beam illumination. Reprinted with permission from Ref. [29], P. Ke and M. Gu, *Opt. Comm.* **171**, 205 (1999). © 1999, Elsevier.

6.6.2 Effect of depolarisation

The degree of polarisation of scattered evanescent waves, γ, can be evaluated according to the following definition:

$$\gamma = \frac{I_\parallel - I_\perp}{I_\parallel + I_\perp},\tag{6.10}$$

where subscripts $\parallel$ and $\perp$ denote the analyser (see Fig. 6.16) parallel or perpendicular to the polarisation state of the illumination He-Ne beam. The larger the γ value, the weaker the effect of depolarisation of the scattered evanescent wave.

The degree of polarisation of the scattered evanescent wave for dielectric particles is plotted in Fig. 6.20. For an s-polarised illumination beam, the degree of polarisation of the scattered evanescent wave is approximately $\gamma = 0.11, 0.17, 0.25, 0.31$ and 0.51 for particles of diameters $\phi = 0.1, 0.2, 0.5, 1$ and 2 μm, respectively. For a p-polarised illumination beam, the degree of polarisation of scattered evanescent waves is approximately $\gamma = 0.06, 0.12, 0.30, 0.65$ and 0.78 for particles of diameters $\phi = 0.1, 0.2, 0.5, 1$ and 2 μm, respectively.

Several features of scattered evanescent waves can be summarised from Fig. 6.20. First, the degree of polarisation of the scattered evanescent wave increases with the particle size for both s- and p-polarised incident beams. This property is similar to that of Mie scattering where the depolarisation of a propagating wave is stronger for a smaller particle [14]. Second, the degree of polarisation is insensitive to the incident angle of the illumination He-Ne laser since the interaction between a particle and an evanescent wave is highly localised at the bottom of the particle [29]. Thirdly, the depolarisation of scattered evanescent waves is more significant for s-polarised light than p-polarised light when the particle size is approximately $\phi \geq 0.5$ μm. However this situation reverses for a smaller particle ($\phi < 0.5$ μm). These features can be modelled using the near-field Mie scattering theory described in Section 6.5.

The depolarisation of the scattered evanescent wave by a gold particle is dependent on the incident angle α [30]. The dependence of the degree of polarisation of the scattered evanescent wave on the size of gold particles is summarised in Fig. 6.21. Here p1, p2, p3, and p4 represent the results measured for p-polarised He-Ne beam illumination

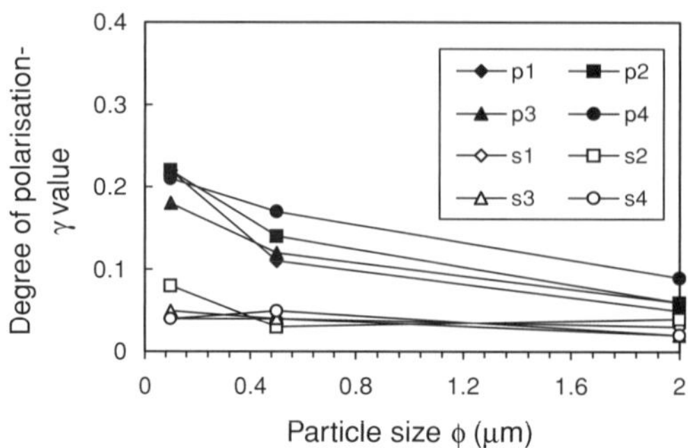

Fig. 6.21 Degree of polarisation as a function of the size of gold particles under s- and p-polarised beam illumination. Reprinted with permission from Ref. [30], M. Gu and P. Ke, *J. App. Phys.* **88**, 5415 (2000). © 2000, American Institute of Physics.

at incident angles of $\alpha = 56°$, $58°$, $60°$ and $62°$, respectively, while s1, s2, s3 and s4 represent those for s-polarised beam illumination at the corresponding incident angles.

Figure 6.21 clearly shows that the degree of polarisation of the scattered evanescent wave decreases with the size of gold particles particularly for p-polarised illumination. For an s-polarised illumination beam, the averaged degree of polarisation is $\gamma = 0.052$, 0.04 and 0.028 for $\phi = 0.1$ μm, 0.5 μm and 2 μm, respectively, while for a p-polarised illumination beam, the averaged degree of polarisation is $\gamma = 0.208$, 0.135 and 0.065 for $\phi = 0.1$ μm, 0.5 μm and 2 μm, respectively. This result differs from that observed in Fig. 6.20 for dielectric particles in which case the degree of polarisation increases monotonically with the size of dielectric particles.

This difference may be related to the generation of surface plasmon resonance [49]. For a metallic particle illuminated with an evanescent wave, the scattered field may be enhanced due to the generation of surface charges. The surface charges form an oscillating distribution under the illumination of evanescent waves and show the characteristics of surface plasmon, which leads to the enhanced scattering for both p- and s-polarised illumination. However, for p-polarised illumination, the enhancement of surface plasmon resonance may occur [49] and becomes stronger for a smaller particle. These physical processes may lead to the feature for gold particles observed in Fig. 6.21.

The impact of depolarisation of near-field Mie scattering on particle-trapped NSOM can be evaluated by measuring the interference evanescent wave pattern with different combinations of the polariser and the analyser shown in Fig. 6.16. I_s and I_p are the intensity measured for s- and p-polarised beam illumination without an analyser. I_{ss} and I_{sp} are measured with an analyser parallel and perpendicular to the incident s-polarised beam and I_{pp} and I_{ps} are measured with an analyser parallel and perpendicular to the incident p-polarised beam, respectively. Images with I_{ss} and I_{pp} are contributed by less-depolarised photons while images with I_{sp} and I_{ps} by strongly-depolarised photons, which is called the polarisation-gating method [28, 30]. Consequently, the effect of depolarised photons on image formation in trapped-particle NSOM can be understood.

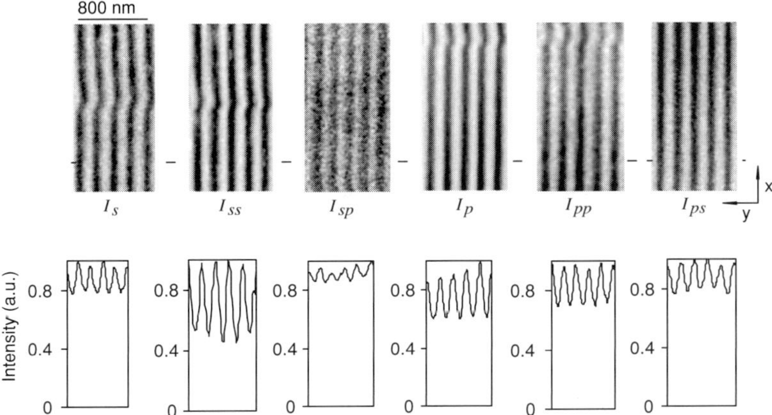

Fig. 6.22 Images (top) and the corresponding cross sections (bottom) of the interference evanescent wave pattern for different polarisation directions of the polariser and the analyser. The images have been normalised by the maximum intensity. The images were recorded with a dielectric particle of $\phi = 1$ μm. Reprinted with permission from Ref. [28], M. Gu and P. Ke, *App. Phys. Lett.* **75**, 175 (1999). © 1999, American Institute of Physics.

For a polystyrene particle of $\phi = 1$ μm, images constructed with I_s, I_{ss}, I_{sp}, I_p, I_{pp} and I_{ps} are shown in Fig. 6.22. The incident angle of the He-Ne laser was $60°$ and the scanning speed was maintained at 1.5 μm/s [28]. According to the intensity cross sections, the contrast for images constructed with I_s, I_{ss}, I_{sp}, I_p, I_{pp} and I_{ps} is approximately 9%, 28%, 6%, 18%, 12% and 14%, respectively. As expected, the image created with I_{ss} shows the best contrast. This result suggests that more strongly depolarised photons carry less information of an object. For p-polarised beam illumination, the image contrast for I_{pp} is slightly poorer than that for I_p because the scattered signal I_p is stronger than I_{pp} and the noise level is increased in the latter case [28].

The effect of polarisation gating on trapped-particle NSOM with a gold particle of $\phi = 0.5$ μm is demonstrated in Fig. 6.23 under the same condition as Fig. 6.22 [30]. The image contrast constructed with I_s, I_{ss}, I_{sp}, I_p, I_{pp} and I_{ps} is approximately 8.3%, 13.4%, 5.0%, 7.0%, 14.1% and 3.0%, respectively. As expected, the image contrast derived with the polarisation gating is better than the others. The improvement in image contrast with I_{pp} is slightly better than that with I_{ss}. This may be caused by the stronger scattered signal of I_{pp}, as suggested by Fig. 6.15.

6.7 A model for conversion of evanescent photons into propagating photons

In Section 6.5, we have shown that scattering of an evanescent wave by a microsphere leads to the propagation of photons in all directions. The aim of this section is to study the imaging process through the evanescent photon conversion by a microscopic particle situated at an interface at which an evanescent wave is generated under total internal reflection conditions.

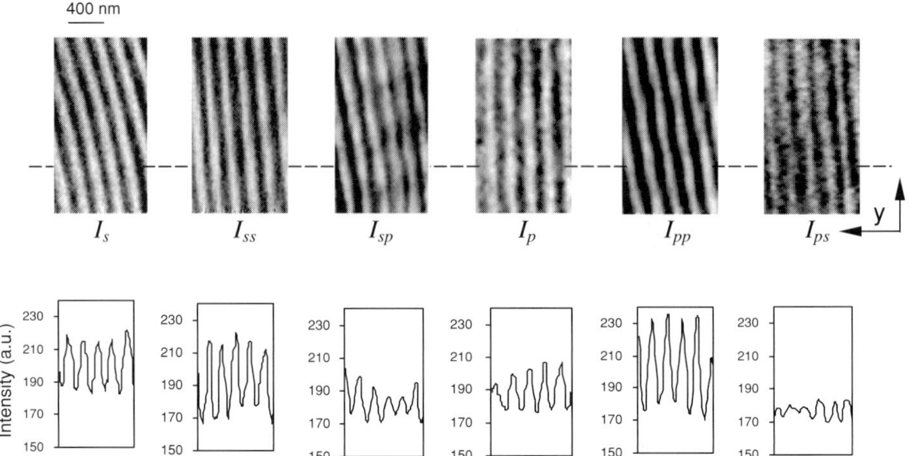

Fig. 6.23 Images (top) and the corresponding cross sections (bottom) of the interference evanescent wave pattern for different polarisation directions of a polariser and an analyser. The images were recorded with a gold particle of $\phi = 0.5$ µm. Reprinted with permission from Ref. [30], M. Gu and P. Ke, *J. App. Phys.* **88**, 5415 (2000). © 2000, American Institute of Physics.

6.7.1 Conversion of evanescent photons to propagation photons

The physical model of the evanescent photon conversion mechanism is based on near-field Mie scattering [50–52] and vectorial diffraction by a high NA lens. To discuss the mathematical treatment of this problem, we first present our analytical expression for the 3D vectorial field distribution around a microscopic particle immersed in an evanescent field. It is derived in the framework of near-field Mie scattering when a particle is situated in a close proximity to the surface on which an evanescent field is generated [50]. The influence of the surface is also taken into account. Subsequently, we include the effect of the trapping/collecting objective by investigating the vectorial diffraction process, to determine the focal intensity distribution (FID) in the image space focal region of the collecting high NA objective. The trapping/collecting objective is one objective which is used both for trapping of a microscopic particle and for collecting the scattered signal. From now on we refer to this objective as the collecting objective only.

Let us consider a microscopic particle in a close proximity of the interface at which an evanescent field is generated by the TIR ($\alpha > \alpha_c$) under either transverse electric (TE) or transverse magnetic (TM) incident illumination (Fig. 6.24(a)). The origin of our coordinate system is located at the centre of the particle with a coordinate system defined in Fig. 6.24(a). The particle is observed by a high NA objective whose focal point coincides with the particle centre. The evanescent wave generated by TIR propagates in the Y_1 direction and decays exponentially in the Z_1 direction, while interacting with the microscopic particle. This interaction can be physically described in terms of superposition of the field scattered by the microscopic particle into upper space (space above the prism surface) and partial reflection of the scattered field into the bottom

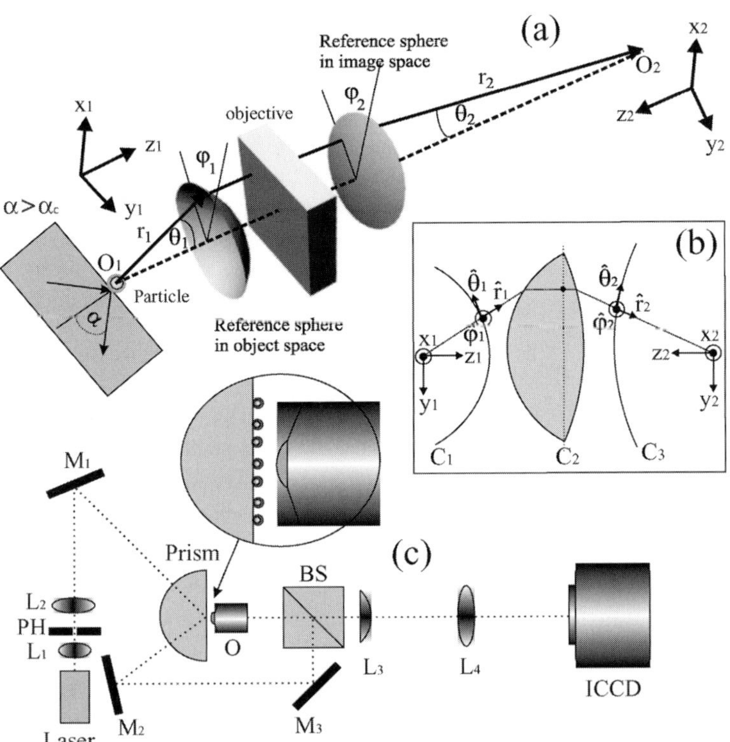

Fig. 6.24 (a) Schematic of our theoretical model for evanescent photon conversion. (b) Representation of the lens transformation process. (c) Experimental setup for recording the FID of converted evanescent photons, collected by a high NA objective. Reprinted with permission from Ref. [52], D. Ganic, X. Gan and M. Gu, *Opt. Express* **12**, 5325 (2004). © 2004, Optical Society of America.

space (space below the prism surface). The scattered field for TE illumination is given by Eq. 6.11, while the full treatment of this physical process is given elsewhere [51]

$$
\mathbf{E}_{\mathrm{SC}}(\mathbf{r}) = \sum_{lm} \left\{ \frac{c\beta_{\mathrm{E}}(l,m)}{n'^2\omega\sqrt{l(l+1)}} \frac{h_l^{(1)}(k'r)}{r\sin\theta} \left[\frac{\partial}{\partial\theta}\left(\frac{\partial Y_{lm}}{\partial\theta}\sin\theta \right) + \frac{1}{\sin\theta}\frac{\partial^2 Y_{lm}}{\partial\varphi^2} \right] \mathbf{r_1} \right.
$$

$$
+ \left[\frac{(-1)\beta_{\mathrm{M}}(l,m)h_l^{(1)}(k'r)}{i\sin\sqrt{l(l+1)}}\frac{\partial Y_{lm}}{\partial\varphi} - \frac{c\beta_{\mathrm{E}}(l,m)}{n'^2\omega\sqrt{l(l+1)}}\frac{1}{r}\frac{\partial Y_{lm}}{\partial\theta}\frac{\partial}{\partial r}(rh_l^{(1)}(k'r)) \right]\theta_1
$$

$$
\left. + \left[\frac{\beta_{\mathrm{M}}(l,m)h_l^{(1)}(k'r)}{i\sqrt{l(l+1)}}\frac{\partial Y_{lm}}{\partial\theta} - \frac{c\beta_{\mathrm{E}}(l,m)}{n'^2\omega\sqrt{l(l+1)}}\frac{1}{r\sin\theta}\frac{\partial Y_{lm}}{\partial\varphi}\frac{\partial}{\partial r}(rh_l^{(1)}(k'r)) \right]\varphi_1 \right\}.
$$

$$(6.11)$$

In Eq. 6.11, $\mathbf{r_1}$, θ_1 and φ_1 are the unit vectors in the spherical coordinate system, $l = 1$ to ∞ and $m = -l$ to $+l$. k' is the wave number and n' is the refractive index of the medium in which a particle probe is immersed. c is the speed of light in vacuum and ω is the angular frequency of the incident light. The functions $\beta_{\mathrm{E}}(l,m)$ and $\beta_{\mathrm{M}}(l,m)$

are the expansion coefficients relating the illumination evanescent field, $h_l^{(1)}(k'r)$ is the spherical Hankel function of the first kind. θ and φ are the variables of the scalar spherical harmonics Y_{lm}. In the case of TM illumination the scattered field is also given by Eq. 6.11 with expansion coefficients $\beta_E(l, m)$ and $\beta_M(l, m)$ substituted by $\tilde{\beta}_E(l, m)$ and $\tilde{\beta}_M(l, m)$, respectively [27].

Using this method one can calculate the superposed scattered field on the reference sphere in object space. The centre of this reference sphere (also known as the entrance pupil of the collecting objective) overlaps with the particle centre, i.e. the origin of the coordinate system O_1. The shadowing effects are neglected in this process. The reason that the shadowing effects can be neglected is because they occur for the reflected portion of the scattered field from the prism surface incident at low angles with respect to the surface normal. At such angles the Fresnel reflection coefficients indicate that only $\sim 4\%$ of incident light is reflected. Furthermore, the field scattered by dielectric particles is most strongly scattered in the direction at a high angle with respect to the surface normal (forward direction) [51]. These two effects indicate that the reflected field contributing to the shadowing effects is 2–3 orders of magnitude weaker than the contribution of the field reflected at a high angle, and therefore can be neglected.

The precise transformation of the field from the reference sphere in object space to the field on the reference sphere in image space requires a detailed model of the lens. We assume that the imaging objective transforms a diverging spherical wave with its origin O_1 in the centre of the particle into a converging spherical wave whose origin O_2 is in the centre of the focal region in image space. Therefore, the lens effect can physically be modelled as a retardation effect affecting the wave traversing two different dielectric media (air and glass). Consider the spherical wavefront C_1 originating from the particle centre O_1 (the origin of the coordinate system $X_1 Y_1 Z_1$), just before the collecting lens (Fig. 6.24(b)). Its curvature corresponds exactly to the curvature of the collecting lens in object space. After traversing the lens front surface, the wavefront becomes the plane wavefront C_2. All points on the spherical wavefront C_1 arrive at the plane wavefront C_2 at the same time. The plane wavefront is then similarly transformed to the converging spherical wavefront C_3, after traversing the lens back surface. The centre of the spherical wavefront C_3 is at O_2 (the origin of the coordinate system $X_2 Y_2 Z_2$). Such transformation further indicates that the lens imparts a scaling effect and a vector rotation. If we consider scattered field vector components, described by its unit vectors $\hat{\mathbf{r}}_1$, $\hat{\theta}_1$ and $\hat{\varphi}_1$ in coordinate system $X_1 Y_1 Z_1$, they are transformed into $-\hat{\mathbf{r}}_2$, $\hat{\theta}_2$ and $\hat{\varphi}_2$ in coordinate system $X_2 Y_2 Z_2$.

Considering such a transformation process of the field from the entrance pupil to the exit pupil, the focal field distribution in image space can be derived by the vectorial diffraction process [35]

$$\mathbf{E}(r_2, \psi, z_2) = \frac{i}{\lambda} \iint_\Omega (-E_{r1}\hat{\mathbf{r}}_2 + E_{\theta 1}\hat{\theta}_2 + E_{\varphi 1}\hat{\varphi}_2) \exp[-ikr_2 \sin\theta_2 \cos(\varphi_2 - \psi)]$$

$$\times \exp(-ikz_2 \cos\theta_2) \sin\theta_2 d\theta_2\, d\varphi_2, \tag{6.12}$$

where E_{r1}, $E_{\theta 1}$ and $E_{\varphi 1}$ are given by Eq. 6.11, while r_2, ψ and z_2 are cylindrical coordinates of a point in image space with a coordinate system shown in Fig. 6.24(a).

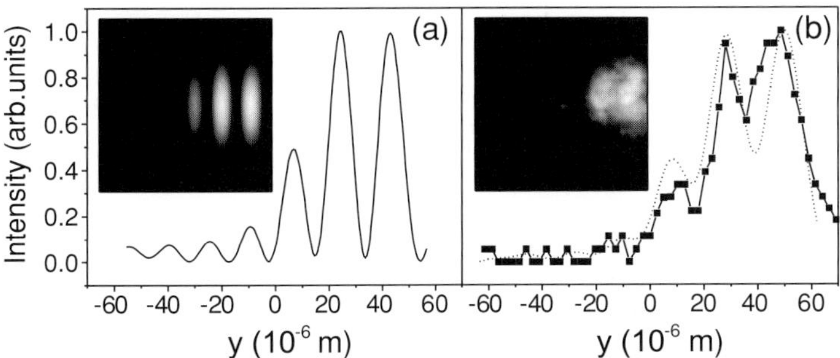

Fig. 6.25 Calculated and observed y-axis scan through $x = 0$, in image focal plane of a 0.8 NA objective collecting propagating photons converted by a 1,000 nm (radius) polystyrene particle under TE incident illumination. (a) Calculated results. (b) Observed results (full line) where the dotted line represents the convolution of the calculated results and the PSF of the imaging lens. Insets show the calculated and observed FID. Reprinted with permission from Ref. [52], D. Ganic, X. Gan and M. Gu, *Opt. Express* **12**, 5325 (2004). © 2004, Optical Society of America.

6.7.2 Theoretical and experimental results

Applied to evanescent photon conversion by a small dielectric particle probe for either TE or TM incident illumination, our model leads to the FID in the far-field of the collecting lens. When the particle radius approaches and exceeds the wavelength of the illuminating light, the FID shows a complex interference-like structure (Fig. 6.25(a)). Furthermore, the result indicates that the conversion and collection of TE evanescent photons are somewhat different from those of TM evanescent photons [51]. The FID in image space of the collecting lens shows a similar interference-like structure for the conversion of either TE or TM localised photons by large particles. However, when the conversion is performed by a small particle, this similarity in the FID is less pronounced [51]. The complex interference-like pattern arises due to the enhancement of morphology-dependent resonance (see Chapter 7), higher multi-poles, scattering properties of large particles and the effects of the interface on which the evanescent field is generated [51].

To confirm the conversion mechanism given by our model we have conducted an experiment. The experimental setup is depicted in Fig. 6.24(c). A helium-neon laser beam was expanded and filtered using lenses L_1, L_2 and a pinhole (PH). It was then directed onto the prism–air surface by mirror M_1 to form an incident angle of $51°$, which was well above the critical angle (α_c). The prism used in the experiment had a refractive index of 1.722 and the particles under investigation were polystyrene particles diluted in water and dried on the prism surface. The particles, immersed into the created localised field, were imaged using a dry 0.8 NA objective and projected onto an intensified CCD camera via lenses L_3 and L_4. The TIR portion of the incident beam was re-routed via mirrors M_2 and M_3 and a beamsplitter (BS) into the back aperture of the collecting objective (O), to enable us to locate the prism surface and thus the centre of the particle under consideration.

The interference-like FID structure for evanescent photon conversion by a large dielectric particle probe can also be experimentally observed. Fig. 6.25(a) shows our

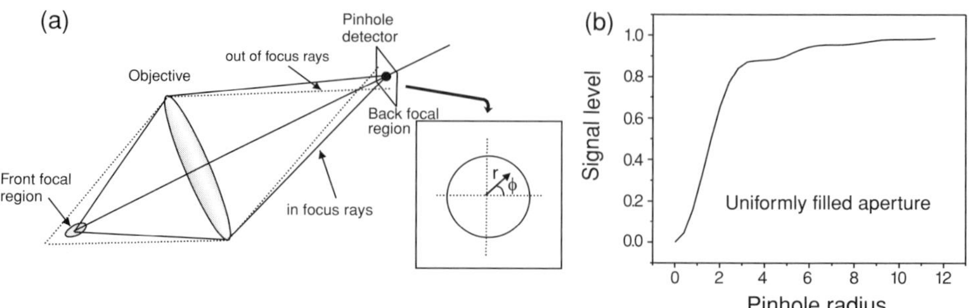

Fig. 6.26 (a) A schematic diagram of a pinhole detection process. Only the rays coming from the front focal region are detected. (b) Detected signal intensity as a function of a pinhole radius, in optical coordinates, for uniformly illuminated objective. Assumed objective NA $=0.8$ in the front focal region, aperture size $\rho_a = 3$ mm and the back focal length of the objective $f = 160$ mm. Reprinted with permission from Ref. [52], D. Ganic, X. Gan and M. Gu, *Opt. Express* **12**, 5325 (2004). © 2004, Optical Society of America.

calculated result of the FID for evanescent photon conversion by a 1 μm (radius) polystyrene particle. The corresponding experimentally observed result is shown in Fig. 6.25(b). Image resolution of the observed result is somewhat degraded due to the imaging properties of lens L_4. The observed structure is a result of the convolution of the point-spread function (PSF) of the imaging lens L_4 and the calculated result shown in Fig. 6.25(a). The agreement between calculated and experimental results confirms that the conversion of evanescent photons is the result of two physical processes, near-field Mie scattering and vectorial diffraction.

6.7.3   Signal level

Now let us turn to the signal level of the scattered near-field signal through a pinhole detector. Such a detector is utilised in the trapped-particle SNOM to discriminate against the out of focus signal. A pinhole detector is essentially a small circular opening of a few to several tens of micrometres, in an otherwise opaque screen, placed perpendicularly to the optical axis at the back focal plane of the imaging objective. The back focal plane focus coincides with the centre of the pinhole. Only the signal that can pass through this opening is detected, and it consists of the signal coming from the front focal region of the imaging objective (Fig. 6.26(a)). Mathematically, the signal level η of a pinhole detector can be expressed as

$$\eta = \frac{\int_0^R \int_0^{2\pi} I(r, \phi) r \, dr \, d\phi}{\int_0^\infty \int_0^{2\pi} I(r, \phi) r \, dr \, d\phi}, \tag{6.13}$$

where R denotes the pinhole radius and $I(r, \phi)$ is the intensity at a point within the pinhole detector determined by distance r from the centre of the pinhole and an angle ϕ. If we express the pinhole radius R in optical coordinates V_R defined as $V_R = 2\pi \rho_a R/(\lambda f)$, where ρ_a denotes objective aperture radius and f is the back focal length of the objective, then the signal level as a function of the pinhole size for a uniformly filled aperture is

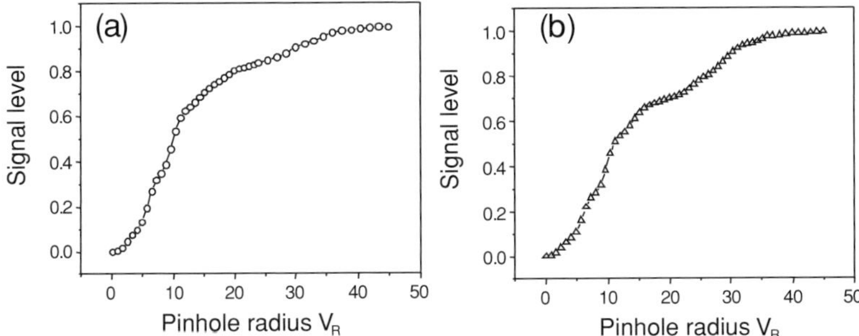

Fig. 6.27 Scattered level as a function of pinhole size (in optical coordinates) of a polystyrene particle for TE illumination (a) and TM illumination (b). Assumed objective NA $=0.8$ in the front focal region, aperture size $\rho_a = 3$ mm and the back focal length of the objective $f = 160$ mm. Particle radius 1.0 μm. Reprinted with permission from Ref. [52], D. Ganic, X. Gan and M. Gu, *Opt. Express* **12**, 5325 (2004). © 2004, Optical Society of America.

shown in Fig. 6.26(b). This result is essentially the fraction of the total energy contained within circles of prescribed radii (varying pinhole size) [14], in the Fraunhofer diffraction pattern of a circular aperture.

Equation 6.13 can be used with the intensity $I(r, \phi)$ determined by our scattering model (Eq. 6.11)) for any point within the pinhole detector. Performing the appropriate integration in Eq. 6.13 determined by the pinhole size, the signal level can be evaluated for the two polarisation states of the incident illumination and a range of pinhole sizes and scattering particles. If we consider a polystyrene scattering particle, the signal level of a pinhole detector is shown in Fig. 6.27. For the wavelength-size particles a large pinhole size is required to collect the signal completely. For the typical conditions given in Fig. 6.27, it can be estimated that a pinhole of a radius of 200 μm is required to collect the total signal for TE and TM incident illumination. Below this pinhole size, the signal level for TE or TM incident illumination is different for a given pinhole.

6.8   Summary

It has been shown that trapped-particle NSOM is capable of converting the localised photons carrying the high-resolution information of an object into a far-field region for detection and imaging. This technique will become a useful nano-metric and nano-imaging tool in many areas including single molecule detection [53] and nanophotonic data storage. However, a few aspects of trapped-particle NSOM should be further improved. The stability of a trapped particle can be improved if a $\text{TEM}_{01}^{\star}$ is used for trapping. To increase the sensitivity of this imaging technique, one can utilise the property of morphology-dependent resonance of a Mie particle. The effect of morphology dependent resonance can be enhanced if a femtosecond laser beam is used, as will be detailed in the next chapter. Such an enhancement may prove the advantage of using a lasing Mie particle in an optical trap [54] to increase the sensitivity of signal collection but also to provide a new mechanism for image contrast.

References

[1] A. Ashkin. Acceleration and trapping of particles by radiation pressure. *Phys. Rev. Lett.*, **24**:1560–1569, 1970.

[2] A. Ashkin, J. M. Dziedzic, J. E. Bjorkholm and S. Chu. Observation of a single-beam gradient force optical trap for dielectric particles. *Opt. Lett.*, **11**:288–290, 1986.

[3] D. Grier. A revolution in optical manipulation. *Nature*, **424**:810–816, 2003.

[4] K. Dholakia, P. Reece and M. Gu. Light takes hold: optical micromanipulation. *Chem. Soc. Rev.*, **37**:42–45, 2008.

[5] P. C. Ke. *Near-field Scanning Optical Microscopy with Laser Trapping*. Ph.D. thesis, Department of Physics, Victoria University of Technology, Australia, 2000.

[6] D. Ganic. *Far-field and Near-field Optical Trapping*. Ph.D. thesis, Faculty of Engineering and Industrial Research, Swinburne University of Technology, Australia, 2005.

[7] D. W. Pohl, W. Denk and M. Lanz. Optical stethoscopy: imaging recording with resolution $\lambda/20$. *App. Phys. Lett.*, **44**:651–653, 1984.

[8] E. Betzig, J. K. Trautman, T. D. Harris, J. S. Weiner and R. L. Kostelak. Breaking the diffraction barrier: optical microscopy on a nanometric scale. *Science*, **251**:1468–1470, 1991.

[9] M. A. Paesler and P. J. Moyer. *Near-field Optics: Theory, Instrumentation and Applications*. New York, John Wiley & Sons, 1996.

[10] L. Malmqvist and M. Hertz. Trapped particle optical microscopy. *Opt. Comm.*, **94**:19–24, 1992.

[11] S. Kawata, Y. Inouye and T. Sugiura. Near-field scanning optical microscopy with a laser-trapped probe. *Jpn. J. Appl. Phys.*, **33**:L1725–L1727, 1994.

[12] T. Sugiura, T. Okada, Y. Inouye, O. Nakamura and S. Kawata. Gold-bead scanning near-field optical microscope with laser-force position control. *Opt. Lett.*, **22**:1663–1665, 1997.

[13] M. Gu and P. Ke. Image enhancement in near-field scanning optical microscopy with laser-trapped metallic particles. *Opt. Lett.*, **24**:74–76, 1999.

[14] M. Born and E. Wolf. *Principles of Optics*. New York, Pergamon, 1980.

[15] P. C. Ke and M. Gu. Characterization of trapping force on metallic Mie particles. *Appl. Opt.*, **38**:160–167, 1999.

[16] C. F. Bohren and D. R. Huffman. *Absorption and Scattering of Light by Small Particles*. New York, John Wiley, 1983.

[17] M. Minsky. Microscopy apparatus. US patent 3013467, Dec 1961. (Filed Nov. 7, 1957).

[18] M. Gu. *Principles of Three-Dimensional Imaging in Confocal Microscopes*. Singapore, World Scientific, 1996.

[19] E. Betzig and J. K. Trautman. Near-field optics: microscopy, spectroscopy and surface modification beyond the diffraction limit. *Science*, **257**:189–195, 1992.

[20] H. A. Bethe. Diffraction by small holes. *Phys. Rev.*, **66**:163–182, 1944.

[21] L. P. Ghislain and W. W. Webb. Scanning force microscope based on an optical trap. *Opt. Lett.*, **18**:1678–1680, 1993.

[22] E. Florin, J. Hörber and E. H. K. Stelzer. High-resolution axial and lateral position sensing using two-photon excitation of fluorophores by continious-wave Nd:YAG laser. *App. Phys. Lett.*, **69**:446–449, 1996.

[23] M. E. Friese, A. G. Truscott, H. Rubinsztein-Dunlop and N.R. Heckenberg. Three-dimensional imaging with optical tweezers. *Appl. Opt.*, **38**:6597–6603, 1999.

[24] M. Gu, P. C. Ke and X. S. Gan. Trapping force by a high numerical aperture microscope objective obeying the sine condition. *Rev. Sci. Instrum.*, **68**:3666–3668, 1997.

[25] P. C. Ke and M. Gu. Characterisation of trapping force in the presence of spherical aberration. *J. Mod. Opt.*, **45**:2159–2168, 1998.

[26] C. Liu, T. Kaiser, S. Lange and G. Schweiger. Structural resonances in a dielectric sphere illuminated by an evanescent wave. *Opt. Comm.*, **117**:521–531, 1995.

[27] H. Chew, D. S. Wang and M. Kerker. Elastic scattering of evanescent electromagnetic waves. *Appl. Opt.*, **18**:2687–2697, 1979.

[28] M. Gu and P. C. Ke. Effect of depolarization of scattered evanescent waves on particle-trapped near-field scanning optical microscopy. *App. Phys. Lett.*, **75**:175–177, 1999.

[29] P. C. Ke and M. Gu. Dependence of strength and depolarization of scattered evanescent waves on the size of laser trapped dielectric particles. *Opt. Comm.*, **171**:205–211, 1999.

[30] M. Gu and P. C. Ke. Depolarization of evanescent waves scattered by laser-trapped gold particles: Effect of particle size. *J. App. Phys.*, **88**:5415–5420, 2000.

[31] A. Ashkin. Forces of a single-beam gradient trap on a dielectric sphere in the ray optics regime. *J. Biophys.*, **61**:569–582, 1992.

[32] W. H. Wright and G. J. Sonek. Radiation trapping forces on microscopheres with optical tweezers. *App. Phys. Lett.*, **63**:715–718, 1993.

[33] D. Ganic, X. Gan and M. Gu. Exact radiation trapping force calculation based on vectorial diffraction theory. *Opt. Express*, **12**:2670–2675, 2004.

[34] D. R. Lide. *CRC Handbook of Chemistry and Physics, 77th edn.* Boca Raton, CRC Press, 1996.

[35] M. Gu. *Advanced Optical Imaging Theory*. Berlin, Springer Verlag, 2000.

[36] J. J. Stamnes. *Waves in Focal Regions*. Bristol, Adam Hilgar, 1986.

[37] S. Sato, Y. Harada and Y. Waseda. Optical trapping of microscopic metal particles. *Opt. Lett.*, **19**:1807–1809, 1994.

[38] H. Furukawa and I. Yamaguchi. Optical trapping of metallic particles by a fixed gaussian beam. *Opt. Lett.*, **23**:216–218, 1998.

[39] M. Gu, D. Morrish and P. C. Ke. Enhancement of transverse trapping efficiency for a metallic particle using an obstructed laser beam. *App. Phys. Lett.*, **77**:34–36, 2000.

[40] M. Gu and D. Morrish. Three-dimensional trapping of Mie metallic particles by the use of obstructed laser beams. *J. Appl. Phys.*, **91**:1606–1612, 2002.

[41] W. H. Wright, G. J. Sonek and M. W. Berns. Parametric study of the forces on microspheres held by optical tweezers. *Appl. Opt.*, **33**:1735–1748, 1994.

[42] D. Day and M. Gu. Effects of refractive-index mismatch on three-dimensional optical data-storage density in a two-photon bleaching polymer. *Appl. Opt.*, **37**:6299–6304, 1998.

[43] D. Ganic, X. Gan and M. Gu. Reduced effects of spherical aberation on penetration depth under two-photon excitation. *Appl. Opt.*, **39**:3945–3947, 2000.

[44] R. Wannemacher, A. Pack and M. Quinten. Resonant absorption and scattering in evanescent fields. *Appl. Phys. B*, **68**:225–232, 1999.

[45] R. Wannemacher, M. Quinten and A. Pack. Evanescent-wave scattering in near-field optical microscopy. *J. Microscopy*, **194**:260–264, 1999.

[46] M. Quinten, A. Pack and R. Wannemacher. Scattering and extinction of evanescent waves by small particles. *Appl. Phys. B*, **68**:87–92, 1999.

[47] P. W. Barber and R. K. Chang. *Optical effects associated with small particles*. World Scientific, Singapore, 1988.

[48] C. Bainier, D. Courjon, F. Baida and C. Girard. Evanescent interferometry by scanning optical tunneling detection. *J. Opt. Soc. Am. A*, **13**:267–275, 1996.

[49] M. Kerker. *The Scattering of Light and Other Electromagnetic Radiation*. New York, Academic Press, 1969.

[50] D. Ganic, X. Gan and M. Gu. Three-dimensional evanescent wave scattering by dielectric particles. *Optik*, **113**:135–141, 2002.

[51] D. Ganic, X. Gan and M. Gu. Parametric study of three-dimensional near-field Mie scattering by dielectric particles. *Opt. Comm.*, **216**:1–10, 2003.

[52] D. Ganic, X. Gan and M. Gu. Near-field imaging by a trapped micro-particle: a model for conversion of localised photons to propagating photons. *Opt. Express*, **12**:5325–5335, 2004.

[53] Y. Ishii and T. Yanagida. Single molecule detection in life science. *Single Mol.*, **1**:5–16, 2000.

[54] K. Sasaki. Microspectroscopy with an optically-manipulated lasing particle. *Mat. Sci. Eng. B Solid*, **48**:147–152, 1997.

7 Femtosecond pulse laser trapping and tweezers

In this chapter, we introduce a new trapping and excitation technique, which utilises a single femtosecond pulse infrared illumination source to simultaneously trap and excite a microsphere probe. The induction of morphology dependent resonance (MDR) in the trapped probe is achieved under two-photon excitation. Monitoring of the MDR in the trapped probe provides a contrast mechanism for imaging and sensing. The experimental measurement of MDR within a laser trapped microsphere excited under two-photon absorption is confirmed in Section 7.2. The effect of the laser power as well as the pulse width on the transverse trapping force is investigated in Section 7.3. The dependence of two-photon induced MDR on the scanning velocity of a trapped particle is then experimentally determined. These parameters are fundamental to the acquisition of images and sensing with femtosecond laser tweezers as described in Section 7.4.

7.1 Introduction

Laser trapping is an ideal method for the remote, non-invasive manipulation of a morphology dependent resonance microcavity. Controlled scanning and manipulation of the microcavity is possible via laser trapping. The microcavity has an enhanced evanescent field at its surface due to the resonant circumferential propagation of radiation at glancing angles greater than the critical angle. Freely suspended in a medium, the cavity becomes increasingly sensitive to its surrounding environment. The interaction of the cavity with its local environment during scanning dynamically alters the coupling to and leakage from the cavity. Monitoring the change in coupling to and leakage from the cavity over time enables imaging and sensing [1].

This scanning particle trapped optical microscopy technique is complementary to that previously described in Chapter 6. The difference being that instead of using a particle to scatter an externally applied evanescent field, in this chapter, the evanescent field is directly generated at the surface of the trapped particle.

Generally, two beams are required for a trapping-MDR system in order to induce MDR in a trapped microparticle. Under this circumstance, the focal spots of the two beams have to be dynamically controlled with high accuracy. Incorporating the trapping and excitation beams into one is advantageous as it greatly simplifies the optical and control systems.

A variety of laser sources have been incorporated into laser trapping systems and used to induce two-photon fluorescence from a trapped dye-doped microsphere [2–4]. It is difficult to achieve simultaneous excitation of MDR and trapping from a single continuous wave (CW) laser beam. The physical reason for this difficulty is that the required power density for particle trapping and MDR is different from that under two-photon femtosecond excitation. For example, under a tight focus of a high numerical aperture (NA) objective, the power required for steady trapping is usually at least one order of magnitude higher than the fluorescence photobleaching level. To overcome this problem, a femtosecond pulse laser beam is advantageous to perform simultaneous trapping and two-photon fluorescence excitation.

As a laser trapped particle experiences the average effect of its trapping laser beam, there is trade-off between the average power and the high peak power of a femtosecond pulse laser beam needed for efficient two-photon absorption. Increasing the power is advantageous for both trapping and two-photon signal. The downside is that too high a power can result in sample damage and even destruction. Single-photon trapping with a femtosecond pulse beam has a similar paradigm in that damage to the sample can occur at a power close to and even below the trapping threshold. MDR induced by two-photon excitation [5] has also been shown to overcome difficulties in separation of excitation and resonant wavelengths and the inability to confine excitation illumination precisely [6].

7.2   Morphology dependent resonance under femtosecond laser illumination

The highly localised spatial nature of two-photon excitation enables the introduction of MDR in a microcavity to be tightly controlled. Therefore fluorescence excitation at various spots within the three-dimensional (3D) space of a microsphere can be investigated in detail.

Morphology dependent resonance arises due to the constructive interference of light rays at near glancing angles by total internal reflection within a microcavity (see Fig. 7.1). A dielectric microsphere possesses natural internal modes of oscillation at characteristic frequencies corresponding to a specific ratio of size to wavelength. These modes of oscillation are also known as whispering gallery modes (WGM) [7]. The MDR effect was first discussed by Purcell [8], who noted that the changes in the final density of states per unit volume and unit frequency would lead to a greatly enhanced probability of spontaneous emission over that normally observed in free space.

7.2.1   Morphology dependent resonance coupling

Both the temporal and spectral coupling of MDR to a spherical microcavity can be investigated by the experimental apparati in Fig. 7.2 and Fig. 7.3, respectively.

There are many basic elements common to both systems. A train of linearly polarised 86 fs pulses of wavelength 870 nm [1] is coupled directly into an inverted microscope so that the back aperture of the objective is filled. A high numerical aperture (NA = 1.2) water immersion objective [1] is used to focus the pulsed laser beam into a sample

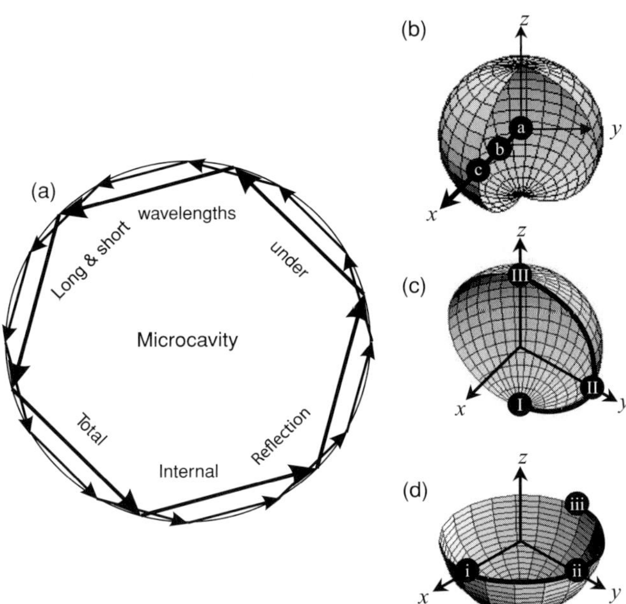

Fig. 7.1 (a) Schematic diagram of a long and short wavelength light ray undergoing total internal reflection, the superposition of which is constructive for a round trip, forming separate MDR modes within a microcavity. (b)–(d) Microcavity planes of interest: radial, meridian and equatorial planes, respectively. Incident polarisation is parallel to the x axis.

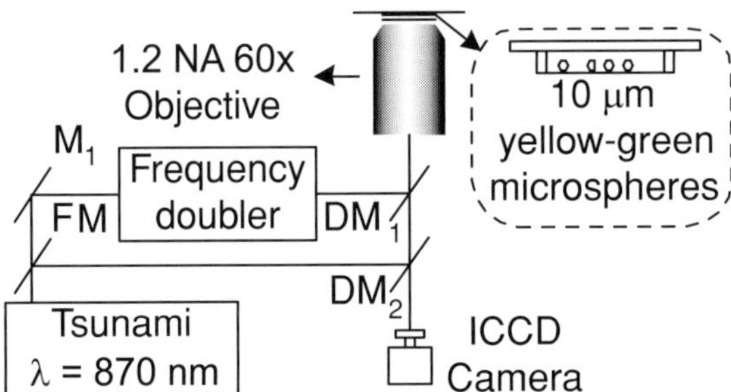

Fig. 7.2 Schematic diagram of the experimental setup for two-photon fluorescence lifetime imaging. Mirror (M1) and flip mirror (FM) enable single- and two-photon excitation via the appropriate dichroic mirrors DM1 and DM2, respectively.

cell. The sample cell consisted of yellow-green fluorescent microspheres of 10 μm in diameter [1], which have an absorption peak close to the laser wavelength for two-photon excitation. The displacement of a particle was achieved by a computer-controlled scanning stage to which the sample cell was attached.

The temporal properties of the fluorescence signal from the microcavity were monitored via an ultrafast intensified CCD camera [1]. The resolution of the ICCD camera was 200 ps and its repetition rate was tuned to synchronise with the repetition rate of the laser, which was 82 MHz.

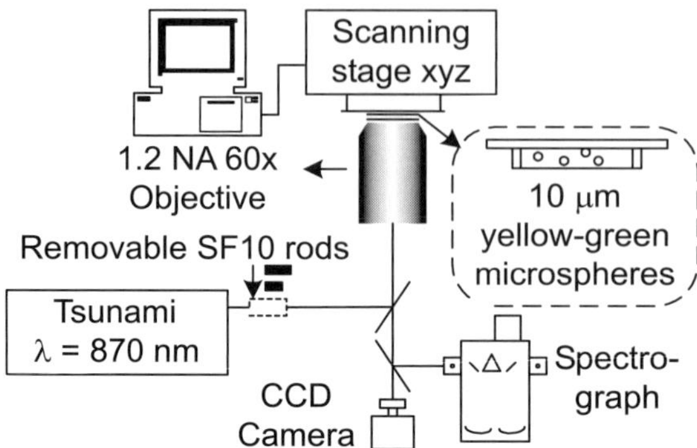

Fig. 7.3 Schematic diagram of the experimental setup for simultaneous single beam trapping and two-photon MDR excitation.

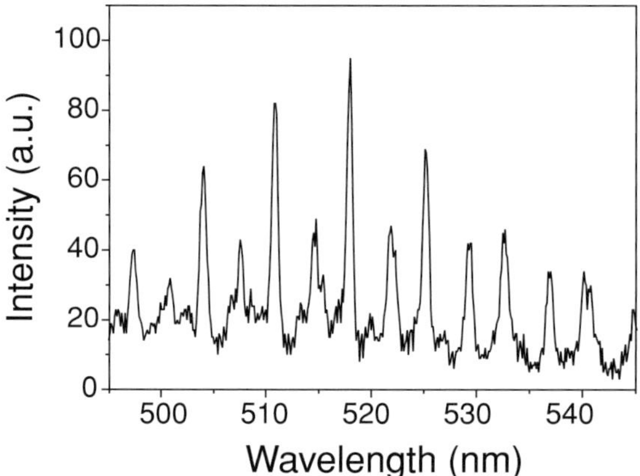

Fig. 7.4 Two-photon fluorescence spectrum ($r = a$, $\phi = 0$ and $\theta = 0$). Reprinted with permission from Ref. [6], D. Morrish, X. Gan and M. Gu, *App. Phys. Lett.* **81**, 5132 (2002). © 2002, American Institute of Physics.

For the temporal measurements and the MDR coupling measurements the microspheres were dried onto the surface of a coverslip. For the laser trapping measurements the microspheres were suspended in water within a sealed sample cell. Throughout this chapter an s-polarised trapping beam is employed, meaning that the polarisation direction of the trapping beam is perpendicular to the direction of the transverse displacement of a trapped particle, unless otherwise specified. The fluorescence emission from an excited microsphere is analysed by a high resolution spectrograph. A series of dispersive pulse-stretch SF-6 glass rods [1] of lengths 25 mm and 65 mm are placed at the output of the laser, so that the laser pulse width incident on a trapped particle can be stretched to 206 fs and 436 fs, respectively.

A typical measured two-photon excited MDR fluorescence spectrum is shown in Fig. 7.4. It is shown that fluorescence emission at specific wavelengths is enhanced due to the MDR effect. The cavity quality factor Q ($Q = \frac{\nu}{\Delta\nu}$, where ν is the

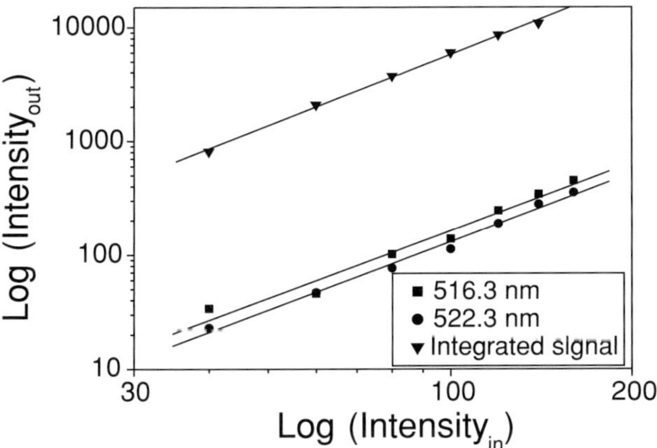

Fig. 7.5 Fluorescence intensity (Intensity$_{out}$) as a function of the input power (Intensity$_{in}$) at $(r = a, \phi = 0$ and $\theta = 0)$. Integrated fluorescence signal (triangles) and the fluorescence MDR peaks at wavelengths 516.3 nm (squares) and 522.3 nm (circles) have linear slope efficiencies of 2.08, 1.98 and 1.97, respectively. Reprinted with permission from Ref. [6], D. Morrish, X. Gan and M. Gu, *App. Phys. Lett.* **81**, 5132 (2002). © 2002, American Institute of Physics.

emission frequency), which can be estimated from the elastic-scattering linewidth based on Lorenz–Mie theory, is approximately 1400 ± 300 for this measurement [9].

In order to verify that the fluorescence emission is truly excited under two-photon absorption, the strength of the fluorescence from an excitation spot focused at the equator of a sphere is measured as a function of input power. It is shown in Fig. 7.5 that the fitted curves to a log–log scale have gradients of 2.08, 1.98 and 1.97, respectively, for the integrated fluorescence signal and the fluorescence peaks at wavelengths 516.3 nm and 522.3 nm. The fluorescence varies proportionally to the square of the input intensity, a characteristic of two-photon process. This measurement confirms that the fluorescence emission is caused by two-photon excitation.

The fluorescence spectra from excitation at different spatial locations in spherical coordinates $(r, \theta, \phi.$ Fig 7.1(b)–(d), respectively) are illustrated in Fig. 7.6. The MDR effect is highly dependent on the spatial location of the excitation spot. Not only does the ratio between fluorescence MDR peaks and background change with the localisation of the excitation position, but also the relative strength between different MDR peaks. If the localised excitation spot is translated along a radial direction in the equatorial plane [Figs. 7.6 (a)–(c), $0 \leq r \leq a, \theta = 0°, \phi = 0°$], the MDR peaks become greatly enhanced when the excitation spot approaches the equator [Fig. 7.6(c), $r = a, \theta = 0°, \phi = 0°$]. The relative strength between the adjacent MDR peaks also varies if the excitation spot is translated around in the equatorial plane [Figs. 7.6 (i)–(iii), $0 \leq \phi \leq 180°, r = a, \theta = 0°$].

Among all the factors that cause changes in fluorescence emission due to different spatial localisation of the excitation, the polarisation nature of the MDR peaks (illustrated in Fig. 7.7) plays an important role. The emission spectrum detected without an analyser from a focal spot localised to the equator of a microsphere $(r = a, \theta = 0°)$, is shown Fig. 7.7(a). Let us take, for example, the two adjacent MDR peaks at wavelengths 509.6

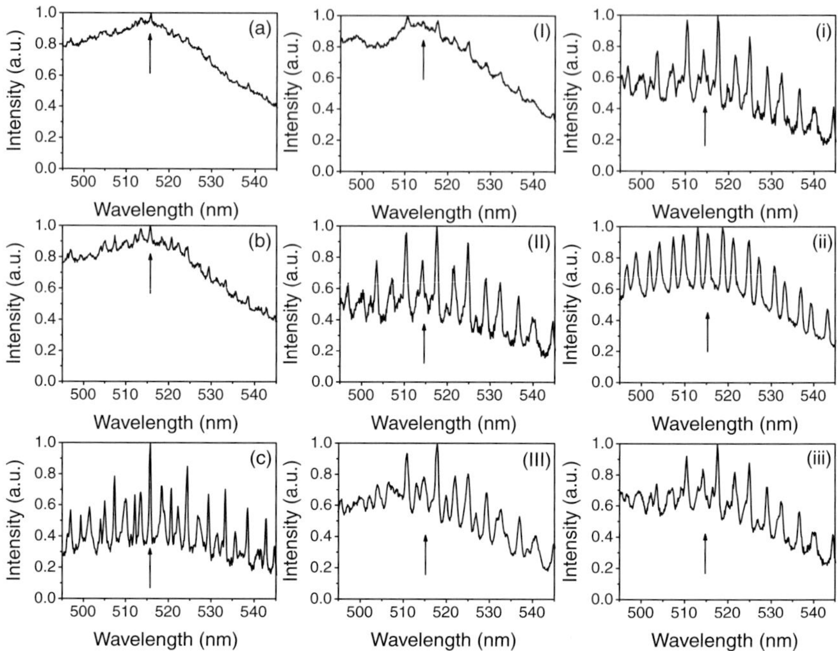

Fig. 7.6 Fluorescence emission spectra for different locations of the excitation spot within a microsphere: (a)–(c) in the radial direction: (a) $r = 0$, (b) $r = 0.5a$, and (c) $r = a$; (I–III) in the meridian plane: (I) $\theta = -90°$, (II) $\theta = 0°$ and (III) $\theta = 90°$; (i)–(iii) in the equatorial plane (i) $\phi = 0°$, (ii) $\phi = 90°$ and (iii) $\phi = 180°$. (Arrow indicates a peak at 514.5 nm). Reprinted with permission from Ref. [6], D. Morrish, X. Gan and M. Gu, *App. Phys. Lett.* **81**, 5132 (2002). © 2002, American Institute of Physics.

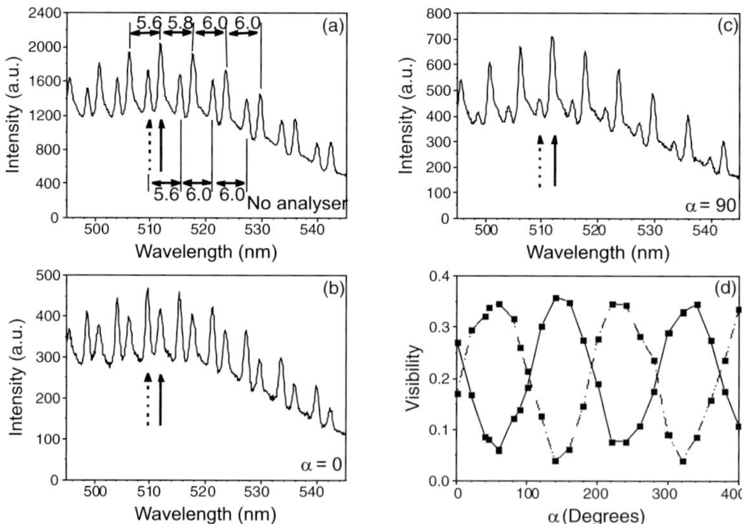

Fig. 7.7 Polarisation dependence of fluorescence spectra: (a) without analyser; (b) analyser angle $\alpha = 0°$; (c) analyser angle $\alpha = 90°$; (d) MDR peak visibility as a function of the analyser angle. Reprinted with permission from Ref. [6], D. Morrish, X. Gan and M. Gu, *App. Phys. Lett.* **81**, 5132 (2002). © 2002, American Institute of Physics.

and 511.9 nm, marked by solid and dashed arrows, respectively. When a polarisation analyser is introduced into the detection path, the relative strength of these two MDR peaks changes (Fig. 7.7(b)).

If the analyser angle, α, is rotated by $90°$, the MDR peak at wavelength 509.6 nm becomes less pronounced compared to the MDR peak at wavelength 511.9 nm (Fig. 7.7(c)). Rotating the analyser by another $90°$ leads to the similar fluorescence spectrum shown in Fig. 7.7(b). In order to quantify the polarisation nature of the MDR peaks, the strength of the two adjacent MDR peaks as a function of the analyser rotation angle is investigated. Figure 7.7(d) shows that MDR peaks periodically change with the analyser angle α; for the two adjacent peaks, one peak is maximised while the other is minimised. Figure 7.7(d) demonstrates the polarisation nature of MDR peaks; the two adjacent peaks representing transverse electric (TE) and transverse magnetic (TM) oscillation modes of a microcavity have orthogonal polarisation states. Furthermore, the measured separation of the two adjacent MDR peaks of same polarisation is approximately $5.9 - 6.0$ nm around the wavelength 510 nm. For a polymer microsphere of refractive index $n = 1.59$, it agrees well with the result of 6.0 nm estimated by

$$\Delta\lambda = \frac{\lambda^2}{2\pi a} \frac{\arctan \sqrt{n^2 - 1}}{\sqrt{n^2 - 1}}, \tag{7.1}$$

based on the plane-wave Mie scattering theory [10].

The characterisation and manipulation of the strength of individual MDR peaks within the 3D space of a microcavity can be achieved due to the highly localised nature of two-photon absorption. In order to quantify the MDR strength and its polarisation nature in relation to the excitation position, the two measurable quantities, visibility

$$V = \frac{(I_{\text{peak}} - I_{\text{background}})}{(I_{\text{peak}} + I_{\text{background}})} \tag{7.2}$$

and the degree of polarisation are introduced:

$$\gamma = \frac{(I_{\alpha=\text{max}} - I_{\alpha=90°})}{(I_{\alpha=\text{max}} + I_{\alpha=90°})}, \tag{7.3}$$

where I_{peak} and $I_{\text{background}}$ are the intensities of a MDR peak and background fluorescence respectively. $I_{\alpha=\text{max}}$ and $I_{\alpha=90°}$ represent the maximum intensity of a MDR peak and the intensity of the peak when the angle of the analyser is rotated by $90°$. The visibility and degree of polarisation are analogous to that in Chapter 6. It should be noted that they have slightly differing definitions when used in this context.

The visibility and the degree of polarisation as a function of polar coordinates (r,θ,ϕ) are demonstrated in Fig. 7.8. Due to the photobleaching effect associated with high power illumination, the results along each spherical coordinate are obtained for a given microsphere. An increase from 3.5% to 48% in visibility between the centre of the cavity and the perimeter is evident (Fig. 7.8(a)) in the radial direction ($0 \leq r \leq a$, $\theta = 0°$, $\phi = 0°$). It is intuitive that the localisation of a focal spot at the microsphere perimeter leads to an increase in the coupling to resonance rays due to glancing angles of incidence with respect to the boundary. Due to the fact that the MDR peaks are

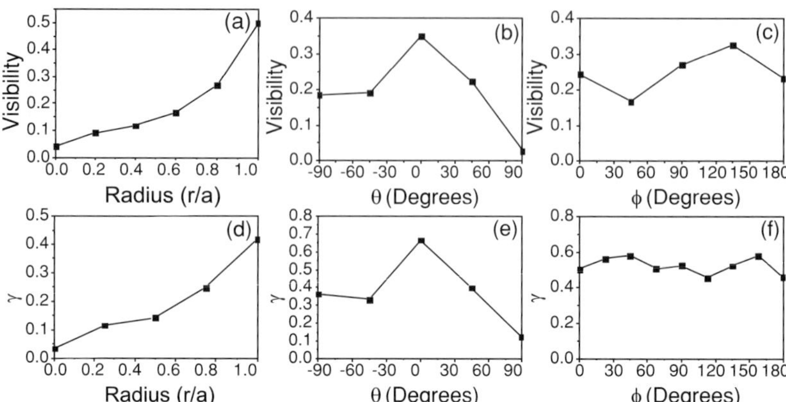

Fig. 7.8 The visibility (a)–(c) and the degree of polarisation (d)–(f) of a MDR peak ($\lambda = 508.6$ nm) as a function of excitation spots in the radial direction (r), the meridian plane (θ) and the equatorial plane (ϕ), respectively. Reprinted with permission from Ref. [6], D. Morrish, X. Gan and M. Gu, *App. Phys. Lett.* **81**, 5132 (2002). © 2002, American Institute of Physics.

highly polarised and the background fluorescence is unpolarised, a similar trend is also observed in the degree of polarisation (Fig. 7.8(d)).

The dependence of the MDR visibility on the localised excitation positions in the meridian plane ($-90° \leq \theta \leq 90°$, $r = a$, $\phi = 0°$) is shown in Fig. 7.8(b). The MDR effect becomes exaggerated when the incident illumination is localised around the equatorial plane of the microcavity, i.e., when $\theta = 0°$. This is because more rays which are coupled into the cavity satisfy the total internal reflection condition if the focal spot is at the equator of a sphere. The break in symmetry between the two hemispheres can be ascribed to the multiple reflection between the cavity and the substrate cover slip when $\theta = -90°$ and to the spherical aberration induced by focusing through the sphere when $\theta = 90°$. With the highly polarised MDR peaks and the unpolarised background fluorescence, a similar trend is observed in terms of degree of polarisation (Fig. 7.8(e)).

The visibility and the degree of polarisation as a function of azimuth angles for excitation locations in the equatorial plane ($0 \leq \phi \leq 360°$, $\theta = 0°$, $r = a$) are shown in Figs. 7.8(c) and (f), respectively. It is found that a periodic variation in the peak visibility is due to the excitation position. When the polarisation state of the incident light is fixed, the strength of the MDR peak can be controlled by the focal position. This occurs due to the fact that the polarisation state of the incident beam with respect to the boundary of the microcavity changes with different values of ϕ. For example, if the excitation beam is linearly polarised in the x-direction (as indicated in Fig. 7.6(c)), the polarisation direction is perpendicular to the surface of the sphere for excitation locations i and iii, and becomes parallel for excitation at location ii. However, it is also noted that the degree of polarisation in the equatorial plane remains constant. The maximum strength of the MDR peaks does not change with different focal positions on the equator, however, the polarisation direction of MDR peaks shifts as the result of varying polarisation orientation of the excitation.

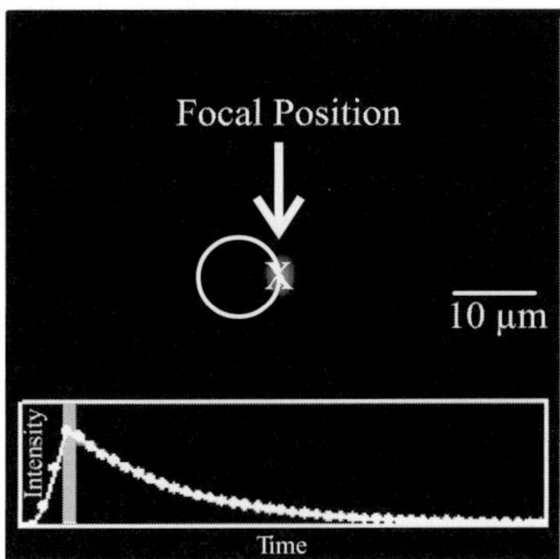

Fig. 7.9 Fluorescence temporal image sequence of a microcavity under two-photon excitation. The location of the excitation is marked by X. The superimposed curve at the bottom represents the fluorescence photon decay at the excitation position.

7.2.2 Cavity photon confinement lifetime

The temporal characteristics of a spherical microcavity under femtosecond pulse laser beam were investigated in the experimental system shown in Fig. 7.2. The fluorescence temporal characteristics are determined by taking a series of images with an exposure time of 200 ps gated over 12.2 ns, which is the time between two successive laser pulses. From this image series, the intensity of the regions of interest can be determined with respect to time. The resultant fluorescence decay curve was then fitted to an exponential decay to determine the fluorescence photon confinement lifetime. Figure 7.9 depicts the evolution of the fluorescence photon confinement lifetime within the microcavity under two-photon excitation. The superimposed curve of the fluorescence photon intensity decay in arbitrary units over the duration of the image sequence is shown for excitation located at the position marked X.

Due to the highly localized nature of two-photon excitation it is possible to examine the fluorescence intensity of different regions within the microcavity. The two fluorescence images in Fig. 7.10 taken over a period of $\Delta\tau = \tau_2 - \tau_1 = 2$ ns in time show the build up of the MDR fluorescence intensity circulating within the region A1 and the excitation region A2. The increase of the intensity ratio between the two regions, $I(A1)$ and $I(A2)$, shows a build up of the MDR fluorescence intensity from 0.24 at τ_1 to 0.32 at τ_2. The circulating MDR fluorescence emission can be observed in the A1 region at τ_2 as a bright spot, which means that the fluorescence photon confinement lifetime is increased due to the confinement of the MDR.

The increase of the measured two-photon fluorescence photon confinement lifetime is further confirmed in Fig. 7.11(a) where the two-photon fluorescence photon confinement lifetime is depicted as a function of the excitation location in the radial direction ($0 < r < a$). The high spatial localisation of the focal region in three dimensions under two-photon excitation within a microcavity results in negligible coupling of fluorescence to the MDR mode profile of the cavity when $0 < r < 0.6$ and thus the corresponding

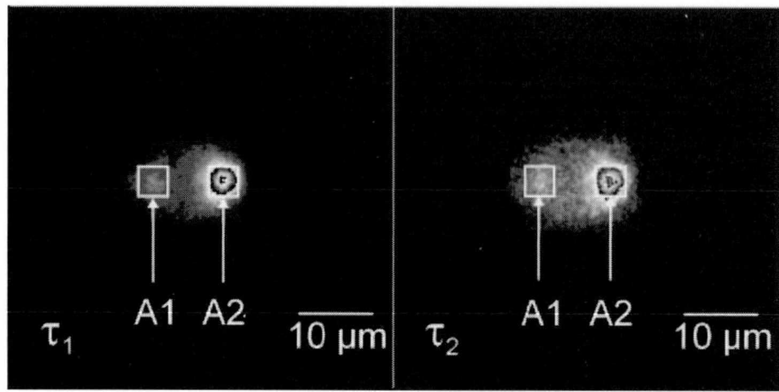

Fig. 7.10 Localised intensity of the two-photon fluorescence temporal images at τ_1 and τ_2. $\Delta\tau = \tau_2 - \tau_1 = 2$ ns. $I(A1)/I(A2) = 0.24$ at τ_1. $I(A1)/I(A2) = 0.32$ at τ_2. Fluorescence at the lifetime of microcavity areas A1 (circulating MDR) and A2 (excitation region) is shown.

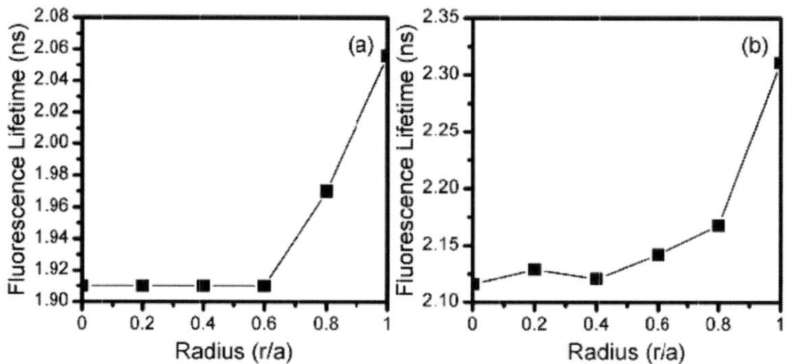

Fig. 7.11 Fluorescence photon lifetime of a microcavity under (a) two-photon and (b) single-photon excitation at various focal positions in the radial direction.

fluorescence photon confinement lifetime is almost constant in this region. As a result, in this region of the microcavity the fluorescence excitation approaches that of a bulk medium [11]. The excited fluorescence and its decay from this bulk is not adversely influenced by its surrounding environment. Only when the excitation is located to the region approaching the perimeter of the microcavity does the fluorescence emission overlap the mode profile of the cavity. When this overlap occurs substantial coupling of the fluorescence to MDR takes place. The fluorescence coupled to MDR remains in the cavity undergoing round trips for an extended period of time before being lost by dissipative or radiative processes. It is therefore expected that the increase of fluorescence photon confinement lifetime is a physical reflection of the enhancement of the MDR visibility [6], demonstrating that as the MDR effect becomes more pronounced, more energy is stored within the cavity.

Comparing the fluorescence photon confinement lifetimes at $r = 0$ and $r = a$ shows an increase by 8%. This increase of the photon lifetime is determined by the Q factor of the microcavity under MDR. According to the relationship between the Q factor and the photon confinement lifetime [12], it can be estimated that the Q factor of the microcavity under the MDR condition is approximately 5×10^3, which is consistent with the observation of the two-photon-induced fluorescence spectra along the radial direction [6]. It seems that the estimated Q factor is lower than that reported elsewhere [13], which may be caused by the leakage of the cover glass slide that is attached to the microsphere under investigation.

Now, let us compare the radial dependence of the fluorescence photon confinement lifetime under two-photon excitation with that under single-photon excitation. For single-photon excitation, the frequency doubled fundamental of wavelength 435 nm was used. The radial dependence of the fluorescence photon confinement lifetime under single-photon excitation is shown in Fig. 7.11(b). It can be seen that instead of the constant region, a gradual increase region exists near the centre. This feature results from the fact that the fluorescence emission under single-photon excitation is not well confined to the focal spot. The fluorescence signal from the out-of-focus regions experiences a weak coupling effect with the MDR mode profile. The coupling of the fluorescence with MDR becomes stronger and approaches a similar situation to two-photon excitation as the excitation is positioned near the perimeter of the microcavity. As such, the increase of the single-photon fluorescence photon confinement lifetime is also approximately 8%. It should be pointed out although the increase of the fluorescence photon confinement lifetimes under two-photon and single-photon is similar, determined by the Q factor of the microcavity, these two excitation mechanisms are physically different.

The fluorescence photon lifetime is increased by 8% in the radial direction of a microsphere cavity as $r \rightarrow a$. Two-photon excitation shows a localisation of the coupling between the fluorescence and the MDR mode profiles near the perimeter of the microcavity, which is consistent with the visibility of the fluorescence spectra [6]. Such a localised coupling condition under two-photon excitation may be advantageous in developing time-resolved particle-trapped near-field optical microscopy (see Section 7.4) [14] and microlasers.

7.3 Simultaneous femtosecond single beam trapping and morphology dependent resonance excitation

The effect of a femtosecond pulse laser beam on laser trapping performance was demonstrated in the experimental system shown in Fig. 7.3. The maximum transverse trapping force [15] as a function of the average laser power is depicted in Fig. 7.12 for an incident pulse width of 86 fs. The maximum transverse trapping force linearly increases with the laser power initially (<2 mW), then undergoes a nonlinear increase and subsequently saturates due to the increased two-photon absorption when the trapping power becomes large. To determine the range within which a particle can be trapped without damage, the trapping performance was monitored at a constant velocity of 10 μm/s as the trapping power increases. The observed power range for the pulse widths of 86 fs, 206 fs and 436 fs is shown in Fig. 7.13. This power range defines a region where the two-photon induced MDR phenomenon can be observed from a particle trapped by a pulsed beam [16].

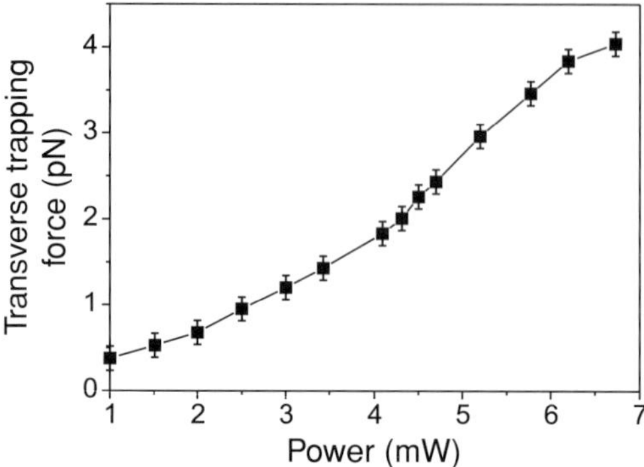

Fig. 7.12 The dependence of the maximum transverse trapping force on laser power. Reprinted with permission from Ref. [16], D. Morrish, X. Gan and M. Gu, *Opt. Express* **12**, 4198 (2004). © 2004, Optical Society of America.

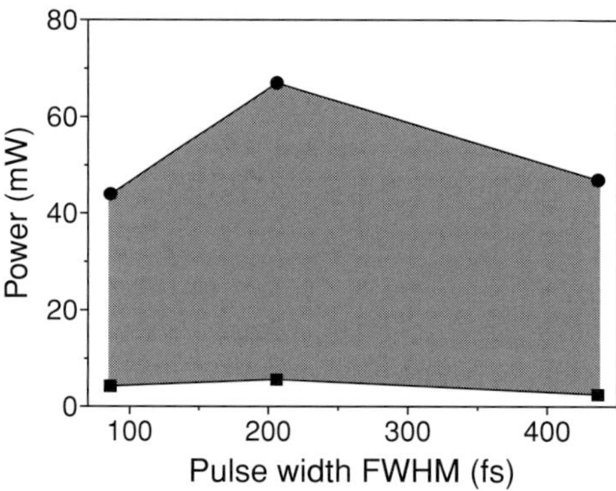

Fig. 7.13 Trapping region bounded by the minimum (squares) and maximum (circles) transverse trapping power limits. Reprinted with permission from Ref. [16], D. Morrish, X. Gan and M. Gu, *Opt. Express* **12**, 4198 (2004). © 2004, Optical Society of America.

Stable laser trapping with femtosecond pulses is achievable with moderate laser powers. The minimum and maximum transverse trapping powers for various femtosecond pulse lengths at a constant scanning velocity of 10 µm/s are given in Fig. 7.13. The minimum and maximum transverse trapping power limits are defined as the points at which trapping occurs and particle destruction, respectively. The trapping power is measured at the back aperture of the trapping objective. The large region of stable trapping provides the ability to tailor trapping conditions to individual applications [2, 17–20]. Trapping with high power can lead to changes within the sample and even explosion [21, 22]. There is a difference between the peak power and the average power of a femtosecond pulse trapping beam. The peak power associated with short pulses is advantageous for the induction of two-photon fluorescence, while an increased pulse width extends the time that the incident radiation is exerting an optical force on the particle. At a fixed laser repetition rate and particle scanning velocity there is a point at which decreasing the pulse width and increasing two-photon absorption does not translate to an increase in trapping efficiency.

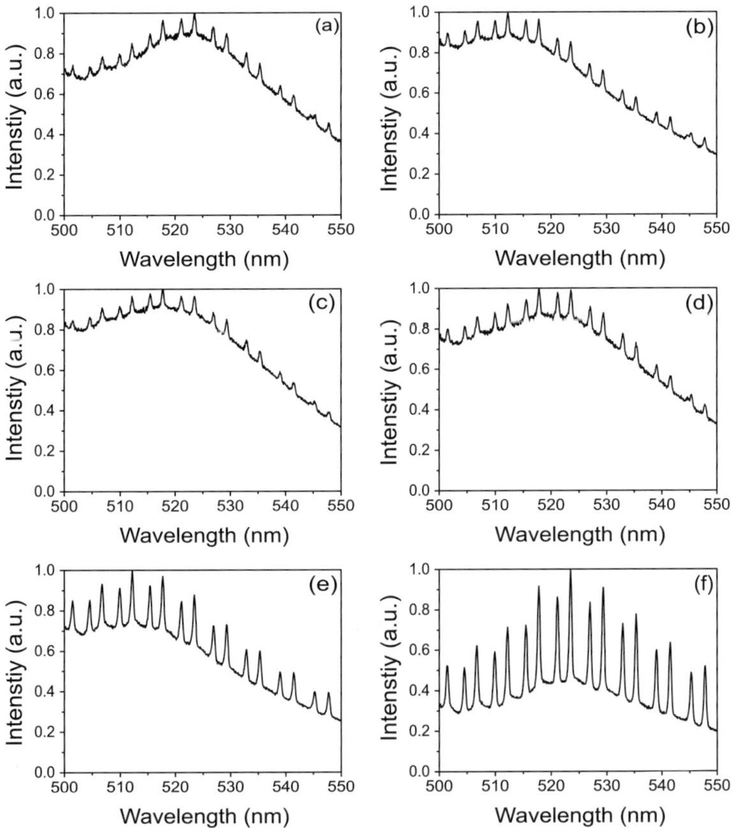

Fig. 7.14 MDR spectra of a laser trapped microsphere at velocities 20, 45, 70, 95, 130 and 150 μm/s (a) to (f), respectively. Reprinted with permission from Ref. [16], D. Morrish, X. Gan and M. Gu, *Opt. Express* **12**, 4198 (2004). © 2004, Optical Society of America.

Figure 7.13 also indicates that a trapped particle can be translated at different velocities when a trapping power level within the marked range is selected, due to the balance between the Stokes force and the transverse trapping force, which depend on the translation velocity of a trapped particle and on the transverse displacement of a trapped particle, respectively. The trapping spot will be located at different transverse positions of a trapped particle. It has been previously demonstrated in Section 7.2 that the strength of the two-photon induced MDR is highly dependent on the location of the excitation spot within a microsphere because of the highly localised nature of two-photon absorption [6]. Therefore, the strength and the visibility of the MDR signal vary with the translation velocity of a trapped particle. In order to quantify the strength of MDR spectral features from the fluorescence background for a trapped microsphere the measurable quantity, visibility (V) defined in Eq. 7.2 is reintroduced.

The MDR signal induced in a stable laser trapped particle for various trapping velocities is given in Fig. 7.14. The MDR effect is greatly enhanced when the translation velocity increases. This phenomenon can be explained in conjunction with Stokes'

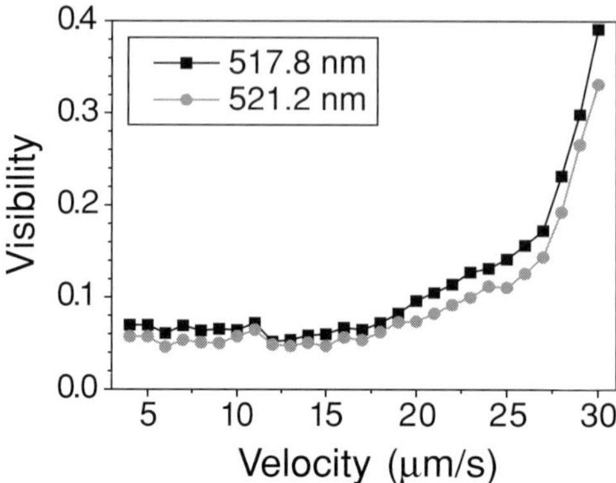

Fig. 7.15 Visibility of peaks at wavelengths 517.8 nm (squares) and 521.2 nm (circles) at trapped microsphere velocities. Reprinted with permission from Ref. [16], D. Morrish, X. Gan and M. Gu, *Opt. Express* **12**, 4198 (2004). © 2004, Optical Society of America.

law [23, 24] and as follows. At a given laser power, the greater the translation velocity, the greater the transverse trapping force is required to balance the viscous drag force. This means that the trapping spot moves towards the perimeter of a trapped particle because the transverse trapping force increases with the displacement of the trapping beam [15]. It has been demonstrated in Fig. 7.8(a) that the two-photon induced MDR effect becomes more significant when the excitation spot moves closer towards the edge of a trapped particle [6]. Therefore, when a particle is scanned fast the visibility of the MDR signal becomes pronounced. Note that the spectral change of the background fluorescence is determined by the collection efficiency of the two-photon fluorescence and the coupling efficiency of the excitation beam at different velocities.

The two adjacent peaks in the MDR fluorescence spectrum shown in Fig. 7.15 represent two cavity modes, the TE and TM modes [6]. Both these modes share a similar increase in the visibility with an increase in the translation velocity. This feature enables the potential for multi-modality imaging or sensing including polarisation dependent signals.

7.4   Resonant particle trapped microscopy

Laser tweezing or laser trapping of a microsphere, based on the radiation pressure [25], has been employed as a sensing probe for particle trapped microscopy [3, 26–33]. In this case, a trapped microsphere is positioned at the sample surface and acts as an imaging probe. Based on this concept, force microscopy and optical microscopy have been the two main imaging modes. The principle of the first imaging mode is the force sensitivity of a trapped particle [3, 26, 27], which can be used for surface profiling and position tracking. The second imaging mode of particle trapped microscopy utilises optical signals generated by various interaction processes between the incident light and a trapped particle. Of these optical interaction processes, scattering of a trapped dielectric [28] or metallic [29, 30] particle with an evanescent field at the sample surface converts the localised photons to a far-field region [31], providing a tool for high resolution near-field optical microscopy. A second optical imaging method is based on the scattering of a trapped particle with the trapping beam [32–34], which is determined by the surface profile and the material of a sample. A third type of optical signal is the radiation induced by the trapping beam in a trapped particle, including second harmonic generation [35] or fluorescence [36].

A fluorescent Mie microsphere is an excellent candidate for use as a laser trapped optical imaging probe rather than the particle acting solely as a scatter. First, a laser trapped imaging probe allows multiple imaging and sensing modes by the excitation of fluorescence resulting from MDR due to the total internal reflection of the internal field within the microcavity [37]. Second, the fluorescence process can be localised if two-photon excitation is employed [2, 3], enhancing the visibility of MDR spectra. Further, the use of a femtosecond laser exhibits an enhancement of the two-photon induced MDR effect [4, 6, 16]. The localised strength and position of the enhanced MDR spectra are highly sensitive to the environment with which a trapped particle interacts. In this section, near-field optical microscopy of a sample surface profile of hundred nanometres in height is investigated based on the two-photon induced MDR effect from a microsphere trapped by a femtosecond laser [6, 16].

The physical principle of this imaging method is the sensitivity of two-photon induced MDR with respect to the position of the trapping focus within a trapped particle [16]. When a trapped particle scanned at a given velocity interacts with a topological feature of a sample, the trapping position initially produced by the balance between trapping force and viscous drag force is displaced further from the centre of the particle to the particle edge. This characteristic leads to the fluorescence imaging contrast resulting from the two-photon excited MDR effect in a trapped particle.

The laser trapped imaging performance of a femtosecond pulse laser beam was demonstrated via the experimental system shown in Fig. 7.3. A train of linearly polarised 86 fs pulses of wavelength 870 nm [1] was directly coupled into an inverted trapping microscope so that the back aperture of the trapping objective was filled. The trapping objective used was a high NA (NA $= 1.2$) water immersion objective [1]. The sample cell consisted of yellow-green fluorescent microspheres of 10 μm in diameter [1], which have an absorption peak close to the laser wavelength for two-photon excitation [16]. The microspheres were suspended in water within a sealed sample cell. The scanning of a trapped particle was achieved by a computer controlled scanning stage on which the sample cell was attached.

An s-polarised trapping beam was employed, such that the polarisation direction of the trapping beam was parallel to the direction of the transverse displacement of a trapped particle. The spectral properties of the fluorescence return signal from the laser trapped probe were monitored via the high resolution spectrograph acting as a monochromator, so that a spectral bandwidth of 1 nm was incident at the detector. This spectral window of detection was then shifted so that the fluorescence intensity from either a single MDR maxima (peak) or minima (valley) was measured (Fig. 7.16(a)). The terms peak, valley and integrated correspond to the signal from the aforementioned MDR maxima, minima and entire fluorescence MDR spectra, respectively.

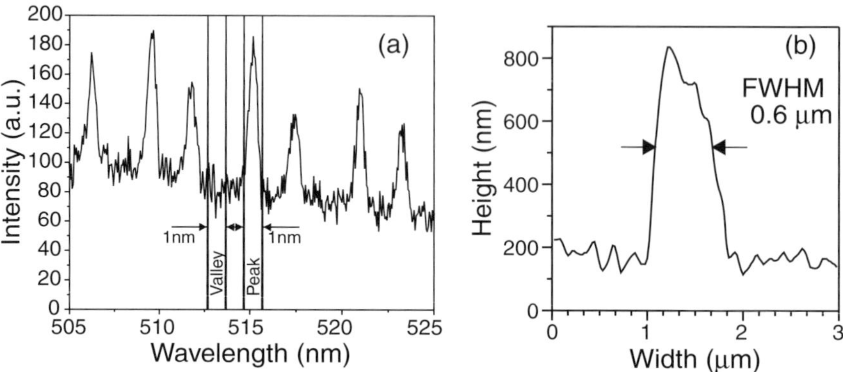

Fig. 7.16 (a) Detection windows of spectral bandwidth 1 nm for measuring the fluorescence intensity from a single MDR peak and valley. (b) One-dimensional AFM trace of the grating structure. Reprinted with permission from Ref. [14], D. Morrish, X. Gan and M. Gu, *App. Phys. Lett.* **88**, 141103 (2006). © 2006, American Institute of Physics.

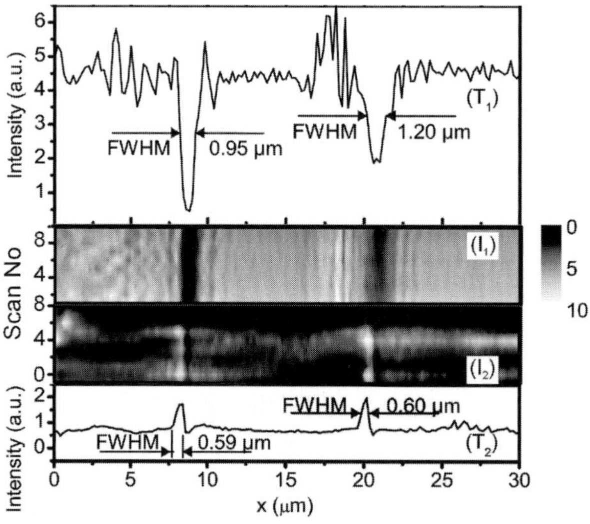

Fig. 7.17 One-dimensional scanning of the grating structure. Image (I_1) is the standard reflection confocal imaging and image (I_2) is the laser trapped probe image. Confocal trace (T_1) and laser trapped probe trace (T_2) show that two elements of the grating have different FWHMs of 0.95, 1.20 and 0.59, 0.60 μm, respectively. Reprinted with permission from Ref. [14], D. Morrish, X. Gan and M. Gu, *App. Phys. Lett.* **88**, 141103 (2006). © 2006, American Institute of Physics.

In order to characterise the imaging potential of a laser trapped MDR probe, a sample with a long period grating profile was used. The sample is a polymer grating produced by the two-photon polymerisation method in NOA63 resin [1] on the surface of a cover slip [38]. It has a spacing of 10 μm, a thickness between 0.1 and 3 μm and a height <1 μm. An atomic force microscope (AFM) [1] in a semi-contact mode was used to take a profile of the grating, showing a typical full width at half maximum (FWHM) of 0.6 μm and a peak height of approximately 600 nm (Fig. 7.16(b)).

Figure 7.17 shows the one-dimensional integrated image obtained by using the laser trapped probe scanning over the grating structure (I_2). A one-dimensional reflection confocal laser scanning image of the grating structure is also shown in Fig. 7.17 (I_1). The respective image cross sections obtained under the laser trapping microscope and the reflection confocal microscope are also illustrated in Fig. 7.17 (T_2 and T_1). A FWHM of approximately 1.00 μm is obtained in the reflection confocal image trace while a FWHM of approximately 0.60 μm is achieved by laser trapped particle microscopy. The laser trapped particle microscopy trace produces a transverse resolution comparable to that of AFM, demonstrating that the resolution is not necessarily limited to the particle size but to the interaction region between the probe and the sample.

The laser trapped particle probe trace shows three regions as the probe scans over the structure (T_2 in Fig. 7.17). The first is an increase in signal as the scanned probe ascends the slope of the structure, forcing the trapping position closer to the perimeter of the trapped particle and hence increasing the coupling of the fluorescence excitation to resonance modes. As the trapped particle crests the top of the structure, and then descends the slope to the substrate, the trapping position moves to the centre, resulting in a sharp decrease in mode coupling and signal intensity. The signal then increases as the scanning of the trapping beam restores the equilibrium of the focal position to the stable transverse trapping position.

The laser trapped image contrast of the long period grating structure under spectral detection of MDR peak, valley and integrated signals is examined over a range of translation velocities of a trapped probe. The average image intensity profiles over the relative displacement of the trapped particle for translation velocities of 8.7, 10.8, 12.9 and 15.0 μm/s for integrated, peak and valley signals are shown in Fig. 7.18. The sharpest image profile was spectrally detected from the MDR peak intensity shown in Fig. 7.18(e)–(h). This is attributed to the fact that MDR is highly dependent on the change in the laser trapping focal position within the scanning microcavity microsphere probe. Alternatively, the MDR valley signal has little modulation as it is essentially a measure of the fluorescence background (Fig. 7.18(i)–(l)). The spectral detection of the integrated MDR fluorescent spectrum shown in Fig. 7.18(a)–(d) approximates the average effect of the peak and valley signals. The respective contrast enhancement of MDR peak and integrated signal images compared with the valley signal images is illustrated in Fig. 7.19(a), showing an approximate contrast increase of 2.7% and 9.3% at a scanning velocity of 8.7 μm/s. At high transverse trapping velocity, the rapid ascent of a structural element can result in the trapped particle breaking contact and leaving the sample surface. This results in a loss of image contrast due to the decreased displacement of the trapping spot within the trapped microcavity.

The spectrally resolved images of the grating can be obtained with a cooled CCD camera attached to the spectrograph for a given exposure time. The resultant spectrally resolved image of a one-dimensional scan in the x-direction is given in Fig. 7.19(b), where the graphic inset along the x-axis schematically depicts the grating structure attached to the cover slip. The spatial resolution of the image is 1 μm every sampling point according to the scanning speed (10 μm/s). It is clear to see from Fig. 7.19(b) that the MDR effect is enhanced at the position of each grating element as the trapping spot is localised further toward the perimeter of the particle, resulting in a stronger MDR signal. The fully spectrally resolved image reveals the variation in coupling of the excitation

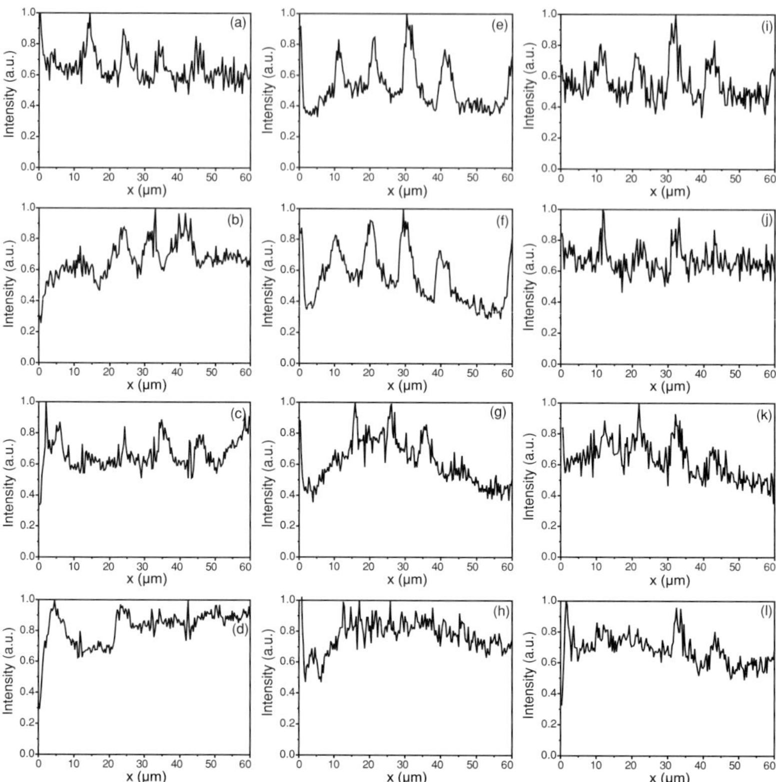

Fig. 7.18 Laser trapped probe profiles: (a)–(d) integrated signal, (e)–(h) peak signal, (i)–(l) valley signal, for scanning velocities of 8.7, 10.8, 12.9 and 15.0 μm/s, respectively. Reprinted with permission from Ref. [14], D. Morrish, X. Gan and M. Gu, *App. Phys. Lett.* **88**, 141103 (2006). © 2006, American Institute of Physics.

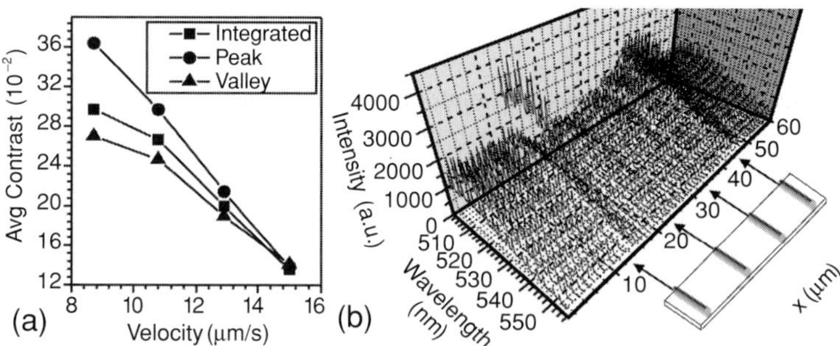

Fig. 7.19 (a) Average contrast of the one-dimensional images of the grating structure with integrated, peak and valley imaging modalities. (b) One-dimensional spectrally-resolved image of the grating structure. Reprinted with permission from Ref. [14], D. Morrish, X. Gan and M. Gu, *App. Phys. Lett.* **88**, 141103 (2006). © 2006, American Institute of Physics.

to the resonance modes as the probe is scanned across the sample, due to the variation of the surface profile of the sample.

7.5 Summary

In summary, we have demonstrated the achievement of simultaneous two-photon induced MDR and trapping of a microsphere by a single femtosecond pulse beam. The use of a femtosecond pulse beam allows for localised two-photon excitation of MDR for a trapped particle scanned at different velocities. The radial trapping position ($0 \leq r \leq a$, $\theta = 0°$, $\phi = 0°$ and $Z = 0$) within the microcavity varies with the transverse trapping velocity. The balance between the optical trapping force and the viscous drag force exerted on the scanned particle results in the trapping position increasing in radius with increasing transverse trapping velocity. The measured dependence of the visibility of the MDR signal on the translation velocity indicates that a high sensitivity and a high scanning velocity of a trapped particle can be achieved simultaneously. The 16% increase in MDR visibility with the transverse scanning velocity is sufficient for sensing applications. This result implies that the two-photon induced MDR signal has the potential to be an invaluable tool for mapping tomography, topography and force.

It has been shown that images in this new system can be constructed with three spectral modalities: integrated, peak, and valley signals. The integrated image trace of a grating structure exhibits transverse resolution comparable to that obtained with AFM, while the contrast from the peak imaging modality shows a considerable enhancement compared with the integrated and valley imaging modalities. A fully spectrally resolved image demonstrates that this technique provides an alternative imaging mechanism for near-field optical microscopy and surface tomography via MDR.

References

[1] D. Morrish. *Morphology Dependent Resonance of a Microsphere and its Application in Near-Field Scanning Optical Microscopy*. Ph.D. thesis, Faculty of Engineering and Industrial Research, Swinburne University of Technology, Australia, 2005.

[2] Y. Liu, G. J. Sonek, M. W. Berns, K. Konig and B. J. Tromberg. Two-photon fluorescence excitation in continuous-wave infrared optical tweezers. *Opt. Lett.*, **20**:2246–2247, 1995.

[3] E. Florin, J. Hörber and E. H. K. Stelzer. High-resolution axial and lateral position sensing using two-photon excitation of fluorophores by continuous-wave Nd:YAG laser. *App. Phys. Lett.*, **69**:446–449, 1996.

[4] B. Agate, C. T. A. Brown, W. Sibbett and K. Dholakia. Femtosecond optical tweezers for in-situ control of two-photon fluorescence. *Opt. Express*, **12**:3011–3017, 2004.

[5] M. Göppert-Mayer. Uber Elementarakte mit zwei Quantensprungen. *Ann. Phys. (Leipzig)*, **5**:273–294, 1931.

[6] D. Morrish, X. Gan and M. Gu. Observation of orthogonally polarized transverse electric and transverse magnetic oscillation modes in a microcavity excited by localized two-photon absorption. *App. Phys. Lett.*, **81**:5132–5134, 2002.

[7] P. Chylek, J. T. Kiehl and M. K. W. Ko. Optical levitation and partial-wave resonances. *Phys. Rev. A*, **18**:2229–2233, 1978.

[8] E. M. Purcell. Spontaneous emission probabilities at radio frequencies. *Phys. Rev.*, **69**:681, 1946.

[9] R. K. Chang and A. J. Campillo. *Optical Processes on Microcavities*. Singapore, World Scientific, 1996.

[10] P. W. Barber and R. K. Chang. *Optical Effects Associated with Small Particles*. Singapore, World Scientific, 1988.

[11] A. Imhof, M. Megens, J. J. Engelberts, D. T. N. de Lang, R. Sprik and W. L. Vos. Spectroscopy of fluorescine (FITC) dyed colloidal silica spheres. *J. Phys. Chem. B*, **103**:1408–1415, 1999.

[12] A. Yariv. *Quantum Electronics*. New York, John Wiley and Sons, 1967.

[13] H. Fujiwara and K. Sasaki. Lasing of a microsphere in dye solution. *Jpn. J. Appl. Phys.*, **38**:5101–5104, 1999.

[14] D. Morrish, X. Gan and M. Gu. Scanning particle trapped optical microscopy based on two-photon-induced morphology-dependent resonance in a trapped microsphere. *App. Phys. Lett.*, **88**:141103, 2006.

[15] A. Ashkin. Forces of a single-beam gradient trap on a dielectric sphere in the ray optics regime. *J. Biophys.*, **61**:569–582, 1992.

[16] D. Morrish, X. Gan and M. Gu. Morphology-dependent resonance induced by two-photon excitation in a micro-sphere trapped by a femtosecond pulsed laser. *Opt. Express*, **12**:4198–4202, 2004.

[17] B. A. Nemet and M. Cronin-Golomb. Microscopic flow measurements using optically trapped microprobes. *Opt. Lett.*, **27**:1357–1359, 2002.

[18] J. Won, T. Inaba, H. Masuhara, H. Fujiwara, K. Sasaki, S. Miyawaki and S. Sato. Photothermal fixation of laser-trapped polymer microparticles on polymer substrates. *App. Phys. Lett.*, **75**:1506–1508, 1999.

[19] H. Furukawa and S. Kawata. Near-field optical microscope images of a dielectric flat substrate with subwavelength strips. *Opt. Comm.*, **196**:93–102, 2001.

[20] H. Ukita and K. Nagatomi. Optical tweezers and fluid characteristics of an optical rotator with slopes on the surface upon which light is incident and a cylindrical body. *Appl. Opt.*, **42**:2708–2715, 2003.

[21] D. Day, M. Gu and A. Smallridge. Use of two-photon excitation for erasable rewritable three-dimensional bit optical data storage in a photorefractive polymer. *Opt. Lett.*, **24**:948–950, 1999.

[22] M. Straub, M. Ventura and M. Gu. Multiple higher-order stop gaps in infrared polymer photonic crystals. *Phys. Rev. Lett.*, **91**:0434901, 2003.

[23] M. Gu, D. Morrish and P. C. Ke. Enhancement of transverse trapping efficiency for a metallic particle using an obstructed laser beam. *App. Phys. Lett.*, **77**:34–36, 2000.

[24] W. H. Wright, G. J. Sonek and M. W. Berns. Parametric study of the forces on microspheres held by optical tweezers. *Appl. Opt.*, **33**:1735–1748, 1994.

[25] A. Ashkin and J. M. Dziedzic. Optical trapping and manipulation of viruses and bacteria. *Science*, **235**:1517–1520, 1987.

[26] L. P. Ghislain and W. W. Webb. Scanning force microscope based on an optical trap. *Opt. Lett.*, **18**:1678–1680, 1993.

[27] M. E. Friese, A. G. Truscott, H. Rubinsztein-Dunlop and N. R. Heckenberg. Three-dimensional imaging with optical tweezers. *Appl. Opt.*, **38**:6597–6603, 1999.

[28] S. Kawata, Y. Inouye and T. Sugiura. Near-field scanning optical microscopy with a laser-trapped probe. *Jpn. J. Appl. Phys.*, **33**:L1725–L1727, 1994.

[29] M. Gu and P. Ke. Image enhancement in near-field scanning optical microscopy with laser-trapped metallic particles. *Opt. Lett.*, **24**:74–76, 1999.

[30] M. Gu and P. Ke. Effects of depolarisation of scattering evanescent waves on near-field imaging with laser-trapped particles. *App. Phys. Lett.*, **75**:175–177, 1999.

[31] D. Ganic, X. Gan and M. Gu. Near-field imaging by a micro-particle: a model for conversion of evanescent photons into propagating photons. *Opt. Express*, **12**:5325–5345, 2004.

[32] L. Malmqvist and M. Hertz. Trapped particle optical microscopy. *Opt. Comm.*, **94**:19–24, 1992.

[33] T. Sugiura, T. Okada, Y. Inouye, O. Nakamura and S. Kawata. Gold-bead scanning near-field optical microscope with laser-force position control. *Opt. Lett.*, **22**:1663–1665, 1997.

[34] T. Konada, H. Sasaki, M. Yamaguchi, Y. Suzuki and Y. Sasaki. Reflection-mode scattering-type scanning near-field optical microscope using a laser trapped gold colloidal particle as a scattering probe. *J. App. Phys.*, **89**:6481–6483, 2001.

[35] L. Malmqvist and M. Hertz. Two-color trapped-particle optical microscopy. *Opt. Lett.*, **19**:853–855, 1994.

[36] K. Sasaki, H. Fujiwara and H. Masuhara. Photon tunneling from an optically manipulated microsphere to a surface by lasing spectral analysis. *App. Phys. Lett.*, **70**:2647–2650, 1997.

[37] R. E. Benner, P. W. Barber, J. F. Owen and R. K. Chang. Observation of structure resonances in the fluorescence spectra from microspheres. *Phys. Rev. Lett.*, **44**:475–478, 1980.

[38] B. Jia, X. Gan and M. Gu. Height/width aspect ratio controllable two-dimensional sub-micron arrays fabricated with two-photon photopolymerization. *Optik*, **115**:358–362, 2004.

8   Near-field optical trapping and tweezers

As discussed in Chapter 6, the trapping volume of a far-field laser trapping geometry is approximately three times larger in the axial direction than that in the transverse direction [1, 2]. Such trapping volume elongation leads to a significant background and poses difficulties in the observations of nano-particle dynamics. In this chapter, we deal with near-field optics using focused evanescent illumination. The recent development of near-field optical tweezers is reviewed in Section 8.1. Section 8.2 introduces the new concept of near-field laser tweezing with a focused evanescent field. This technology is characterised both experimentally and theoretically in Section 8.3. Section 8.4 presents the utilisation of a femtosecond laser beam in a near-field optical trap. Finally, some discussions on this new method are given in Section 8.5.

8.1   Near-field optical tweezers

Near-field laser trapping or tweezers means that radiation force that is used for trapping and manipulating a micro-object results from the interaction with an evanescent wave. Recently, a new trapping modality based on the evanescent wave illumination, also called near-field illumination, has been proposed [3–9] and demonstrated [7]. This trapping technique results in a significantly reduced trapping volume due to the fact that the strength of an evanescent wave decays rapidly with the distance from the surface at which the field is generated. In this section, the near-field trapping mechanism based on the different ways to generate a localised near-field is reviewed.

8.1.1   Near-field trapping with wide-field evanescent illumination

An evanescent field can be generated at a surface of a prism under the total internal reflection condition. It is called wide-field because the area at which evanescent field is generated is much larger than the microspheres (Fig. 8.1). If a small dielectric microsphere is in the vicinity of the surface it can convert evanescent photons into propagating photons through a photon tunnelling process. The tunnelled photons scatter from the particle through the process of multiple internal reflections inside the particle. Consequently, a part of the momentum of photons of the incident laser beam is transferred to the particle, driving the moving particle.

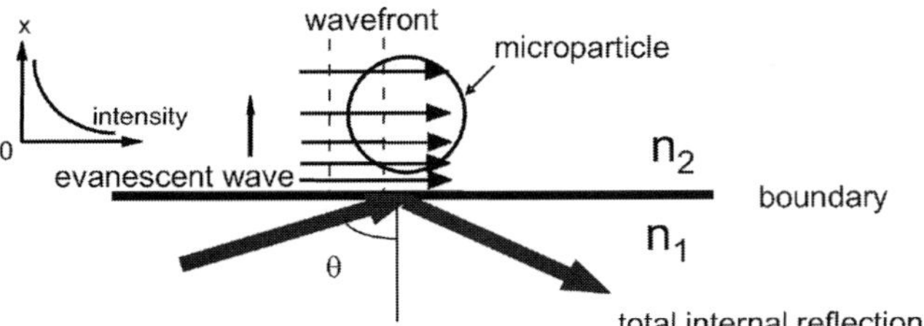

Fig. 8.1 Near-field trapping under wide-field evanescent illumination.

The concept of using this wide-field evanescent illumination for trapping was first demonstrated by Kawata and his colleagues [3]. Microspheres have been successfully guided along the surface. Later they demonstrated the collection and guiding of particles into a single-mode channelled waveguide by the use of the evanescent field along the channel [10]. The dielectric and metallic particles are dispersed in water on the surface of the glass substrate and a channelled waveguide is created on the surface. The evanescent field generated along the waveguide could drive the microscopic particles with speeds of up to 14 μm/s. Later, it was demonstrated that the propulsion of small particles in the evanescent field generated in the multi-mode waveguides can also be achieved [11].

Optically induced organisation of microparticles over an extended area has been reported recently using wide-field evanescent illumination generated at the surface of a prism by total international reflection [12]. A thousand microspheres have been trapped over macroscopic areas, and have also been guided along the surface of the prism at controllable speeds. The system provides easy switching between trapping and guiding, thus creating an optical conveyor belt for large scale assembly of microspheres. The strength of the evanescent field on the prism, i.e. the trapping performance, can be enhanced by using a resonant microcavity [13, 14]. However, because the evanescent wave field on a prism is not laterally localised, it is difficult to achieve laser manipulation and tweezing in this case.

8.1.2 Near-field trapping using a metallic tip

To overcome the drawback of the wide-field near-field trapping method, a metallic tip illuminated by a laser beam has been proposed to produce a localised evanescent field by the surface plasmon effect so that a particle of a few nanometres in size can be trapped. Theoretically, Novotny *et al.* [4] have investigated the possibility of trapping a nano-particle using the field enhancements close to a laser illuminated sharply pointed metal tip. The near-field close to the tip mainly consists of evanescent components which decay rapidly with distance from the tip. The analysis is based on characterising the field enhancements near the tip using a numerical method. The effect of the particle in close proximity of the tip is also included. The intensity at the foremost part of the tip is approximately 3000 times stronger than the illuminating intensity. Once the field distribution on the surface of the particle is determined, the Maxwell stress tensor approach is used to determine the near-field force exerted on the particle. Utilising this methodology, it has been shown that nano-particles can be trapped using the field enhancements near a metallic tip when the polarisation of the incident light is directed along the tip axis.

Another novel approach to producing a localised near-field is based on the use of a combination of evanescent illumination and light scattering at the probe apex, which can shape the optical field into a localised three-dimensional optical trap [5]. Two counter-propagating evanescent waves are generated by the total internal reflection of plane waves at the substrate/air interface. A tungsten probe is placed in close proximity to the dielectric microsphere that is above the surface. As the probe tip moves closer to the sphere, a force is exerted on the particle. The coupled-dipole method and the Maxwell stress tensor are used to determine the near-field force exerted on the particle. It is shown that small objects can be selectively captured and manipulated by the near-field force generated using an apertureless probe. Furthermore, the magnitude and direction of the trapping force greatly depend on the polarisation state of the incident illumination.

8.1.3   Near-field trapping using a nano-aperture

Okamoto and Kawata [6] have proposed a near-field trapping technique that utilises a circular nano-aperture as a localised field source. The trapping is achieved by interaction between the aperture and the dielectric sphere via evanescent photons. They have made a series of numerical calculations of the radiation force exerted on such a particle. The particle position near the aperture is constantly changed in this calculation, thus a spatial distribution of the radiation force, also known as a force mapping, is determined. The electromagnetic field distribution is calculated using the finite-difference time-domain (FDTD) method. From each calculated electromagnetic field distribution, the radiation force was obtained from the Maxwell stress tensor on the surface of the sphere. Using this methodology, it was confirmed that optical near-field trapping using a nano-aperture configuration can be achieved. The result indicates that a particle is attracted towards the aperture. The near-field radiation force is found to be larger than the forces due to thermal fluctuations and to gravity. Furthermore, they have found that if two particles are near the aperture, the first particle is trapped and the second one is also attracted to the first one.

8.2   Near-field optical tweezers under focused evanescent wave illumination

Although the localised near-field techniques proposed in the last section are exciting and important in single molecule manipulation and detection, it is difficult to implement them in practice. First, it is difficult to control the distance between the probe and samples, which is in the range of tens of nanometres due to the evanescent nature of the illumination. Thus, the use of a sharp tip or a nano-aperture hampers the nano-manipulation operation such as rotation. Second, although a metallic tip leads to an

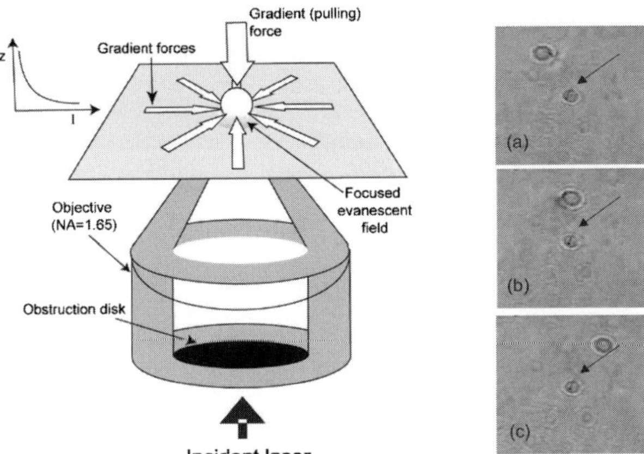

Fig. 8.2 Concept of evanescent-field laser trapping under focused evanescent wave illumination. The strength of the evanescent wave decays rapidly with the distance. (a)–(c) Experimental images of a trapped particle of diameter 2 μm (indicated by an arrow) while the sample cell is translated. Trapping power: 10 mW; $\varepsilon = 0.8$; translating speed: 3 μm/s. Reprinted with permission from Ref. [7], M. Gu, J. Haumonte, Y. Micheau, J. Chon and X. Gan, *App. Phys. Lett.* **84**, 4236 (2004). © 2004, American Institute of Physics.

enhanced evanescent wave, the heating effect caused by surface plasmons associated with a metallic tip significantly reduces the stability of the trapping and tweezing. Third, the light throughput of near-field tips or apertures is quite low. In this section, we propose and demonstrate a novel near-field trapping technique that is based on focused evanescent wave illumination [7, 15].

8.2.1 Near-field optical tweezing of a microsphere

The focused evanescent field is generated by the use of a ring beam produced by a high numerical aperture objective that is centrally obstructed, as shown in Fig. 8.2. The opaque disk has such a size that the minimum angle of convergence of a ray is larger than the critical angle determined by the refractive index difference between the two media. As a result, each incident ray results in total internal reflection and thus an evanescent wave on the interface. Because of the circularly symmetric nature of the illumination, the resulting evanescent wave constructively interferes at the centre of the focus, enhancing the strength of the evanescent field and reducing the lateral trapping size. The transverse profile of the evanescent focal spot leads to a gradient force, as shown in Fig. 8.2, pushing a small particle toward the centre of the focus. Because of the fast decaying nature of the evanescent field, there exists a downward gradient force with a reduced trapping depth. Therefore, optical trapping of a microsphere can be achieved under evanescent wave illumination. In fact, the axial pulling force can prove advantageous for holding a particle in three dimensions if a trapping beam propagates downward.

A detailed characterisation of the focused evanescent wave and the resultant trapping force will be given in the next section. As an example of the calculated results, the

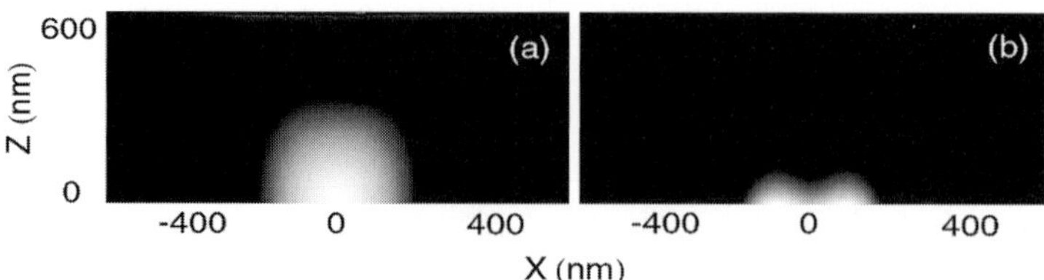

Fig. 8.3 Calculated density plots (a) and (b) represent the modulus squared of the electric field at wavelength 532 nm in the focal region of an objective of numerical aperture 1.65 at the interface between the cover slip ($n = 1.78$) and water ($n = 1.33$). (a) Result given by propagating components ($\varepsilon = 0$, ε is defined the obstruction radius normalised by the radius of the back aperture of the trapping objective). (b) Result with an obstruction whose size satisfies the total internal reflection condition ($\varepsilon = 0.8$). The evanescent focal spot exhibits two peaks that eventually reduce the axial size of the trapping volume significantly. Reprinted with permission from Ref. [7], M. Gu, J. Haumonte, Y. Micheau, J. Chon and X. Gan, *App. Phys. Lett.* **84**, 4236 (2004). © 2004, American Institute of Physics.

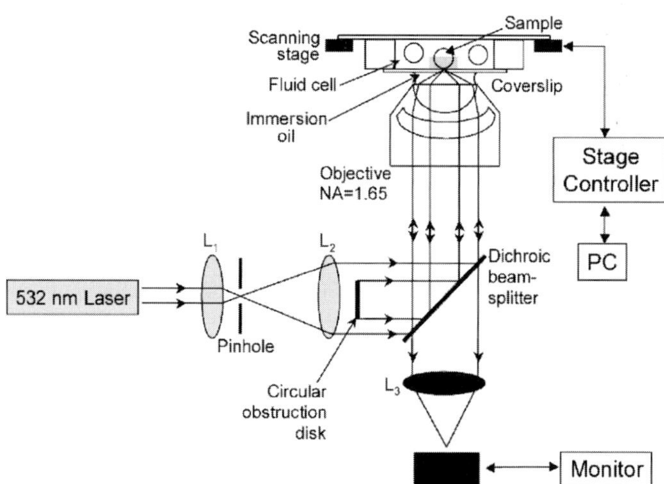

Fig. 8.4 A schematic diagram of an evanescent-field trapping system. Reprinted with permission from Ref. [7], M. Gu, J. Haumonte, Y. Micheau, J. Chon and X. Gan, *App. Phys. Lett.* **84**, 4236 (2004). © 2004, American Institute of Physics.

trapping spots for an objective with and without an opaque disk are shown in Fig. 8.3. The axial size of the trapping volume, defined by the position where the intensity drops to 50% of that at the interface, is reduced to approximately 60 nm, while the lateral trapping size is reduced by 10%. The other advantage associated with this method is that there is no heating problem and that the distance between the trapping site and the objective is sufficiently large for micromanipulation.

In Fig. 8.4, the schematic diagram of the experimental setup for near-field laser trapping is illustrated. The illumination light beam from a continuous-wave laser of wavelength 532 nm or 1064 nm is filtered by a small pinhole and is expanded to a parallel beam by lens L_2. The beam is then directed into a high NA objective lens

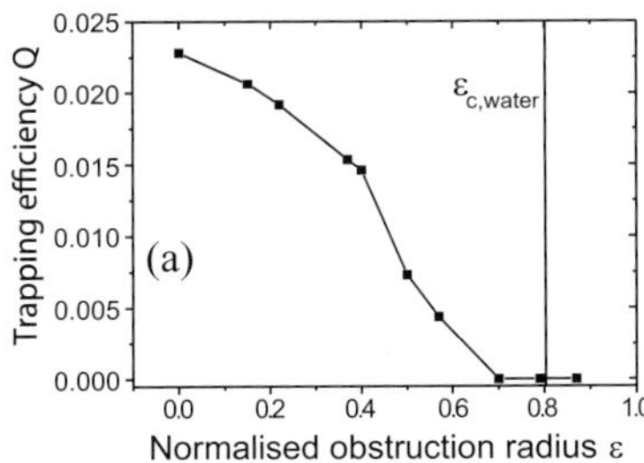

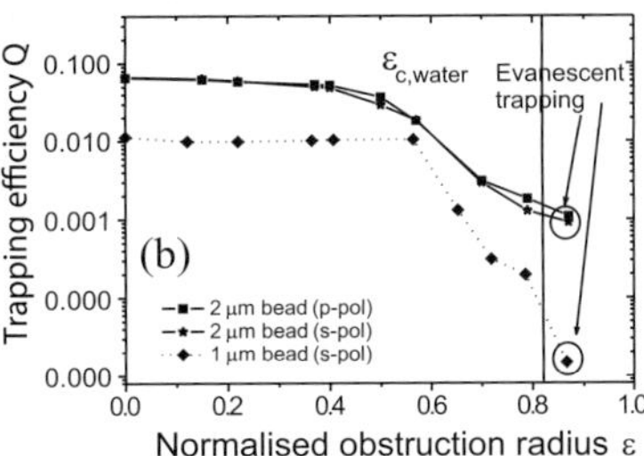

Fig. 8.5 Measured trapping efficiency as a function of the normalised obstruction radius ε. (a) Axial trapping efficiency Q_a; (b) transverse trapping efficiency Q_t. The solid and dotted curves correspond to particles of 2 µm (wavelength 532 nm) and 1 µm (wavelength 1064 nm) in diameter, respectively. The x-direction is along the polarisation direction of the trapping beam. Reprinted with permission from Ref. [7], M. Gu, J. Haumonte, Y. Micheau, J. Chon and X. Gan, *App. Phys. Lett.* **84**, 4236 (2004). © 2004, American Institute of Physics.

(NA $= 1.65$) to focus light on to a sample cell containing polystyrene microspheres of diameter 2 µm or 1 µm, suspended in water. A cover slip ($n = 1.78$) used as a bottom wall of the cell has a refractive index of 1.78 that matches the refractive index of the immersion solution of the objective. The cell is sealed by a cover slide mounted on a computer controlled scanning stage and the trapping process is monitored by a CCD camera. To achieve total internal reflection illumination, we coaxially insert a circular obstruction disk in the beam path and the size of the obstruction disk is carefully selected to block low angle rays that do not satisfy the total internal reflection condition ($\varepsilon = 0.8$). In order to keep the consistency of the experimental results, the laser power after the objective lens is kept at 10 mW throughout the measurement.

To demonstrate near-field laser trapping under focused evanescent wave illumination, we measured the trapping efficiency along the transverse and axial directions, respectively. The trapping efficiency Q, a parameter independent of trapping power, for the evaluation of trapping force is defined as $Q = Fc/nP$, where n is the refractive index relative to that of the surrounding medium, P is the laser power, and c is the speed of light in vacuum. The trapping force F is derived from the measured maximum translating velocity at which a trapping particle falls out of the trap and estimated by the Stokes law

$$F = 6\pi R \eta v. \tag{8.1}$$

Here R is the radius of a trapped particle, v is the maximum translation speed, and η is the viscosity of the surrounding medium [16].

In this experiment, the NA of the objective is larger than that corresponding to the critical angle. The electric field includes the propagating and evanescent components. The propagating component leads to attractive force in the focal region [16] while the evanescent field results in a downward force due to its fast decaying nature. The combined effect of the two forces is large enough to overcome the weight of a trapped microsphere, so that the microsphere can be translated in the axial direction. This feature is confirmed in Fig. 8.5(a) for $\varepsilon = 0$. When the obstruction size increases, the axial trapping efficiency

Q_a decreases, as shown in Fig. 8.5(a). This is because, for a large obstruction disk, the weighting of the propagating electric field decreases, therefore resulting in a decrease in the axial trapping efficiency. It is noticed that axial translation cannot be achieved if the normalised size of the obstruction disk is larger than 0.6. Although this size is less than the 0.8 that corresponds to the critical angle in the experiment, the corresponding propagating component of the trapping beam contributes less than 55% to the total electric field, and thus the resultant attractive axial force is inadequate to overcome the weight of the trapped particle. Beyond this point, the fast decaying nature of the evanescent field results in downward force only and hence it is impossible to translate a particle in the axial direction.

However, laser trapping is achieved for any size of the obstruction in the transverse direction. In Fig. 8.5(b), the transverse trapping efficiency as a function of normalised obstruction size is illustrated. Similar to the axial trapping efficiency, the transverse trapping efficiency Q_t decreases with increasing size of the beam obstruction. There are two physical reasons for this feature. First, increasing the size of the obstruction reduces the contribution of the propagating component to the transverse force. Second, as demonstrated in the earlier research, the higher angle rays of convergence of a trapping objective are less efficient in transverse trapping of a dielectric particle [17]. However, it is noticed that transverse trapping is successfully achieved at an obstruction size larger than 0.8, in which case all rays satisfy the total internal reflection condition and there exists no propagating component. Under such a circumstance, the transverse trapping efficiency is reduced by two orders of magnitude due to the increased friction force between the particle and the interface caused by the downward gradient force, compared with that for no obstruction. However, transverse evanescent-field trapping can be achieved when the polarisation direction of the trapping beam is parallel or perpendicular to the translation direction of the trapping beam. As may be expected from the far-field laser trapping performance, the trapping efficiency in the latter case is slightly less than that in the former (Fig. 8.5(b)).

Figures 8.2(a)–(c) show a trapped particle under focused evanescent wave illumination while the sample cell is translated. As shown in Fig. 8.5(b), the phenomenon described in Fig. 8.5 can be also achieved for a smaller particle or using a near-infrared trapping beam at wavelength 1.06 μm although the trapping efficiency is reduced.

8.2.2   Near-field optical tweezing of a red blood cell

As will be studied in the next section, near-field tweezing of a microsphere can be explained by the multiple reflection and transmission on a rigid interface of the microsphere. This trapping mechanism does not necessarily hold for a soft object such as a red blood cell (RBC), in which case the shape and size dynamically varies under the radiation pressure. Since the inception of laser tweezers based on a single highly focused laser beam [18], this technology has been used to manipulate erythrocytes or RBCs [19] because they play important roles in drug therapy [20] and disease diagnostics [21].

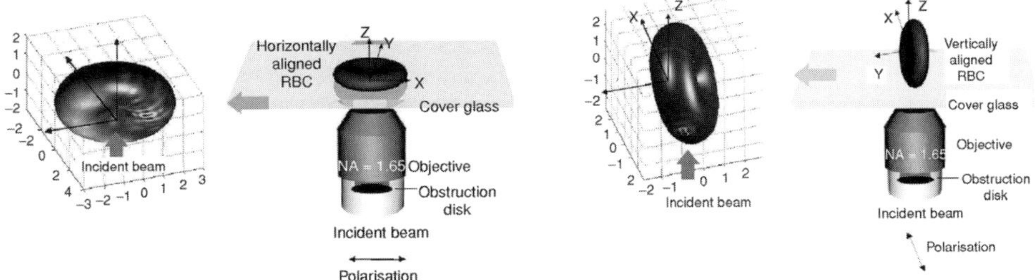

Fig. 8.6 (a) Stress profile on a biconcave RBC oriented horizontally. (b) Stress profile on a biconcave RBC oriented vertically. Reprinted with permission from Ref. [15], M. Gu, S. Kuriakose and X. Gan, *Opt. Express* **15**, 1369 (2007). © 2007, Optical Society of America.

The field distribution in an evanescent focal region of an objective of NA $= 1.65$ is depicted in Fig. 8.3(b), showing the two peaks produced due to the strong depolarisation effect in a ring beam illumination under the total internal reflection condition, as will be shown in the next section. These two localised evanescent wave peaks form a directional trap and can be used to trap an RBC both horizontally and vertically. In other words, a near-field trap produces an axially asymmetrical force distribution and can effectively 'stick' an object in various orientations on the surface where total internal reflection occurs. This situation is confirmed in Fig. 8.6 [15]. Distributions of optical stress on a biconcave RBC oriented in horizontal and vertical planes in the focal region of the objective (NA $= 1.65$) at wavelength 1.06 μm, calculated using the FDTD [22] and Maxwell stress tensor [23] methods, are shown in the insets of Figs. 8.6(a) and (b), respectively. The simulation reveals that both horizontal (Fig. 8.6(a)) and vertical (Fig. 8.6(b)) orientations are stable trapping positions. This feature is fundamentally different from far-field laser tweezers in which case an RBC can be trapped only in the vertical plane [24, 25].

The experimental setup of the near-field laser tweezing of an RBC under focused evanescent wave illumination is the same as that shown in Fig. 8.2. The red blood cells were diluted (1:10000) and suspended in phosphate buffered saline which ensured the isotonicity of the suspension [15, 26]. The bottom central portion of a Petri dish was drilled away, and the special high index cover glass was attached. One drop of RBC suspension was carefully pippetted onto the high index cover glass, and the Petri dish was sealed partly so as to ensure the complete illumination of the sample with white light. The Petri dish was then attached to the computer controlled piezoelectric scanning stage, and the entire sample stage was aligned tilt free horizontally using the vertically aligned laser beam. Figure 8.7 shows that RBCs are trapped using the focused evanescent illumination; Figs. 8.7(a)–(c) show the successive frames from a movie showing an RBC trapped in the horizontal position, and Figs. 8.7(d)–(f) show the successive frames from a movie showing an RBC trapped in the vertical position, while the neighbouring red blood cells are shown to be scanned perpendicular to the direction of polarisation. Once an RBC is trapped, one can perform cell rotation and folding as well as high ratio optical stretching/squeezing at a power level of at least one order of magnitude less than that by other methods [15, 26].

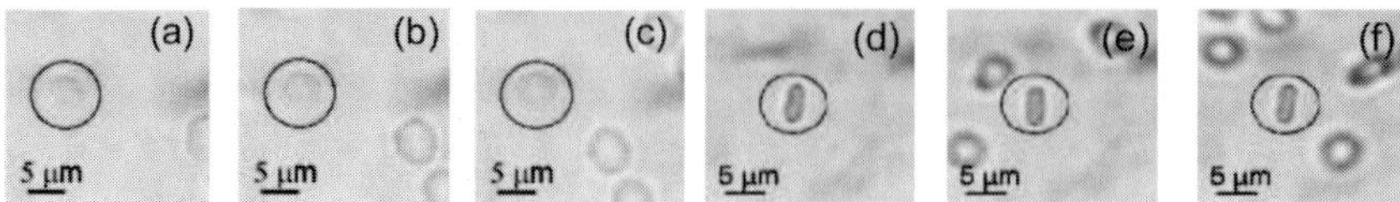

Fig. 8.7 (a)–(c) Successive frames taken from the movie showing near-field trapping of the RBC in the horizontal orientation. (d)–(f) Successive frames taken from the movie showing near-field trapping of the RBC in the vertical orientation.

8.3 Characterisation of near-field tweezing

In this section, the electromagnetic field in a focused evanescent focus and the resultant trapping force on a microsphere are characterised both experimentally and theoretically.

8.3.1 Characterisation of the electromagnetic field in a focused evanescent focus

Let us first demonstrate the direct mapping of a pure focused evanescent field generated by an annular beam illuminated high NA objective (NA $=1.65$) using a scanning near-field optical microscope (SNOM) with a metallic coated fibre tip [27, 28]. A schematic diagram of the experimental setup is shown in Fig. 8.8(a). A linearly polarised He-Ne laser beam is coupled into a high NA objective (NA $=1.65$). A confocal pinhole, located in front of a photomultiplier tube (PMT) at the detection arm provides optical sectioning properties and is used to ensure that the centre of the focal spot is placed at the glass–air interface. An SNOM head with a fibre probe vertical to the interface is placed on top of a cover glass. The tightly focused field is directly mapped with the fibre probe scanning in a transverse plane parallel to the interface (Fig. 8.8(b)). An aluminium coated probe with an aperture of 30–100 nm in diameter is used for measurements. During the experiment, the distance between the tip and the cover glass surface is controlled by the shear force mechanism.

In order to obtain a thorough understanding of a pure focused evanescent field, it is required to isolate it from the propagating component. For this purpose, an obstruction disk with a size larger than the critical radius ε_c (corresponding to the critical angle and normalised by the back aperture radius of the objective, for example, $\varepsilon_c = 0.6$ for a glass–air interface) is inserted in the illumination path as close as possible to the back aperture of the objective, as shown in Fig. 8.8(b), to produce a pure focused evanescent field. In the experiments, the obstruction disk has a normalised radius of $\varepsilon = 0.803$. The mapping of the intensity distributions in the transverse plane close to the interface is shown in Fig. 8.9. It can be clearly seen that the focused evanescent intensity has two identifiable peaks in the direction of the input polarisation (Fig. 8.9(a)). The cross section of the distribution along the polarisation direction is depicted in Fig. 8.9(c), showing a pronounced dip at the centre of the focus. It should be pointed out that one of the advantages of using a fibre probe is that the electric field can be mapped three dimensionally [27, 28]. It has been demonstrated that the pure focused evanescent field decays exponentially and the decay constant is estimated to be 54 nm [27, 28].

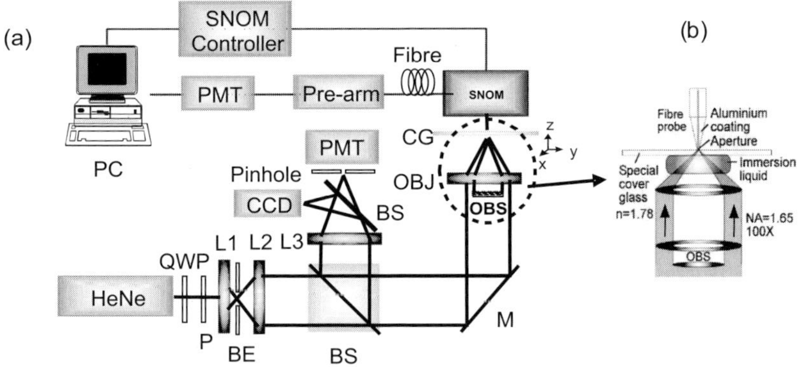

Fig. 8.8 (a) A schematic diagram of the experimental setup for mapping a focused evanescent wave using a SNOM. QWP: quarter wave plate, P: polariser, BE: beam expansion system, BS: beamsplitter, M: mirror, L1, L2, L3: lenses, OBJ: objective, NA $=1.65$, OBS: obstruction. (b) Schematic of the generation of a focused evanescent field and the detection using a SNOM probe. Reprinted with permission from Ref. [27], B. Jia, X. Gan and M. Gu, *App. Phys. Lett.* **86**, 131110 (2005). © 2005, American Institute of Physics.

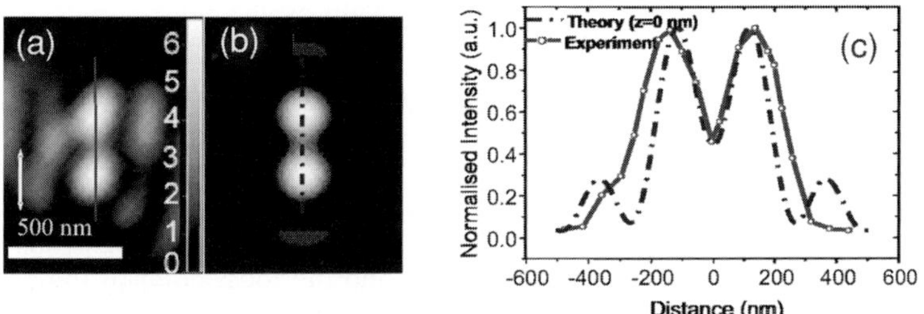

Fig. 8.9 Intensity distributions of pure evanescent focuses at the horizontal plane close to the glass–air interface ($\varepsilon = 0.803$). The incident polarisation direction is vertical. (a) Experimental measurement. (b) Theoretical calculation. (c) Cross sections corresponding to (a) and (b). Reprinted with permission from Ref. [27], B. Jia, X. Gan and M. Gu, *App. Phys. Lett.* **86**, 131110 (2005). © 2005, American Institute of Physics.

To confirm the experimental mapping result, we can calculate the electromagnetic field distribution on the interface using the vectorial diffraction theory [29]. For a linearly polarised monochromatic plane wave in the coordinates system shown in Fig. 8.8, the 3D electric field distribution in the focal region of the objective can be calculated according to the following expression:

$$E(r_2, \phi, z_2) = \frac{\pi i}{\lambda} \{[I_o + \cos(2\phi)I_2]i + \sin(2\phi)I_2 j - 2i \cos\phi I_1 k\}. \tag{8.2}$$

Here the unit vectors in the x, y and z directions are represented by i, j and k, respectively. Variables r_2, ϕ and z_2 are the cylindrical coordinates of an observation point. Due to the focusing process, E is depolarised and has three components, E_x, E_y and E_z, which are determined by the three variables I_0, I_1 and I_2 given by

$$I_0 = \int_\beta^\alpha \sqrt{\cos\theta_1} \sin\theta_1 (t_s + t_p \cos\theta_2) \exp[-ik_0\Phi(\theta_1)] J_0(k_1 r_2 \sin\theta_1)$$

$$\times \exp(-ik_2 z_2 \cos\theta_2) d\theta_1, \tag{8.3}$$

$$I_1 = \int_\beta^\alpha \sqrt{\cos\theta_1} \sin\theta_1 (t_p \sin\theta_2) \exp[-ik_0\Phi(\theta_1)] J_1(k_1 r_2 \sin\theta_1)$$

$$\times \exp(-ik_2 z_2 \cos\theta_2) d\theta_1, \tag{8.4}$$

$$I_2 = \int_\beta^\alpha \sqrt{\cos\theta_1} \sin\theta_1 (t_s - t_p \cos\theta_2) \exp[-ik_0\Phi(\theta_1)] J_2(k_1 r_2 \sin\theta_1)$$

$$\times \exp(-ik_2 z_2 \cos\theta_2) d\theta_1, \tag{8.5}$$

where t_s and t_p are the Fresnel amplitude transmission coefficients for different polarisation states [29]. $J_0(x)$, $J_1(x)$ and $J_2(x)$ are the zero-order, first-order and second-order Bessel functions of the first kind, α and β are the convergence angles of waves corresponding to the outer and inner radii of a ring beam, respectively.

$\Phi(\theta_1)$ in Eqs. 8.3–8.5 is called the spherical aberration function and is physically caused by the mismatch of the refractive indices at the interface, which can be expressed as

$$\Phi(\theta_1) = -d(n_1 \cos\theta_1 - n_2 \cos\theta_2), \tag{8.6}$$

where d is the distance between the interface and the focal point of the objective as if there were no interface. For $d = 0$, i.e. at the interface, this term disappears.

The intensity distribution is proportional to the modulus squared of Eq. 8.2, $|E|^2$. At the interface, the intensity distribution under the experimental condition in Fig. 8.8 is depicted in Fig. 8.9(b), revealing a good agreement with the experimentally measured result with an SNOM. The calculated cross sections along the illumination polarisation direction of the intensity distribution are shown in Fig. 8.9(c) for a further comparison with the experiment. After taking the coupling efficiencies and the finite probe sizes (60 nm on average) into consideration [27, 28], the theoretical cross sectional intensity distribution has a good agreement with the experimental measurement (Fig. 8.9(a)).

8.3.2   Characterisation of the trapping force in a near-field trap

To characterise the dependence of the trapping force of the central obstruction size, shown in Fig. 8.5, one can adopt the vectorial diffraction approach to represent the highly focused beam and its interaction with a microsphere, while the Maxwell stress tensor method is used to calculate the trapping forces on a microsphere [30, 31]. Such an approach enables one to consider the vectorial properties of the electromagnetic field distribution in the focal region of a high NA microscope objective. Effects such as the complex phase modulation, the refractive index mismatch, i.e. spherical aberration, the polarisation dependence and the objective apodisation can be considered. Therefore, one

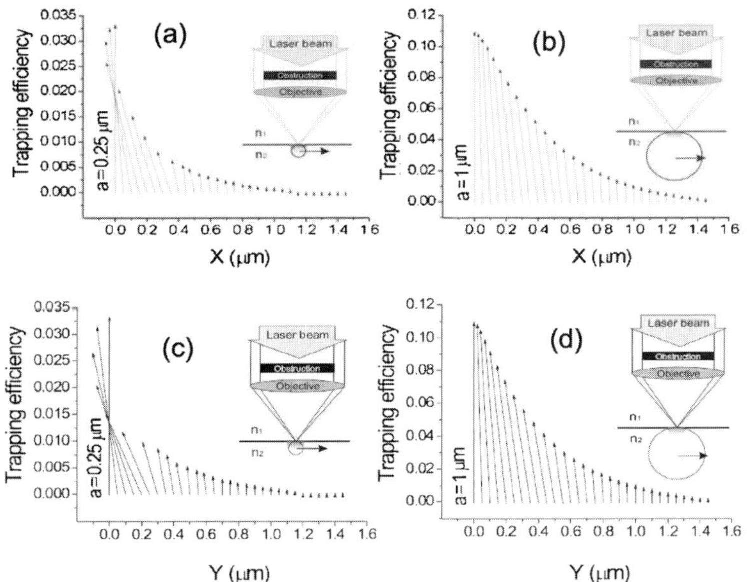

Fig. 8.10 (a) and (b) Trapping efficiency mapping for a small and a large polystyrene particle of radius a, scanned in the X direction (parallel to the polarisation direction) across the focused evanescent field. (c) and (d) Trapping efficiency mapping for a small and a large polystyrene particle of radius a, scanned in the Y direction (perpendicular to the polarisation direction) across the focused evanescent field. NA $= 1.65$, $\lambda = 532$ nm, $\varepsilon = 0.85$, $n_1 = 1.78$ and $n_2 = 1.33$. Reprinted with permission from Ref. [31], D. Ganic, X. Gan and M. Gu, *Opt. Express* **12**, 5533 (2004). © 2004, Optical Society of America.

can express the electric field distribution in the focal region of a high NA objective by Eqs. 8.2–8.5, while the magnetic field distribution can be derived in a similar way [29].

Now let us consider a homogeneous microsphere situated in the second medium. The net radiation force on the microsphere according to the steady-state Maxwell stress tensor analysis is given by [32]

$$\langle \mathbf{F} \rangle = \frac{1}{4\pi} \int_0^{2\pi} \int_0^{\pi} \left\langle \left(\varepsilon_2 E_r \mathbf{E} + H_r \mathbf{H} - \frac{1}{2}(\varepsilon_2 E^2 + H^2)\hat{\mathbf{r}} \right) \right\rangle r^2 \sin\phi \, d\phi \, d\theta, \quad (8.7)$$

where r, ϕ and θ are spherical polar coordinates. Equation 8.7 can be further expressed as a series over the incident and scattered field coefficients [32].

The trapping efficiency defined as a dimensionless factor Q is given by $Q = cF/n_2 P$, where c denotes the speed of light in vacuum, F is the trapping force and P is the incident laser power at the focus. When Q is evaluated in the transverse direction it is known as the transverse trapping efficiency (TTE), while when evaluated in the axial direction it is known as the axial trapping efficiency (ATE). We can distinguish between two ATEs, forward ATE (negative value) corresponds to the inverted microscope configuration and backward ATE (positive value) corresponds to the upright microscope configuration.

Consider an upright trapping system. Figure 8.10 shows the trapping efficiency mapping, when a small and a large polystyrene particle are transversally scanned in the X (also called P) and Y (also called S) directions, respectively. The focused evanescent field distribution is generated by placing a central obstruction ($\varepsilon = 0.85$) perpendicularly to

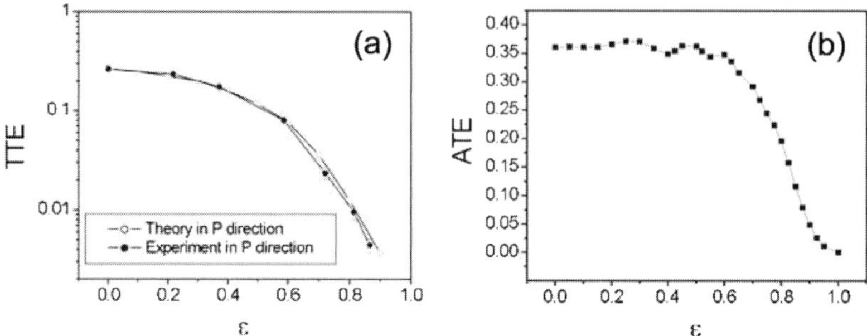

Fig. 8.11 (a) The calculated and measured maximal TTE of a polystyrene particle of 1 μm in radius as a function of the obstruction size ε. (b) The maximal ATE of a polystyrene particle of 1 μm in radius as a function of the obstruction size ε. Reprinted with permission from Ref. [31], D. Ganic, X. Gan and M. Gu, *Opt. Express* **12**, 5533 (2004). © 2004, Optical Society of America.

the path of an incoming laser beam. ε is defined as the obstruction radius normalised by the radius of the back aperture of the trapping objective, and it produces the focused evanescent field for $\varepsilon > 0.8$ (the refractive indices of the coverslip and immersion water are $n_1 = 1.78$ and $n_2 = 1.33$. The laser beam ($\lambda = 532$ nm) propagates in the Z direction.

For both small and large particles the axial trapping efficiency is positive, meaning that the particles can be lifted up. Not only is the ATE for large particles stronger, the force mapping structure in the case of large particles is markedly different from the small particle case due to the larger interaction cross section of the small particle with the focused evanescent field. The maximal transverse trapping efficiency for small particles is much larger relative to the ATE than that for large particles. The maximal TTE constitutes 16.8% of the maximal ATE for small particles, compared to 7.4% for large particles. Further, the ATE in the X direction is slightly larger than that in the Y direction.

The dependence of the maximal TTE and ATE on the obstruction size is critical to capture a particle in the evanescent focal spot and is shown in Fig. 8.11 for polystyrene particles of radius 1 μm. The maximal TTE decreases with an increase in the size of the beam obstruction due to the reduced contribution of the propagating component to the transverse force and because the high angle rays are less efficient in the transverse trapping of a dielectric particle [17]. The theoretical dependence shown in Fig. 8.11(a) agrees well with the experimental results obtained in the near-field trapping system, which confirms the validity of the model. The dependence of the maximal ATE on the obstruction size is at first relatively unchanged until $\varepsilon \sim 0.6$, at which point the maximal ATE decreases for increasing ε (Fig. 8.11(b)). At the focused evanescent field condition, i.e. when $\varepsilon > 0.8$, the maximal ATE is still approximately 43% of the far-field case when no obstruction is present.

8.4   Near-field laser tweezers under femtosecond laser illumination

The tight confinement of the evanescent field achieved by focusing the ring beam illumination at the interface with a high numerical aperture objective is advantageous to

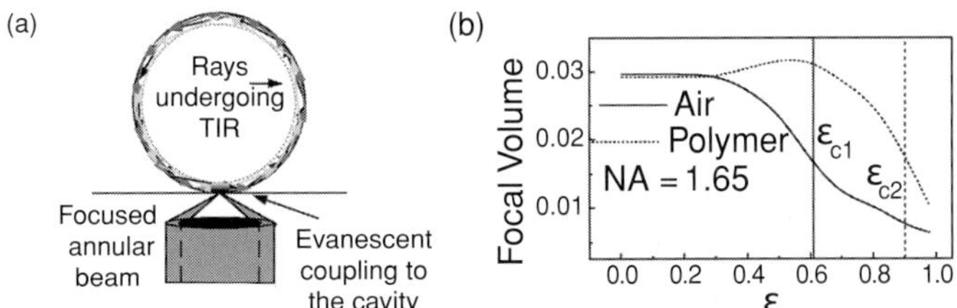

Fig. 8.12 (a) Generation of a focused evanescent wave using a circular obstruction disk. (b) The focal volume ΔV under two-photon excitation at a wavelength of 800 nm as a function of ε (for NA $= 1.65$). The solid and dashed curves represent the focal volumes calculated for the glass–air and glass–polymer interfaces respectively. The refractive indices of glass, polymer and air are 1.78, 1.5 and 1.0, respectively. Reprinted with permission from Ref. [33], S. Kuriakose, D. Morrish, X. Gan, J. Chon and *M. Gu, App. Phys. Lett.* **89**, 1211125 (2006). © 2006, American Institute of Physics.

inducing a nonlinear effect such as two-photon absorption at the focus if a femtosecond laser beam is used for trapping. This section presents two-photon-induced near-field morphology dependent resonance [33] and femtosecond near-field tweezing [34].

8.4.1　Two-photon-induced near-field morphology dependent resonance

Morphology dependent resonance (MDR) or whispering gallery modes (WGM) occurs when light within a dielectric microsphere resonator, which has a higher refractive index than its surroundings, gets total internally reflected and interferes [35]. Coupling the light into WGM can lead to many useful applications [36] including enhanced spontaneous emission [36], structure fluorescence resonance [37], microcavity lasing [38] and trapped-particle microscopy [39, 40]. In these applications, an effective coupling condition, that is, the effective overlapping of the excitation beam and the region where the WGM are generated, has been achieved by a focused beam or a prism.

As shown in Chapter 7, in the case of focused beam coupling to a microsphere, the 3D confining nature of the focal spot is a critical factor for effective coupling. Such a confinement can be enhanced if a two-photon excitation method is used. As a result, it has been shown that the MDR effects such as the strength and polarisation of the MDR peaks are highly dependent on the excitation position [41]. However, the axial dimension of the focal spot is still approximately three times larger than that in the transverse direction. This problem does not exist in a prism coupling method, in which case an evanescent field generated by total internal reflection on a prism significantly reduces the axial dimensional of the excitation. However, the evanescent wave on a prism is not confined in the transverse direction. Since the strength of the focused evanescent focus is strong, it can be used to induce two-photon excitation, as shown in the last section, and thus to excite WGM for enhancing the coupling of the fluorescence to WGM.

The geometry of focused evanescent wave coupling to WGM is shown in Fig. 8.12(a). In this geometry, a tightly confined and enhanced evanescent field is produced, with which it is possible to effectively induce two-photon absorption in a fluorescent microsphere on the cover glass. In fact, the volume of the focused evanescent spot is also dependent on the NA of the focusing objective.

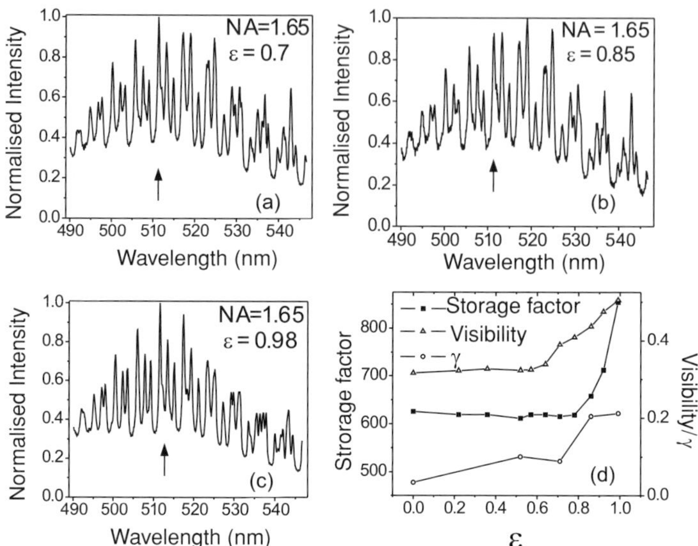

Fig. 8.13 (a), (b) and (c) The two-photon fluorescence spectra for $\varepsilon = 0.7$, $\varepsilon = 0.85$ and $\varepsilon = 0.98$ respectively. The arrow indicates the WGM peak at 513 nm. (d) Spectrum visibility (V), photon storage factor and degree of polarisation (γ) of the peak at 513 nm as a function of the obstruction disk size ε. Reprinted with permission from Ref. [33], S. Kuriakose, D. Morrish, X. Gan, J. Chon and M. Gu, *App. Phys. Lett.* **89**, 1211125 (2006). © 2006, American Institute of Physics.

Figure 8.12(b) shows the focal volume ΔV under two-photon excitation as a function of ε for both the glass–air and glass–polymer interfaces, as calculated by the vectorial theory for a high NA objective [42]. For an objective of NA $= 1.65$, ΔV decreases drastically as soon as ε is larger than the critical value of $\varepsilon_{c1} = 0.61$, determined by the critical angle at the glass–air interface, and at $\varepsilon_{c2} = 0.9$, determined by the critical angle at the glass–polymer interface. The slight increase in the focal volume before the critical obstruction disk size ε_{c2} is caused by the stronger effect of the propagating component of the wave, after the glass–polymer interface, than the evanescent one. The reduction of ΔV between $\varepsilon = 0$ and $\varepsilon = 0.98$ is as large as six times in both cases because the illumination field becomes purely evanescent.

In an experimental setup similar to Fig. 8.13 [33], a Ti:sapphire ultrashort pulse laser at a wavelength of 800 nm and at a repetition rate of 80 MHz was used to induce two-photon absorption in polymer microspheres. Fluorescent polystyrene microspheres of diameter 10 μm with an excitation peak at wavelength 486 nm were used so as to ensure efficient two-photon absorption. The pulsed laser beam was obstructed in the centre using a suitable obstruction disk. A normalised obstruction disk size, ε, of 0.61 corresponded to the critical size to cut off the cone of rays below the critical angle at the cover-glass–air interface. Such an annular beam was tightly focused using a high NA objective (NA $=1.65$) and made to undergo total internal reflection generating a focused evanescent field. WGM were excited for ε of values ranging from 0 to 0.98 so that the relative coupling efficiency of the evanescent field with respect to far-field illumination could be analysed.

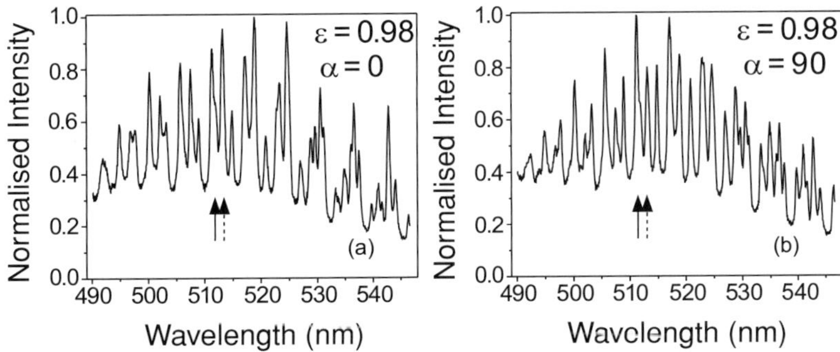

Fig. 8.14 Two-photon fluorescence spectra for NA $= 1.65$ and $\varepsilon = 0.98$ and corresponding to analyser angles of (a) $\alpha = 0°$ and (b) $\alpha = 90°$. The double arrow indicates the two WGM peaks at 513 nm and 514 nm, respectively. Reprinted with permission from Ref. [33], S. Kuriakose, D. Morrish, X. Gan, J. Chon and M. Gu, *App. Phys. Lett.* **89**, 1211125 (2006). © 2006, American Institute of Physics.

The effective coupling between the fluorescence and the WGM can be enhanced if the obstruction size ε is larger than the critical size, as shown in Figs. 8.13(a)–(c). The two-photon fluorescence spectra from the WGM induced by the pure evanescent field illumination demonstrate that when ε increases the MDR becomes more pronounced, which is not only reflected by the increase in the ratio between the resonance peak intensity and the background but also by the decrease in the width of the MDR peaks. Such an enhancement can be described by the visibility of the spectra

$$V = (I_{\text{peak}} - I_{\text{background}})/(I_{\text{peak}} + I_{\text{background}}), \qquad (8.8)$$

where I_{peak} and $I_{\text{background}}$ are the intensity of the fluorescence peak and background and the photon storage factor ($Q = \nu/\Delta\nu$, where ν is the emission frequency), which is a measure of the photon storage time in the cavity.

The visibility of the spectra (V) and the storage factor Q measured as a function of ε are shown in Fig. 8.13(d), showing a significant increase when ε is larger than the critical size. For an extremely thin annulus ($\varepsilon = 0.98$), an increase of 60% in the visibility and 37% in the Q factor is observed as compared with those for $\varepsilon = 0$. The physical reasons for this enhancement are of two folds. First, the focal depth, which is approximately 80 nm for $\varepsilon = 0.98$, is much smaller than the cavity thickness of WGM, which is approximately 0.8 μm for a microsphere of diameter 10 μm. The excitation focal volume of the fluorescence as shown in Fig. 8.12(b) is well within the WGM cavity. Therefore, enhanced effective coupling is expected. Further, the component of the wave vector parallel to the interface, i.e. k_x, for a beam linearly polarised along the x-direction, should contribute to stronger coupling between the fluorescence and WGM within the microsphere.

The degree of polarisation of the individual peaks can be investigated by placing an analyser in the detection arm. Figures 8.14(a) and (b) show the evanescent wave excited spectra with an analyser at angles of $\alpha = 0$ and 90 degrees with respect to the incident polarisation direction. It is noted that the relative strengths of the two peaks indicated by the two arrows change with the analyser position. When the analyser was rotated by 90 degrees the peak at 514 nm (indicated by the dashed arrow) became less pronounced while the peak at 513 nm (indicated by the solid arrow) became stronger. This demonstrates that the MDR peaks are polarised and that the peaks at 513 nm and 514 nm have orthogonal polarisation states representing the TE and TM modes in a microcavity, which is consistent with the discussion in Chapter 7. To quantify this phenomenon, we measured the degree of polarisation (defined as $\gamma = (I_{\alpha=\mathrm{max}} - I_{\alpha=90^\circ})/(I_{\alpha=\mathrm{max}} + I_{\alpha=90^\circ})$, where $I_{\alpha=\mathrm{max}}$ and $I_{\alpha=90^\circ}$ represent the maximum intensity when the analyser is placed by 0° and 90°, respectively) for $\varepsilon = 0$ and $\varepsilon = 0.98$, showing an increase in γ by five times. This feature further indicates that the highly localised evanescent field provides strong coupling of the fluorescence to WGM.

8.4.2 Near-field optical trapping with a femtosecond pulse laser beam

As has been demonstrated, one of the main features of the focused near-field trapping geometry is that the strength of the focused evanescent wave is so strong that nonlinear excitation becomes possible if an ultrashort pulse laser beam is applied. Therefore, using a pulsed laser beam to perform focused evanescent trapping of fluorescent polymer microspheres should enable simultaneous nonlinear excitation in the near-field tweezed sample. Such functionality can be potentially useful to manipulate, diagnose and sort samples of interest with increased detectability in micromanipulation and sensing.

For this purpose, the microspheres are suspended in a glass cell of thickness approximately 600 μm, which is scanned transversely in directions either parallel (P) or perpendicular (S) to the laser polarisation direction while a microsphere of interest is trapped using the focused evanescent field. The maximum scanning speed at which the particle falls out of the trap is determined, and is used to calculate the trapping force using the drag force method, in accordance with Faxen's law for the approximate drag on a spherical particle near an interface [16]. The trapping efficiency η is calculated from the drag force F [34].

In order to determine the maximum near-field transverse trapping efficiency (η) using the femtosecond laser beam, non-fluorescent polystyrene microspheres of diameters (ϕ) ranging from 1 μm to 3.2 μm are used. As the size of the particles decreases, it becomes increasingly difficult to trap them as the Brownian force increases with the particle size. The size of the obstruction disk has been changed for values of $0 \leq \varepsilon \leq 0.86$, so that the relative evanescent contribution to the total trapping illumination is gradually increased until it is all evanescent for $\varepsilon_\mathrm{c} = 0.86$ for the given cover-glass–water interface.

Figures 8.15(a)–(d) show the dependence of η on ε in trapping polystyrene microspheres with ϕ ranging from 1 to 3.2 μm using pulsed illumination. In general, η decreases with ε as is expected. For a given size of the microsphere, the maximum trapping efficiencies along the S (perpendicular) direction η_s are smaller than those in the P (parallel) direction η_p for $\varepsilon = 0$, while $\eta_\mathrm{s} > \eta_\mathrm{p}$ for $\varepsilon = 0.86$. The trapping performance for $\varepsilon = 0$ is similar to the far-field case shown in Chapter 7. According to the dependence of η on the particle size for near-field trapping, we observe that

$$\eta_\mathrm{s} > \eta_\mathrm{p} \quad \text{for} \quad \phi \gg \lambda, \quad \eta_\mathrm{s} \approx \eta_\mathrm{p} \quad \text{for} \quad \phi \leq \lambda. \tag{8.9}$$

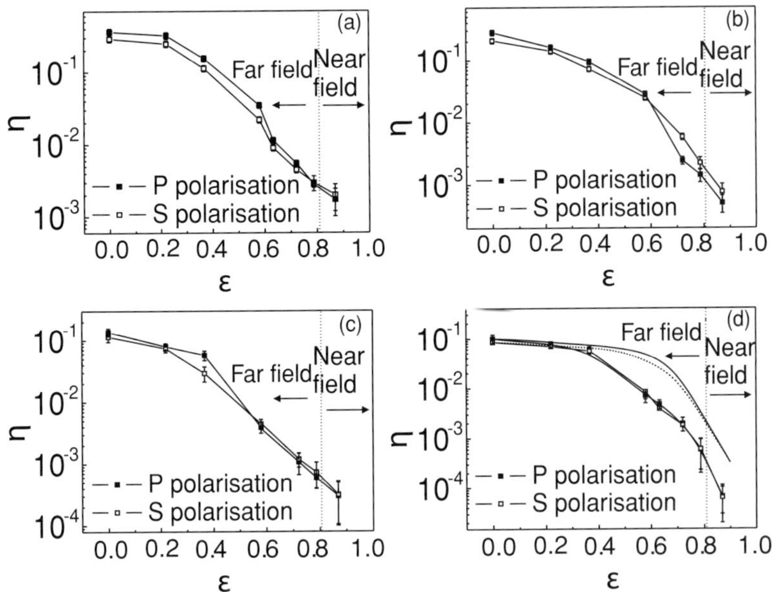

Fig. 8.15 The logarithmic values of the maximum transverse trapping efficiency η measured as a function of ε for femtosecond laser trapping ($\lambda = 780$ nm), for (a) $\phi = 3.2$ μm, (b) $\phi = 2$ μm, (c) $\phi = 1.5$ μm and (d) $\phi = 1$ μm. Reprinted with permission from Ref. [34], S. Kuriakose, D. Morrish, X. Gan, J. Chon, K. Dholakia and M. Gu, *App. Phys. Lett.* **92**, 081108 (2008). © 2008, American Institute of Physics.

To calculate the near-field trapping efficiency under femtosecond illumination, we use the model shown in Section 8.3. The trapping efficiency as a function of ε is shown in Fig. 8.15(d), which qualitatively confirms the observed change in the values of trapping efficiency with ε. The experimentally measured near-field trapping efficiency is lower than that predicted by the theory, which could be attributed to the increased Brownian motion for smaller particles and also to the strong axial trapping force and multiple scattering from the interface.

In the case of near-field trapping, i.e. for $\varepsilon = 0.86$, the dependence of the maximum transverse trapping efficiency on the particle size along the S and P directions is shown in Fig. 8.16, confirming the observation shown in Eq. 8.9. The values of η_p for $\phi = 1$, 1.5, 2 and 3.2 μm are 6.43×10^{-5}, 2.98×10^{-4}, 5.11×10^{-4} and 1.73×10^{-3} respectively, while the values of η_s are 6.77×10^{-5}, 3.16×10^{-4}, 7.7×10^{-4} and 2.02×10^{-3} respectively. Defining $\beta = \eta_s - \eta_p$, we can see that the value of β increases with ϕ and is equal to 0.5 for $\phi = 2$ μm.

Based on the observed trapping performance under pulsed illumination, fluorescent polystyrene microspheres with $\phi = 6$ μm and $\phi = 10$ μm were used to study the excitation of WGM in such near-field tweezed microspheres. The successive frames of a movie showing the focused evanescent trapping of a fluorescent microsphere of $\phi = 6$ μm are shown in Figs. 8.17(a)–(c). The near-field tweezed particle is shown by a box while the neighbouring particle is scanned along the s direction at a speed of $v = 1 \pm 0.05$ μm/s. The maximum near-field trapping efficiency has been calculated to be 1.04×10^{-3},

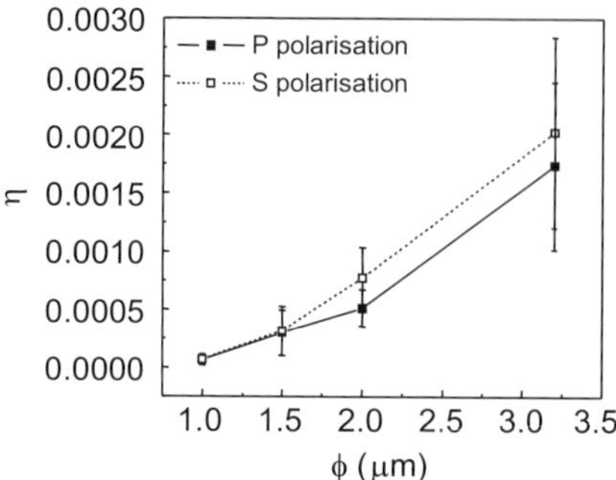

Fig. 8.16 Maximum near-field trapping efficiency η as a function of the particle diameter ϕ for femtosecond laser illumination ($\lambda = 780$ nm, $\varepsilon = 0.86$). Reprinted with permission from Ref. [34], S. Kuriakose, D. Morrish, X. Gan, J. Chon, K. Dholakia and M. Gu, *App. Phys. Lett.* **92**, 081108 (2008). © 2008, American Institute of Physics.

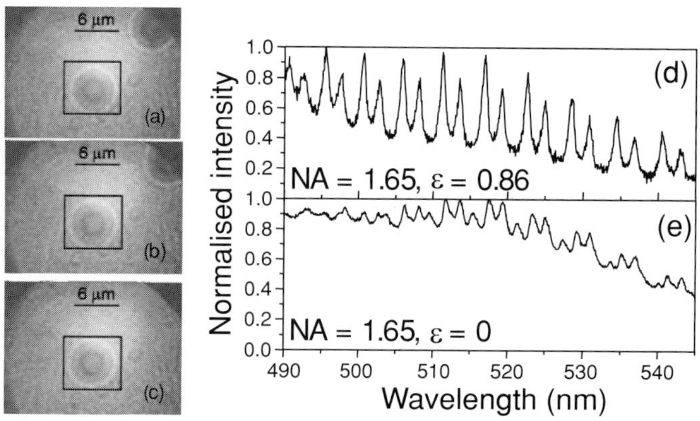

Fig. 8.17 (a)–(c) Successive frames taken from a movie showing femtosecond near-field trapping ($\varepsilon = 0.86$) of a microsphere of $\phi = 6$ μm. The trapped microsphere is marked by a black box. (d) and (e) The two-photon fluorescence spectra obtained from the microsphere of $\phi = 10$ μm excited using by near-field ($\varepsilon = 0.86$) and far-field traps ($\varepsilon = 0$), respectively. Reprinted with permission from Ref. [34], S. Kuriakose, D. Morrish, X. Gan, J. Chon, K. Dholakia and M. Gu, *App. Phys. Lett.* **92**, 081108 (2008). © 2008, American Institute of Physics.

which corresponds to a trapping force of approximately 1 femtonewton at a power of 10 mW in the trapping plane. The trapping efficiency value calculated for a microsphere of $\phi = 6$ μm from the size dependence graph in Fig. 8.16 is 6.25×10^{-3}. It is seen from Figs. 8.17(a)–(c) that there appears to be a bright ring around the surface of the trapped fluorescent particle. Using the spectral measurement method, we have observed that the fluorescence spectra in the bright region exhibit the excitation of WGM which become pronounced for larger microsphere diameters.

The two-photon fluorescence emission from a microsphere of $\phi = 10$ μm is analysed using a spectrograph of resolution 0.1 nm at a power of 9.15 mW and at $\varepsilon = 0.86$, and is shown in Fig. 8.17(d). The visibility of the spectra [40] is calculated to be 0.45, which is facilitated by the strong suppression of the background intensity. This visibility value is four times higher than that of the spectrum acquired using two-photon excitation in far-field tweezed microspheres (Fig. 8.17(e)) at a scanning velocity of 4 µm/s and also higher by 18% than that of the highest visibility achieved at a maximum translational speed of 29 µm/s [40].

It is shown that the resonance under the near-field trapping is significantly stronger than that in the far-field case. The calculated mode photon storage factor, defined as $Q = \nu/\Delta\nu$, for one of the dominant resonant peaks at wavelength 511 nm is 706, which is 46% higher compared to the photon storage factor of 483 for a resonant spectra from a microsphere excited by a far-field trap. Though the photon storage factor of a microsphere is improved in near-field trapping, the absolute value is not so high. The factors contributing towards lowering the photon storage factor could be the small size of the cavity resulting in radiation losses [43], coupling losses due to the presence of the dielectric substrate [44] and the lower refractive index contrast available, as the microsphere is suspended in water [45].

8.5   Summary

One of the future developments in focused evanescent wave tweezers is to reduce the lateral size of the trapping volume. As near-field optics are not constrained by the diffraction limit, they may also offer the potential for creating optical trapping volumes that are significantly smaller than those that can be achieved in conventional optical tweezers. The use of a radially polarised beam illumination can lead to a confined evanescent focal spot of 150 nm or less [28, 46]. The reduction of the trapping spot may be also achieved by using a doughnut beam [8, 47].

Another development aspect is the enhancement of the strength of an evanescent wave. Many near-field geometries are currently under investigation that may be used to enhance the effects of evanescent waves, including surface plasmons [48], dielectric resonators such as a double-layer dielectric stack [14] and multi-layer stacks [49]. For example, the use of a dielectric double-layer structure or a metallic layer, coated on the cover slip, can result in the enhancement of the evanescent field by approximately two to three orders of magnitude.

The use of an evanescent light field can potentially create a large number of trapping sites over a large area on a substrate [12], which is ideal for extended studies of colloidal and biological particles. The creation of patterned evanescent fields using holographic or other light-patterning techniques could also be used to address colloids individually in a large aggregation, enabling simultaneous and independent micromanipulation. Further, the use of near-field optics may lead to nanometric trapping sites for manipulation of nano-scale objects. Therefore, near-field laser tweezers will become a novel tool in single molecule detection and manipulation. The enhanced electric field within the focused evanescent spot may also prove advantageous for simultaneous laser tweezing and nonlinear excitation within a trapped object.

References

[1] A. Ashkin. Applications of laser radiation pressure. *Science*, **210**:1081–1088, 1980.

[2] P. C. Ke. *Near-field Scanning Optical Microscopy with Laser Trapping*. Ph.D. thesis, Department of Physics, Victoria University of Technology, Australia, 2000.

[3] S. Kawata and T. Sugiura. Movement of micrometer-sized particles in the evanescent field of a laser beam. *Opt. Express*, **17**:772–774, 1992.

[4] L. Novotny, R. X. Bian, and X. S. Xie. Theory of nanometric optical tweezers. *Phys. Rev. Lett.*, **79**:645–648, 1997.

[5] P. C. Chaumet, A. Rahmani and M. Nieto-Vesperinas. Optical trapping and manipulation of nano-objects with an apertureless probe. *Phys. Rev. Lett.*, **88**:1236011–1236014, 2002.

[6] K. Okamoto and S. Kawata. Radiation force exerted on subwavelength particles near a nanoaperture. *Phys. Rev. Lett.*, **83**:4534–4537, 1999.

[7] M. Gu, J.-B. Haumonte, Y. Micheau, J. Chon and X. Gan. Laser trapping and manipulation under focused evanescent wave illumination. *App. Phys. Lett.*, **84**:4236–4238, 2004.

[8] D. Ganic. *Far-field and Near-Field Optical Trapping*. Ph.D. thesis, Faculty of Engineering and Industrial Research, Swinburne University of Technology, Australia, 2005.

[9] K. Dholakia, P. Reece and M. Gu. Light takes hold: optical micromanipulation. *Chem. Soc. Rev.*, **37**:42–45, 2008.

[10] S. Kawata and T. Tani. Optically driven Mie particles in an evanescent field along a channeled waveguide. *Opt. Lett.*, **21**:1768–1770, 1996.

[11] T. Tanaka and S. Yamamoto. Optically induced propulsion of small particles in an evanescent field of higher propagation mode in a multimode, channeled waveguide. *App. Phys. Lett.*, **77**:3131–3133, 2000.

[12] V. Garcés-Chávez, K. Dholakia and G. C. Spalding. Extended-area optical induced organization of microparticles on a surface. *App. Phys. Lett.*, **86**:1–3, 2005.

[13] P. J. Reece, V. Garcés-Chávez and K. Dholakia. Near-field optical micromanipulation with cavity enhanced evanescent waves. *App. Phys. Lett.*, **88**:221116, 2006.

[14] P. C. Ke, X. S. Gan, J. Szajman, S. Schilders and M. Gu. Optimising the strength of an evanescent wave generated from a prism coated with a double-layer thin-film stack. *Bioimaging*, **5**:1–8, 1997.

[15] M. Gu, S. Kuriakose and X. Gan. A single-beam near-field laser trap for optical rotation, stretching and folding of erythrocytes. *Opt. Express*, **15**:1369–1375, 2007.

[16] W. H. Wright and G. J. Sonek. Radiation trapping forces on microscophers with optical tweezers. *App. Phys. Lett.*, **63**:715–718, 1993.

[17] M. Gu, D. Morrish and P. C. Ke. Enhancement of transverse trapping efficiency for a metallic particle using an obstructed laser beam. *App. Phys. Lett.*, **77**:34–36, 2000.

[18] A. Ashkin, J. M. Dziedzic, J. E. Bjorkholm and S. Chu. Observation of a single-beam gradient force optical trap for dielectric particles. *Opt. Lett.*, **11**:288–290, 1986.

[19] A. Ashkin, J. M. Dziedzic and T. Yamane. Optical trapping and manipulation of single cells using infrared laser beams. *Nature*, **330**:769–771, 1987.

[20] A. Krantz. Red-cell mediated therapy: opportunities and challenges. *Blood Cells, Molecules and Diseases*, **23**:58–68, 1997.

[21] A. Elgsaeter, B. T. Stokke, A. Mikkelsen and D. Branton. The molecular basis of erythrocyte shape. *Science*, **234**:1217–1223, 1986.

[22] J. Yu, J. Chen, Z. Lin, L. Xu, P. Wang and M. Gu. Surface stress on the erythrocyte under laser irradiation with finite-difference time-domain calculation. *J. Biomed. Opt.*, **10**:064013, 2005.

[23] J. P. Gordon. Radiation forces and momenta in dielectric media. *Phys. Rev. A*, **8**:14–21, 1973.

[24] S. C. Grover, R. C. Gauthier and A. G. Skirtach. Analysis of the behaviour of erythrocytes in an optical trapping system. *Opt. Express*, **7**:533–539, 2005.

[25] S. K. Mohanty, K. S. Mohanty and P. K. Gupta. Dynamics of interaction of RBC with optical tweezers. *Opt. Express*, **13**:4745–4751, 2005.

[26] S. Kuriakose (Varghese). *Characterisation of Near-Field Optical Trapping and Biological Applications*. Ph.D. thesis, Faculty of Engineering and Industrial Research, Swinburne University of Technology, Australia, 2007.

[27] B. Jia, X. Gan and M. Gu. Direct observation of pure focused evanescent wave of a high numerical aperture objective lens by scanning near-field optical microscopy. *App. Phys. Lett.*, **86**:131110, 2005.

[28] B. Jia. *A Study on the Complex Evanescent Focal Region of a High Numerical Aperture Objective and its Applications*. Ph.D. thesis, Faculty of Engineering and Industrial Research, Swinburne University of Technology, Australia, 2006.

[29] M. Gu. *Advanced Optical Imaging Theory*. Berlin, Springer Verlag, 2000.

[30] D. Ganic, X. Gan and M. Gu. Exact radiation trapping force calculation based on vectorial diffraction theory. *Opt. Express*, **12**:2670–2675, 2004.

[31] D. Ganic, X. Gan and M. Gu. Trapping force and optical lifting under focused evanescent wave illumination. *Opt. Express*, **12**:5325–5345, 2004.

[32] J. P. Barton, D. R. Alexander and S. A. Schaub. Theoretical determination of net radiation force and torque for a spherical particle illumination by a focused laser beam. *J. App. Phy.*, **66**:4594–4602, 1989.

[33] S. Kuriakose, D. Morrish, X. Gan, J. Chon and M. Gu. Enhanced coupling to whispering gallery modes by two-photon absorption induced by focused evanescent field. *App. Phys. Lett.*, **89**:211125, 2006.

[34] S. Kuriakose, D. Morrish, X. Gan, J. Chon, K. Dholakia and M. Gu. Near-field optical trapping with an ultrashort pulsed laser beam. *App. Phys. Lett.*, **92**:081108, 2008.

[35] L. Rayleigh. Further applications of Bessel's functions of high order to the whispering gallery and allied problems. *Phil. Mag.*, **27**:100–109, 1914.

[36] K. J. Vahala. Optical microcavities. *Nature*, **424**:839–846, 2003.

[37] R. E. Benner, P. W. Barber, J. F. Owen and R. K. Chang. Observation of structure resonances in the fluorescence spectra from microspheres. *Phys. Rev. Lett.*, **44**:475–478, 1980.

[38] A. J. Campillo, J. D. Eversole and H.-B. Lin. Cavity quantum electrodynamic enhancement of stimulated-emission in microdroplets. *Phys. Rev. Lett.*, **67**:437–440, 1991.

[39] D. Morrish, X. Gan and M. Gu. Scanning particle trapped optical microscopy based on two-photon-induced morphology-dependent resonance in a trapped microsphere. *App. Phys. Lett.*, **88**:141103, 2006.

[40] D. Morrish, X. Gan and M. Gu. Morphology-dependent resonance induced by two-photon excitation in a micro-sphere trapped by a femtosecond pulsed laser. *Opt. Express*, **12**:4198–4202, 2004.

[41] D. Morrish, X. Gan and M. Gu. Observation of orthogonally polarized transverse electric and transverse magnetic oscillation modes in a microcavity excited by localized two-photon absorption. *App. Phys. Lett.*, **81**:5132–5134, 2002.

[42] J. Chon and Min Gu. Scanning total internal reflection fluorescence microscopy under one-photon and two-photon excitation: image formation. *Appl. Opt.*, **43**:1063–1071, 2004.

[43] J. R. Buck and H. J. Kimble. Optimal sizes of dielectric microspheres for cavity QED with strong coupling. *Phys. Rev. A*, **55**:033806–033818, 2003.

[44] H. Ishikawa, H. Tamaru and K. Miyano. Microsphere resonators strongly coupled to a plane dielectric substrate: coupling via the optical near field. *J. Opt. Soc. Am. A*, **17**:802–813, 2000.

[45] P. Chylek, J. T. Kiehl and M. K. W. Ko. Optical levitation and partial-wave resonances. *Phys. Rev. A*, **18**:2229–2233, 1978.

[46] B. Jia, X. Gan and M. Gu. Direct measurement of a radially polarized focused evanescent field facilitated by a single lcds. *Opt. Express*, **13**:6821–6827, 2005.

[47] D. Ganic, X. Gan and M. Gu. Optical trapping force with annular and doughnut laser beams based on vectorial diffraction. *Opt. Express*, **13**:1260–1265, 2005.

[48] D. Sarid. Long-range surface plasma wave on very thin metal films. *Phys. Rev. Lett.*, **47**:1927–1930, 1981.

[49] R. C. Nesnidal and T. G. Walker. Multilayer dielectric structure for enhancement of evanescent waves. *Appl. Opt.*, **35**:2226–2229, 1996.

9 Femtosecond cell engineering

Continued development of optical systems for simultaneous observation and manipulation of live biological specimens has produced advances in understanding cell physiology. Traditional optical microscopes have given way to multi-functional, multi-laser based observation platforms that provide us with the opportunity to interact with the specimen on a subcellular level.

This chapter gives a brief review on the development of advanced photonics technologies for biological applications including the use of femtosecond pulse lasers to interact with target cells for the stimulation of cellular responses (Section 9.1). Section 9.2 is focused on the technology of femtosecond pulse laser based microfabrication to develop microfluidic devices for applications in biology, while Section 9.3 demonstrates the use of femtosecond laser fabricated microenvironments for advanced live cell imaging of T cells. Section 9.4 discusses the use of an integrated sensor for optical sensing in microfluidic devices.

9.1 Femtosecond cell stimulation

Fluorescence signals of cells can be linked to the overall health and integrity of those cells [1], with fluctuations in the signals indicating effects such as changes in dye loading, fluorescence resonance energy transfer (FRET), fluorescence lifetime imaging (FLIM), fluorescence recovery after photobleaching (FRAP), fluorescence loss in photobleaching (FLIP), cell activation and cell destruction. Monitoring the integrity of biological specimens that are being altered due to focused femtosecond irradiation is important to ensure no damage is being caused by such illumination. While limited exposure to high peak intensity laser pulses has been demonstrated as a successful tool in microsurgical applications [2–5] and to a lesser extent cell trapping [6, 7], continued monitoring of cell vitality during and post-exposure will help explain the intracellular and extracellular processes that occur as a result of the exposure. There are numerous advantages to using femtosecond lasers in the near infrared for manipulating cells as compared with typical single-photon induced photoeffects, such as highly localised nonlinear photodamage, increased penetration depth and limited heat transfer to sample. The finite interaction period of a femtosecond pulse with a cell provides a mechanism for altering particular cell characteristics without leading to the destruction of the cell. Recently femtosecond lasers have been demonstrated as a method for non-invasive procedures like dissection [8], photodisruption [9,10], microinjection [11] and cell transfection [12]. The use of femtosecond irradiation has also been used to generate a calcium ion response in HeLa cells [13] and neocortical neurons in rat brain slices [10].

9.1.1   Experimental systems

9.1.1.1   Real time imaging system

Imaging samples in real time (typically 25–30 frames per second) has a couple of advantages over standard beam scanning based modalities. The Nipkow disk [14] reduces the high levels of photobleaching associated with beam scanning configurations whilst allowing millisecond monitoring of fluorescence intensity and/or spectrum across the desired samples.

The use of a real time single-photon fluorescence monitoring of cells that are activated using a femtosecond laser beam allows for novel applications in cell biology. Compared with standard commercial fluorescence microscopes that use a beam scanning method coupled with photomultiplier tubes, the time resolution of the real time system allows detection of cellular response to femtosecond irradiation in previously undetectable time periods. Traditional beam scanning microscopes can take up to one second to produce an image, which could lead to distorted images of relatively high-speed cellular activity. A Nipkow disk scanning microscope can produce up to hundreds of frames a second depending on the rotational speed of the Nipkow disk and the data transfer rate of the CCD camera. The fast imaging times afforded by this system enable the monitoring of the speed that calcium waves traverse a cell following a short femtosecond laser pulse. Also, examples of calcium ion (Ca^{2+}) oscillations being instigated as a result of femtosecond laser pulses can be observed.

The schematic diagram of the modified Nipkow disk real time microscope is illustrated in Fig. 9.1, developed by Day *et al.* in 2005 [15]. The real time scanning unit is connected to a modified inverted microscope. The excitation source utilised in the real time system is a fibre coupled argon:krypton (Ar:Kr) laser with excitation bandpass filters at the wavelengths 488 nm, 547 nm and 647 nm on a stepper motor controlled filter wheel (ExF). The excitation beam is then passed through two matched, rotating disks. The first disk contains an array of microlenses which focuses the light through the array of pinholes on the second disk (Nipkow disk). From there the light enters the IX71 microscope and is focused into the sample with a 0.6 numerical aperture (NA) objective (Obj 1). The fluorescence light is collected with the same objective and passed back through the rotating pinhole array where the dichroic mirror (DM) reflects the fluorescence into the charge coupled detector (CCD). Another stepper motor controlled filter wheel (EmF) located before the CCD contains the emission filters.

The femtosecond pulse laser excitation beam is a Ti:sapphire femtosecond pulse laser which produces 80 fs pulses at a repetition rate of 80 MHz and an average power of 950 mW. The laser has a tunable wavelength range from 730 nm to 870 nm, which allows for multi-photon excitation of most biological dyes and specimens.

The femtosecond pulse laser beam is passed through a mechanical shutter that is used to control the exposure time from ms to minutes. The beam is then expanded and

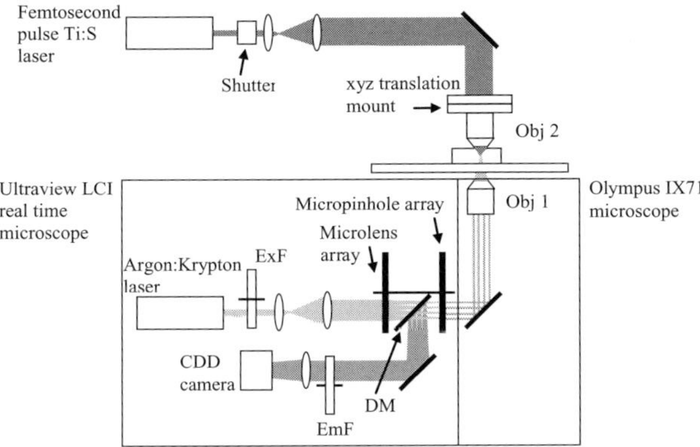

Fig. 9.1 Schematic diagram of the modified real time scanning system with counter-propagating focused femtosecond irradiation.

directed through the modified transmission head of the microscope where it is directed down through a 0.9 NA water objective (Obj 2) into the sample. An over expanded top-hat beam profile is used so that when the objective is translated there is no change in the profile of the focus spot. The objective is mounted in a computer-controlled x–y stepper motor scanning stage which is used to control the focus spot within the field of view of the imaging objective. A piezo-scanner attached to the objective is used to position the focus spot in the z-axis.

The system can also be used to directly image and monitor the two-photon signal coming from the femtosecond focus spot in a cell by blocking the single-photon excitation beam with an appropriate beam block in the excitation filter wheel.

9.1.1.2 Laser scanning confocal imaging system

Delivery of the femtosecond pulse laser to cells was carried out using an adapted confocal microscope. The shutter-controlled femtosecond pulse laser was expanded to exceed the width of the back aperture of a 1.25 NA objective, and directed into the back port of an inverted microscope, where a short pass dichroic mirror was installed. This dichroic allowed for simultaneous fluorescence excitation using the scanned 488 nm line of the Ar:Kr ion laser of the confocal microscope whilst simultaneously allowing for the femtosecond pulse laser to target the sample at a fixed point.

The source for the femtosecond pulse laser beam was the Ti:sapphire femtosecond pulse laser. The femtosecond beam was passed through a mechanical shutter with a fixed shutter time of 500 ms for the experiments. The beam then passed through a neutral density filter wheel before it was expanded and directed by the use of lenses and mirrors through the rear of an inverted microscope, where it was then directed through a 1.25 NA oil objective into the sample. The power at the back aperture of the objective was measured to be between 8 and 15 mW. In order to visualise the transmitted image whilst laser exposure was occurring, a short pass filter was placed in the transmission path of the confocal microscope. The system used for simultaneous imaging and femtosecond irradiation is similar to that shown in Fig. 9.1, except that the real time imaging system is replaced with the confocal scanning system.

Imaging at 37°C was performed with the assistance of a Focht Chamber System. This allowed cells to be imaged on the inverted microscope, whilst simultaneously being perfused with a medium that is kept at a constant 37°C (±0.1°C). The perfusion medium consisted of Dubelco's modified eagle medium (DMEM) with HEPES (20 mM), but without any other supplements. The medium was continually bubbled with carbon dioxide gas to maintain a stable pH. Cells in regions of high cell confluence were chosen as targets for laser irradiation.

9.1.1.3   Cell culture

The effectiveness of the modified system for imaging cellular activation using a counter-propagating femtosecond laser beam, with respect to the imaging system, was conducted on two types of cells: GH3 rat pituitary cells and COS cells, a simian fibroblast cell line. The cell types were grown in 35 mm diameter culture dishes with a 170 μm coverglass bottom. The cells were cultured in DMEM with 20 mM HEPES, 10% fetal bovine serum, 2% penicillin-streptomycin and 200 mM L-glutamine solution in a CO_2 incubator at 37°C. The cells were loaded with 2 μm Fluo-3 AM for 30 minutes at room temperature in DMEM without supplements. After loading the cells were washed and finally immersed in 2 ml of DMEM solution with 20 mM HEPES again without the supplements.

MC3T3-E1 (3T3) murine osteoblast-like cells were seeded onto autoclaved round 40 mm diameter coverslips at a density of 100 000 cells per coverslip. The 3T3 cells were grown in DMEM supplemented with HEPES (20 mM), 10% fetal bovine serum, L-glutamine (200 mM), and penicillin-streptomycin (1%) for 24 hours prior to imaging.

Prior to laser stimulation and imaging, cells were incubated with 1–2 μm fluo-3 AM for 1 hour at room temperature in DMEM and 20 mM HEPES without supplements for Ca^{2+} monitoring. Cells were kept out of the light for the duration of fluoro-3 incubation. After fluoro-3 dye incubation, the cells were washed and immersed in fresh DMEM (with no supplements) ready for imaging.

9.1.2   Femtosecond laser induced calcium oscillations

Femtosecond laser irradiation of GH3 cells can be used to induce Ca^{2+} oscillations. The oscillations in the targeted cell (1) seen in Fig. 9.2(a) were initiated in direct response to a 15 ms exposure at 800 nm. The average power of the laser in the focus spot was 8 mW. To image the cells they were excited with the 488 nm laser line from the Ar:Kr laser and the corresponding fluorescence was collected after it passed through the 525 nm filter. The cells were imaged for 54 seconds before cell (1) was exposed to the femtosecond laser beam and then imaged a further 146 seconds to monitor the oscillations. For such an extended imaging period, the frequency of image capturing was decreased to two frames per second to reduce potential photobleaching. The oscillating fluorescence intensity levels measured in Fig. 9.2(b)–(e) was an average of the fluorescence intensity over the whole cell.

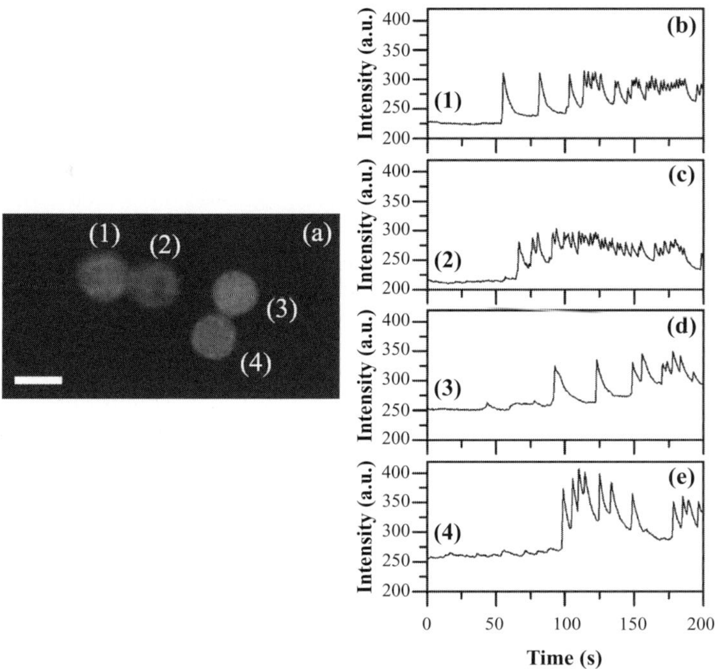

Fig. 9.2 Fluorescence intensity profiles of femtosecond laser induced Ca^{2+} oscillations in GH3 cells versus time. (a) Single-photon fluorescence image of GH3 cells. (b), (c), (d) and (e) Intensity profiles of cells (1), (2), (3) and (4), respectively. Cell (1) was exposed to the femtosecond laser at the time indicated by the dashed line. The scale bar is 10 μm [15].

The oscillations generated in cell (1) as a result of the laser stimulus, triggered oscillations in the neighbouring cells (2), (3) and (4) in response to the possible release of chemical factors. The exact chemical mechanism that triggers the oscillations in neighbouring cells is not fully understood but it is possibly due to the release of extracellular chemical factors [16] (e.g. growth hormone). The exposure to the femtosecond laser beam results in no permanent damage to the cells, including the targeted cells, during the period of monitoring, as evident by the continued fluctuations in Ca^{2+} concentrations. However, higher power levels or longer exposure can be used to disrupt the cell membrane and effectively kill the cell.

9.1.3   Femtosecond laser induced calcium wave

The same process used to induce oscillations in GH3 cells can be used to increase the Ca^{2+} levels in COS cells. The focused femtosecond pulse laser beam induces a localised increase in the fluorescence signal which is followed by a Ca^{2+} wave that travels the length of the cell, increasing the overall cell fluorescence with it. The Ca^{2+} wave expands radially away from the point of interaction, slowing gradually the further it travels. Figures 9.3(a)–(d) show the increase in fluorescence signal over time.

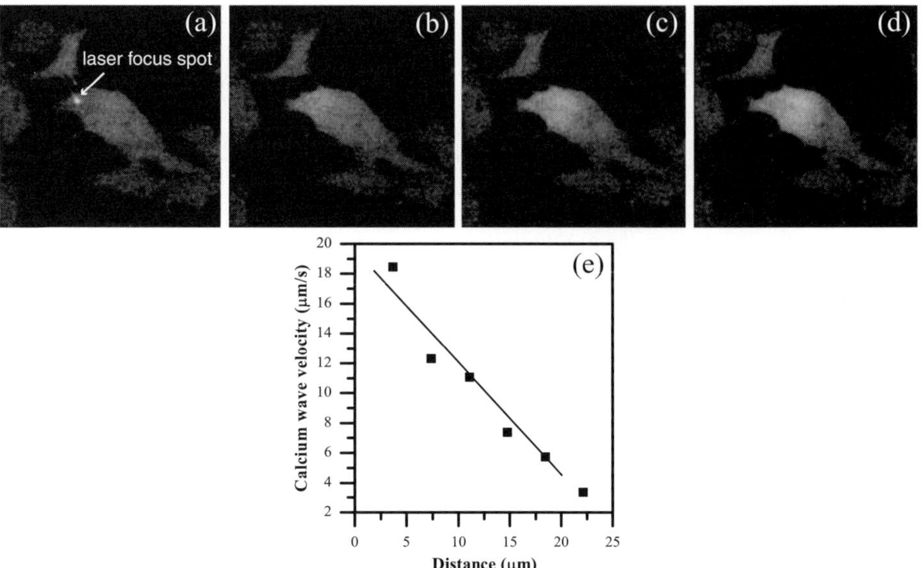

Fig. 9.3 Femtosecond laser induced Ca^{2+} wave expanding across a COS cell. Selected images represent images taken at (a) 0 s, (b) 0.4 s, (c) 0.8 s and (d) 1.4 s after exposure. (e) The measured velocity profile of the Ca^{2+} wave as it expands across the COS cell. The scale bar is 10 μm [15].

Due to the high frame rate of the imaging system it is possible to measure the velocity of the Ca^{2+} wave as it expands across the cell. Figure 9.3(e) shows the decrease in the measured velocity of the Ca^{2+} wave as it travels along the distance of the cell. The cell was exposed to the femtosecond irradiation for 25 ms with an average power of 15 mW in the focus. The capture rate of the images was increased to 10 frames per second in order to image the travelling Ca^{2+} wave. The limitation on the frame rate is the strength of the fluorescence signal of the cell. A weaker fluorescence signal requires a longer exposure time and therefore a slower image capture rate.

The combined use of a real time imaging system and targeted femtosecond pulse laser irradiation can now be used to study various aspects of cell physiology on a timescale not previously achieved. The use of the counter-propagating femtosecond beam provides the freedom to manipulate the sample in three dimensions while maintaining a simplified imaging system.

In this section it has been demonstrated that femtosecond pulse lasers can be used to initiate intracellular Ca^{2+} oscillations, and that this effect might also be useful in initiating oscillations in adjacent cells. It is postulated that this trigger of oscillations in neighbouring cells is due to extracellular chemical factors being released from the GH3 cell as a result of the femtosecond pulse. The release of these chemical factors then affects the neighbouring cells, inducing comparable oscillations in them. It should be noted that cells do not have to be touching for this to occur. Caution should be used in interpreting these results, however, as GH3 cells can show spontaneous oscillations. This ability to induce oscillations in adjacent cells, when used in conjunction with various chemical antagonists, could open the way for novel methods of micropharmacological research.

The rapid propagation of a calcium wave across a cell often requires real time imaging in order to observe it fully. This system can follow this propagation from the instant the cell was irradiated with the femtosecond laser. This system's ability to micromanipulate the counter propagating objective for the femtosecond pulse laser allows for cells to be trapped and moved. The added ability to monitor cells in real time means that the vitality of the cells can be monitored simultaneously.

9.1.4 Femtosecond induced mechanical strain

One of the areas in which stimulation of cells with femtosecond pulse laser irradiation might find extended applications is orthopaedic tissue engineering. Tissue engineering aims to create biologically functional tissue substitutes. Generally this is done by seeding cells on a biocompatible scaffold and cultivating within a bioreactor. The bioreactor provides the appropriate environment and stimuli to the cellular construct so that tissue integrity is optimised prior to implantation. It is widely recognised that successful tissue engineering requires not only coordinate biochemical stimulus of the cell, but also physical stimulus of the cell. In particular, engineering of load-bearing tissues such as bones requires mechanical stimulation of the constructs in order to optimise the engineered scaffold's mechanical properties [17].

A variety of methods have been used to mechanically stimulate bone cells in artificial tissue scaffolds. These methods include compression rigs [18] and shear flow chambers [19]. Although these methods are extremely useful, one of their limitations is that they apply a gross mechanical strain to the entire structure. Thus it is difficult to control the stimulation of single cells within specific regions of the construct. This kind of stimulation is desirable as it allows for control of tissue substitute properties on a highly localised scale. There is evidence that suggests that focused femtosecond laser radiation can be used to stimulate single osteoblastic cells in a monolayer, since laser radiation can be controlled with high spatial and temporal resolution, which might be useful for bone tissue engineering. This technique has the potential to be incorporated into a sophisticated bioreactor for the culturing of bone tissue [20].

Cellular strains and displacement were computed using digital image correlation (DIC), which is a pattern matching technique that allows measurement of displacements with sub-pixel resolution from sequences of images. DIC has previously been used for measuring deformation in articular cartilage [21], trabecular bone [22], compressed chondrocytes [23] and intracellular strains in mechanically-stimulated smooth muscle cells [24]. Thus DIC has become an established technique for biomechanical measurements at the tissue and cellular levels. The ability to measure displacements and strains on a highly localised level can provide a quantitative measure for evaluating local mechanical properties in engineered tissue constructs. Thus this information could be used to locate regions within the construct in which cellular stimulation is required to modify the local construct properties. This study used algorithms described previously [23, 25, 26]. The algorithm was realised using Matlab 7.0, and was applied to the sequences of fluorescent images associated with the fluo-3 fluorophore. Prior to applying DIC, a Wiener adaptive noise reduction filter with a kernel size of

10×10 pixels was applied to all images in a sequence. Next, approximately 700 regions of interest (ROI) of size 49×49 pixels were randomly placed throughout the image. The DIC algorithms were then applied to measure the displacement of the centre of each ROI throughout the sequence of images. This was accomplished by comparing the first image to all other images in the sequence. In order to calculate the strain components from the displacement data, a thin plate smooth spline was fitted to the measured displacements. Then Delaunay triangulation was applied to draw the smallest possible set of triangles to connect the centres of all ROIs. The displacements of the vertices of these triangles as measured from the thin plate spline defined a set of three linear equations for each triangle. These equations were solved to yield the average deformation tensor within each triangle

$$
F = \begin{bmatrix} \dfrac{\mathrm{d}x}{\mathrm{d}X} & \dfrac{\mathrm{d}x}{\mathrm{d}Y} \\ \dfrac{\mathrm{d}y}{\mathrm{d}X} & \dfrac{\mathrm{d}y}{\mathrm{d}Y} \end{bmatrix}. \tag{9.1}
$$

In Eq. 9.1 (X, Y) denote coordinates in the reference image (first image in the sequence) and (x, y) denote coordinates in the deformed image. The average Lagrangian strain, $\mathbf{E}$, within each triangle was then calculated as

$$
\mathbf{E} = 0.5(\mathbf{F}^{\mathrm{T}}\mathbf{F} - \mathbf{I}), \tag{9.2}
$$

where $\mathbf{F}^{\mathrm{T}}$ is the transpose of $\mathbf{F}$ and $\mathbf{I}$ is the unity tensor. Next, the eigenvalues of the strain tensor were found to yield the principal strains ($E1$, $E2$) within each triangle.

The principal strains were interpolated to yield an estimate of the strain fields throughout the field of view. The strain fields were smoothed with a moving average filter with a kernel size of 20×20 pixels to filter noise. Visualisation of the distribution of strains throughout the field of view was obtained by creating intensity-maps depicting the spatial distribution of strains.

In order to demonstrate that targeting cells with the femtosecond pulse laser was having a visible effect it was necessary to load the cells with a marker that could rapidly detect cellular changes. One of the most effective methods in accomplishing this was to use calcium ion fluorophores that can rapidly detect any alterations in local calcium levels caused by the laser. Targeting of individual 3T3 cells with femtosecond pulse laser irradiation caused an instant transient rise in intracellular Ca^{2+} levels. In almost all cases it was possible to target cells with the femtosecond pulse laser multiple times to get a repeated increase in intracellular Ca^{2+}, as shown in Fig. 9.4. Ca^{2+} would normally return to baseline levels approximately 2 min after point laser irradiation. As the calcium load returned to resting levels a cellular contraction was observed in nearly all cases ($n = 11$). The displacements of the individual target cell then induced a radial displacement pattern from the surrounding cells as indicated by the direction of the arrows in Fig. 9.5. These displacements were measured using DIC. Calculation of the strain field from the measured displacements was performed using the procedure described

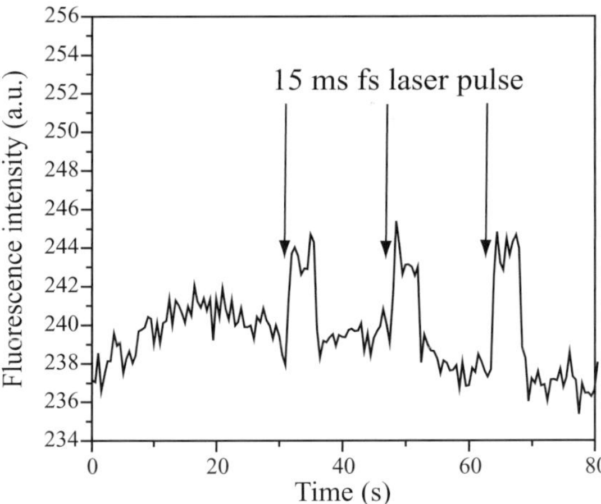

Fig. 9.4 Graphical representation of the change in Ca^{2+} levels as measured within a region of interest surrounding a typical target cell. Repeated targeting of the cell induces repeated calcium spikes. Arrows indicate the time-points of the exposure to 500 ms of femtosecond pulse laser radiation [20].

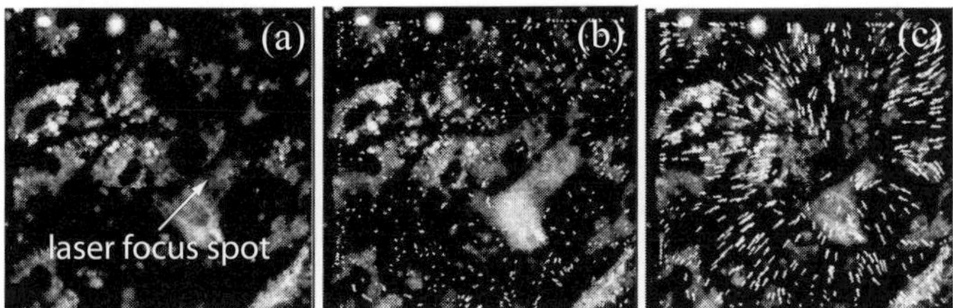

Fig. 9.5 Images of 3T3 cells loaded with Fluo-3 AM (a) before femtosecond pulse laser targeting ($t = -150$ s); (b) 20 s after femtosecond pulse laser targeting ($t = 20$); and 2 min after that ($t = 140$). Images (b) and (c) are overlaid with arrows which represent cell displacements brought about by the target cell contracting. Arrows are three times larger than actual displacements. A clear rise in intracellular fluorescence due to increased Ca^{2+} can be seen in the targeted cell at $t = 20$ (b), which has subsided by $t = 140$ (c). Image scale: 160 μm × 160 μm [20].

in [25]. The calculations revealed a heterogeneous strain field within the monolayer. The principal strains around the targeted cell were of the order of 20% (Fig. 9.6). The magnitude of the strains decreased with distance from the targeted cell. However, even at a distance of about 5 cell lengths ($\sim$50 μm) significant strains with a magnitude of more than 5% were measured. This level of deformation is larger than strain levels reported to stimulate bone mineralisation *in vitro* [24].

This shows that focused femtosecond pulse laser irradiation can stimulate murine 3T3 osteoblast-like cells. Femtosecond pulse laser irradiation of cells loaded with fluoro-3 AM causes a transient rise in intracellular Ca^{2+} levels, which is accompanied by contraction of the irradiated cell. This contraction causes a heterogeneous strain field in the surrounding monolayer. The resulting strains are of the order of 20% near

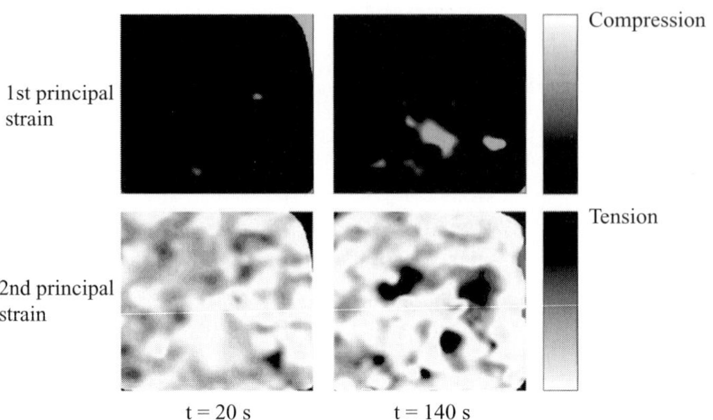

Fig. 9.6 A comparison of the first and second principal strains comparing $t = 20$ and $t = 140$ using the same data as that of Figure 9.3. Image scale: 160 µm $\times$ 160 µm [20].

the target cell, and gradually decrease with distance from the cell. Cell strains of as little as 0.8–1% have been reported to up-regulate bone mineralisation related transcriptional activity [27]. Osteoblasts have also been shown to up-regulate osteopontin production 2.8 fold after a tensile strain application of up to just 10% [28]. Furthermore, it has been shown that an increase in intracellular Ca^{2+} in osteoblasts also has effects on bone cell stimulation [29]. Thus femtosecond pulse lasers might be useful for stimulating bone mineralisation in tissue constructs by artificially inducing increases in intracellular Ca^{2+}. In the past, continuous wave lasers have been used to enhance bone repair and bone stimulation [29, 30], however, using femtosecond pulse lasers in the near infrared for manipulating cells might be preferable because of the increased penetration depth, highly localised nonlinear photodamage and limited heat transfer to samples.

The monitoring of the alterations in Ca^{2+} as a result of laser stimulation showed that calcium ions are not the only intracellular messenger that can be monitored for mechanotransduction processes using this system; up-regulation of nitric oxide can be monitored using commercial fluorophores such as DAF-FM, and alterations in structural proteins as a result of laser induced mechanotransduction can be assessed using appropriate green fluorescent protein type targeted vectors. Using the femtosecond pulse laser to target the extracellular matrix, individual microtubules, or even microfilaments inside cells, thereby inducing localised displacements, would help in the understanding of the tensegrity [31] of cellular structure, and how this might be affecting the mechanotransduction signals of cells.

9.2   Femtosecond microfabrication

Fabrication of microstructures in dielectrics using femtosecond lasers has gained momentum recently due to the benefits associated with multi-photon excitation. When femtosecond pulses are focused with sufficient energy within the volume of a substrate, nonlinear excitation can lead to physical processes such as avalanche ionisation, electron plasma formation and shock-wave induced microexplosions. Considerable research has been conducted into the effects of machining or microstructuring transparent materials based on shock-wave induced microexplosions [32–34]. The short pulse width of femtosecond lasers provides a means of micromachining metals and dielectrics without defects and debris that are typically associated with a build up of heat in the focal region. The energy density that is created from focusing mJ femtosecond pulses ($> 10^{15}$ W cm^{-2}) typically results in the ionisation of the material in the focal region. The creation of a plasma in the focal region transfers the energy and heat away from the surrounding material, resulting in a heat affected zone (HAZ) that is limited to the extent of the focal region.

With respect to these nonlinear processes a range of applications have been investigated, including three-dimensional optical data storage [35], fabrication of optical waveguides [36], microstructuring of optical components [37, 38] and fabrication of microchannels [39]. Fabrication of microchannels in dielectrics for the development of complex three-dimensional microfluidic devices has also been thoroughly investigated [40, 41]. In all those applications amplified femtosecond pulse lasers producing microjoules of energy per pulse were used to overcome the optical breakdown threshold of the substrate, typically glass or fused silica.

Polymers as substrates have advantages over their glass counterparts as their properties are more easily tailored for specific applications and are cheaper and easier to manufacture. One of the advantages of polymers is their lower threshold for optical breakdown, which can be reached using non-amplified nanojoule femtosecond pulses. Considerable research has also been conducted into the different applications involving fabrication by femtosecond pulses in polymers, such as three-dimensional data storage [42, 43] and fabrication of photonic band gap structures [44].

The significant benefit from amplified femtosecond pulse laser systems is the ability to overcome the breakdown threshold with a single femtosecond pulse, thereby reducing any possible effects due to heating. However, amplified laser systems typically have repetition rates in the kHz, which could limit the machining speed, whereas the non-amplified systems have repetition rates in the MHz.

9.2.1 Experimental systems

9.2.1.1 Non-amplified femtosecond laser fabrication system

The setup used in the experiment is illustrated in Fig. 9.7. The femtosecond pulse laser producing 80 fs pulses at a repetition rate of 80 MHz was used as the fabrication laser. The per pulse energy in the experiments was maintained at 0.9 nJ, which is below the energy threshold required to ablate poly(methyl methacrylate) (PMMA). The laser beam was focused into the sample by a long working distance, water immersion objective with 0.9 NA. The objective is designed to be immersed in water and as such has no cover-slip correction built in. The samples were polished blocks of commercial PMMA mounted

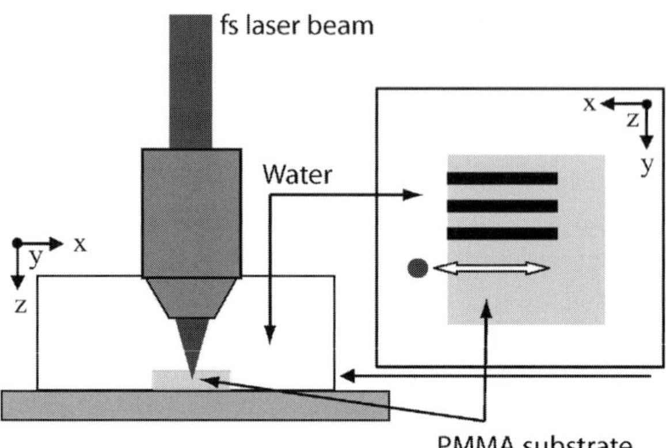

Fig. 9.7 Schematic diagram of the experimental setup for the fabrication of microchannels in a PMMA substrate. Reprinted with permission from Ref. [45], D. Day and M. Gu, *Opt. Express* **13**, 5939 (2005). © 2005, Optical Society of America.

in a glass dish filled with de-ionised water. The samples were immersed in water in order to reduce the aberrations resulting from the type of objective used and the sample–immersion medium interface. The sample holder was mounted on x–y stepper motor translation stages, with the objective mounted on the z-axis stepper motor translation stage. Monitoring of the fabrication process was via a CCD camera positioned behind the objective lens, allowing for the viewing of the x–y plane. In the experiment, fabrication of the microchannels was in the x–y plane at a depth of approximately 75 μm below the sample surface.

The focus spot was initially focused in the water beside the sample and then translated laterally through the water–sample interface and into the sample. As the sample was translated perpendicular to the direction of propagation of the laser beam it was expected that the microchannels would reflect the elongated shape of the focal region along the z direction. Water as an immersion medium for femtosecond drilling has been demonstrated to improve the fabrication process by assisting in the removal of debris from inside and near the edge of the channel [41], which can otherwise impede the process. However, in this research water immersion was used to reduce aberrations and improve the performance and consistency of the fabrication method.

9.2.1.2   Amplified femtosecond laser fabrication system

The photomask substrate consists of a PMMA substrate cut to the size of a glass slide (75 mm × 25 mm) with a thickness of 2 mm. A metal layer of aluminium (Al) was coated onto one surface of the PMMA using evaporation vacuum deposition. The thickness of the Al layer was controlled in order to limit the transmission of the ultraviolet light from the mercury lamp used to polymerise the photoresist during fabrication of the master mould. A layer thickness of 250 nm was found to remove sufficient ultraviolet light from a focused 100 W mercury lamp.

Fabrication of the photomask was achieved with an amplified Ti:sapphire femtosecond pulse laser that produced 100 fs pulses at a repetition rate of 1 kHz at a wavelength of

800 nm. A 0.25 NA objective lens was used to focus the laser onto the sample using a fabrication system similar to that illustrated in Fig. 9.7. A white light source and CCD camera were mounted to monitor the fabrication *in situ*. Mounting the white light source in transmission mode allowed real time inspection of the quality of the metal layer removal, by observing the quality of the light being transmitted through the newly etched regions.

9.2.2　Microchannel fabrication

This section presents a study of the characteristics of microchannels fabricated in a water-immersed PMMA substrate under high repetition rate, nanojoule, femtosecond laser pulses. Using transmission optical microscopy the dependence of channel properties on the different fabrication parameters is investigated. Based on the fabrication method demonstrated, cylindrical microfluidic channels can be fabricated in PMMA with diameters ranging from 8 μm to 20 μm.

The fabrication process in PMMA under high repetition rate nanojoule femtosecond laser pulses described in this section does not have the required energy per pulse to achieve optical breakdown through direct ionisation of the substrate. Instead, it is proposed that the absorption of multiple pulses results in a significant increase in the temperature for a localised region surrounding the focal spot, which becomes the dominant fabrication process [46]. Located within the region affected by the fabrication process is modified material caused by the decomposition of the polymer substrate due to the high temperatures. Subsequent post fabrication processing is used to develop the microchannel.

9.2.2.1　Fabrication mechanism

In order to fabricate microchannels in PMMA under nanojoule high repetition rate femtosecond pulse conditions a two-step procedure was utilised.

- **Laser exposure** The PMMA samples are exposed to 0.9 nJ in the focal region at a wavelength of 750 nm, while the focal spot is translated through the sample at speeds between 100 μm/s and 1 mm/s. The absorption cutoff for the PMMA samples is 370 nm, which indicates that the dominant absorption process of the focused laser is two-photon absorption. Given the repetition rate of the laser of 80 MHz, in the time it takes the sample to move a distance equal to the size of the focus spot at a speed of 1 mm/s, 3.3×10^4 pulses will have been absorbed by the sample. As the excited electrons relax back to the ground state over a period of picoseconds [47] and in doing so transfer energy to the polymer matrix, the successive pulses will lead to a localised increase in temperature. The delay between successive repeated laser scans allows the temperature in the irradiated region to decrease further by diffusion. The density of the material in the irradiated region is reduced due to the heating, which is reduced even further with every repeated laser scan. As the scans are repeated the region of less dense material is extended uniformly, radially away from the focal region. As the energy is low enough not to induce optical breakdown there is no debris created during the laser exposure.

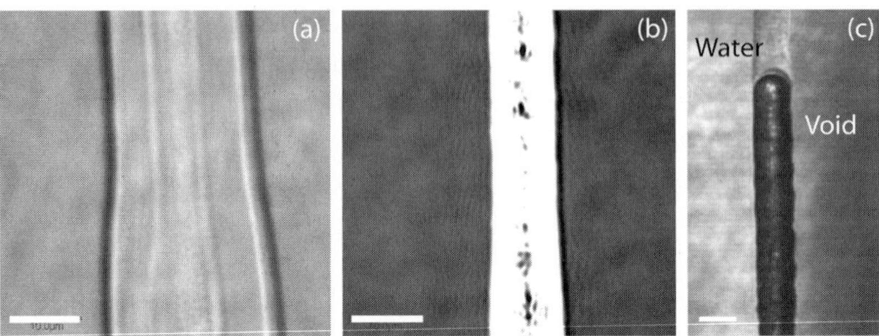

Fig. 9.8 Transmission images of a microchannel (a) before and (b) after annealing at 200°C for 30 seconds. (c) Illustrates that hollow channels are formed as water enters the channel via capillary action. The scale bars are 10 μm. Reprinted with permission from Ref. [45], D. Day and M. Gu, *Opt. Express* **13**, 5939 (2005). © 2005, Optical Society of America.

- **Post-exposure annealing** After the laser exposure the samples are annealed on a hotplate at a temperature above the glass transition temperature of the material. As the laser irradiated region has a different density to the bulk material it is affected slightly differently than the bulk material. This in effect results in differing amounts of expansion as the temperature is increased, thereby producing a microchannel.

9.2.2.2   Effect of annealing

During and after the fabrication process it can be seen that there is a region of material with modified optical properties compared with the bulk substrate, as shown in Fig. 9.8(a). The width of the central region is 6.5 μm, which is surrounded by another region of modified material with a width of 17.5 μm. The substrate was exposed to femtosecond pulses with energy of 0.9 nJ at a wavelength of 750 nm. With an energy of 0.9 nJ per pulse there is no direct optical breakdown of the material, as evident by the lack of visible radiation typically emitted during plasma generation. The sample was translated at 800 μm/s with the same region being repeatedly irradiated 10 times in succession. After the irradiation the sample was placed on a hotplate at a temperature of 200°C for 30 seconds. Annealing the sample at a temperature above the glass transition temperature of the material formed a microchannel in the substrate, as shown in Fig. 9.8(b). The width of the channel in Fig. 9.8(b) is 9 μm. Variation of the channel diameters is observed near the interface between the sample and the water immersion medium as boundary effects dominate. Even at the lower energies used ablation can occur at the surface of the sample as the focal spot moves into the sample. Confirmation of a hollow channel structure can be seen in Fig. 9.8(c) where water has entered the channel via capillary action.

Such large channel diameters with respect to the focal spot size may be as a result of heating of the polymer through multiple passes, which allows additional energy transfer to the polymer matrix, increasing the region of interaction and thereby resulting in larger diameter channels after the post-exposure annealing.

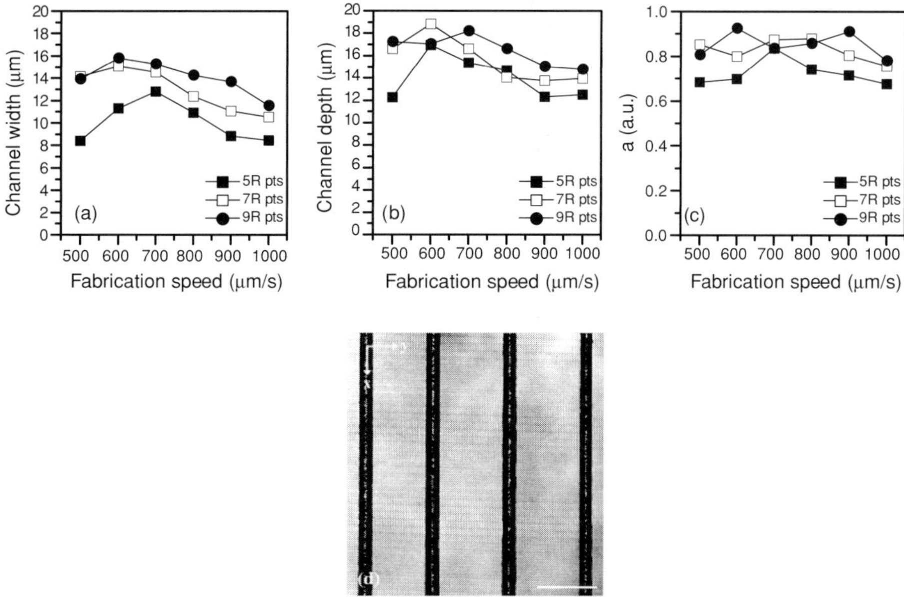

Fig. 9.9 Measured microchannel characteristics as a function of the fabrication speed. (a) Width, (b) depth and (c) ratio of width to depth. (d) Transmission image of microchannels. The scale bars are 50 μm. Reprinted with permission from Ref. [45], D. Day and M. Gu, *Opt. Express* **13**, 5939 (2005). © 2005, Optical Society of America.

9.2.2.3 Effect of fabrication speed

Microchannels were fabricated in a PMMA substrate at speeds of 500 μm/s to 1 mm/s with energy of 0.9 nJ in the focus at a wavelength of 750 nm. Figure 9.9 shows the change in channel width and depth as the speed at which the sample is translated is increased. Here the depth of the channel refers to the axial dimension of the channel, as the fabrication geometry used will create elliptically shaped channels. It can be seen from Figs. 9.9(a) and (b) that both the width and depth of the channels decrease as the speed of fabrication is increased. In order to measure the width and depth of the channels the edge of the substrate is polished to remove any ablation effects near the substrate and immersion water interface. The channel dimensions are then measured directly from the channel cross sections. Due to the large number of pulses absorbed by the sample on a millisecond timescale and with a 10 s delay between repeated scans, it is proposed that the diffusion of heat through the matrix results in a channel with a cross section 650 times that of the focal region.

In order to characterise the shape of the channel, the function α is defined as the ratio of the width to the depth, where 1 represents a circle. The value of α for the focal region of the objective used is 0.16. Figure 9.9(c) shows that the cross section of the channel is considerably more circular than the elliptical profile of the focal spot. Based on the shape and size of the channel with respect to the focal spot it would indicate that some uniformly diffusive process begins in the centre of the focal spot and expands radially.

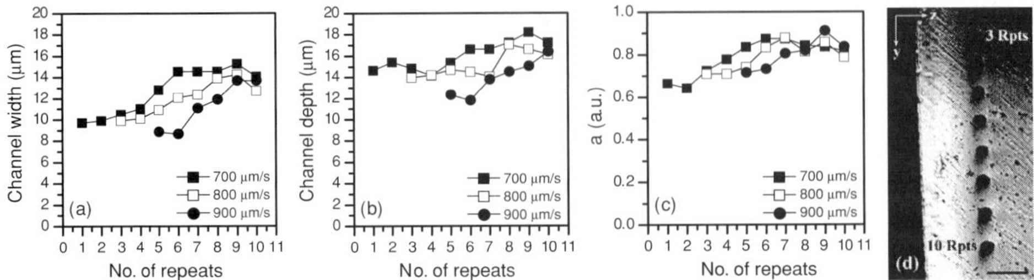

Fig. 9.10 Measured microchannel characteristics as a function of the number of fabrication repeats: (a) width, (b) depth, (c) ratio of width to depth. (d) Transmission image of microchannel cross sections. The scale bars are 50 μm. Reprinted with permission from Ref. [45], D. Day and M. Gu, *Opt. Express* **13**, 5939 (2005). © 2005, Optical Society of America.

The microchannels shown in Fig. 9.9(d) were fabricated at 900 μm/s using ten repeats with a repeat delay of 10 s. An advantage of this method of channel fabrication is that it produces channels with relatively smooth channel walls.

9.2.2.4   Effect of repeated fabrication

The nature of the repeated fabrication process is such that the energy required to reach plasma expansion and microexplosions and thus optical breakdown is not achieved when the sample is translated at speeds greater than 100 μm/s. As a result repeated scans over the same region are employed in order to create the conditions required to produce a channel after annealing, where the delay between repeats is fixed at 10 s. However, at fabrication speeds between 100 and 700 μm/s channels can be formed from a single scan after annealing. This indicates that while the temperature increase wasn't great enough to induce optical breakdown it was enough to modify the density of the material in the interaction region.

The effect of repeated scans over the same region can be seen in Figs. 9.10(a), (b) and (c). For any given fabrication speed an increase in both the width and depth of the channel is associated with an increase in the number of repeats. From Figs. 9.10(a) and (b) there is a limit below which a channel cannot be formed; this occurs at increasingly higher energy levels for faster fabrication speeds. As the number of repeated fabrications is increased, the subsequent reheating of the irradiated region produces a more uniform heating of the surrounding medium, resulting in an almost circular channel cross section, as seen in Fig. 9.10(c).

The channels shown in Fig. 9.10(d) were fabricated at 800 μm/s with the number of repeats varying from 3 to 10, with a repeat delay of 10 s. As can be seen from the figure, all of the channels were fabricated at a depth of approximately 75 μm below the surface of the polymer.

9.2.2.5   Effect of delay between repeated fabrication

The repeated energy absorption is shown to affect both the size and shape of the channels due to the localised heating and diffusion of the fabrication process. In Fig. 9.11 it is shown that for channels fabricated at 800 μm/s with ten repeats there is an increase in the width of the channel as the delay is increased. At 800 μm/s, the energy deposited from

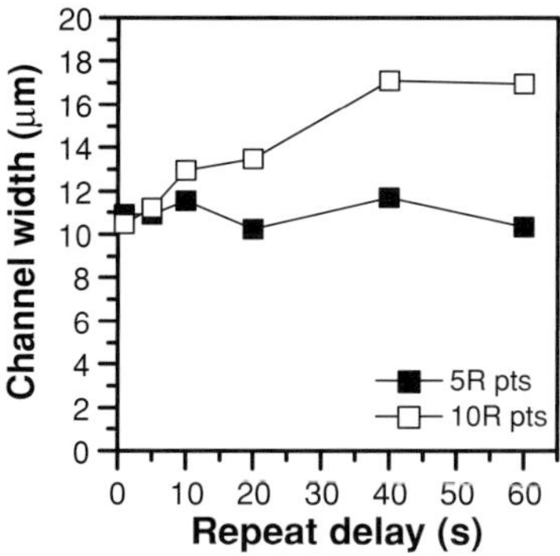

Fig. 9.11 Measured microchannel characteristics as a function of the delay between repeats. Reprinted with permission from Ref. [45], D. Day and M. Gu, *Opt. Express* **13**, 5939 (2005). © 2005, Optical Society of America.

five repeats is just above the threshold required to modify the density in the interaction region and therefore there is not enough subsequent heating of the substrate to cause the same broadening effect as in the ten repeat case.

Femtosecond laser pulses with energy of 0.9 nJ per pulse and a 80 MHz repetition rate at a wavelength of 750 nm were used to fabricate straight microchannels in a PMMA substrate. The size and shape of the microchannels can be controlled by changing the fabrication parameters of speed, the number of fabrication repeats and delay between fabrication repeats. It has been proposed that the absorption of energy in the focal region modifies the density of the polymer matrix which after annealing the sample above the glass transition temperature results in the formation of the microchannels. Diffusion of heat through the substrate is a uniform process which has the effect of creating symmetrically shaped channels. This fabrication method is expected to have applications in the fabrication of microstructures or microfluidic devices in polymer substrates.

9.2.3 Microfluidic photomask fabrication

Rapid prototyping of microfluidic devices is growing in importance, both in terms of time and cost as the microfluidic industry continues to generate new avenues for biological and chemical research. The fabrication of photomasks is still a fundamental step in the development of microfluidic devices, regardless of the method used for device replication. Typically high resolution photomasks are fabricated using a photolithographic or e-beam lithographic process [48, 49] in order to achieve features with dimensions less than a micrometre. However, both lithographic fabrication methods rely on the use of a photoresist for transfering an image to a photomask. When a photoresist is used to generate a photomask, multiple steps are required just to produce the photomask, which then in turn can be used to replicate the designed structure. The disadvantage of using conventional photolithography or e-beam lithography is the time and expense required to produce a new photomask. This becomes apparent during the development cycle of a device, where the frequent redesign of structures or geometries requires repeated fabrication of the photomask. Recently we have seen the emergence of high resolution printing as an alternative method for generating a photomask for microfluidic device manufacturing [50]. Generally microfluidic devices require feature resolutions on the order of a micrometre with total combined channel lengths of at least centimetres.

A microfluidic device can be fabricated using any number of methods other than lithography. Micromachining with excimer and Nd:YAG lasers is the most common method for micromachining; however, the nanosecond pulse width generated by those lasers results in the formation of a large HAZ. Depending on the material being machined, fabrication defects such as cracks can appear in the HAZ and ejected debris can be found around the interaction region. Moreover, the selective etching of a metal layer or layers from the surface of a transparent substrate is not possible with excimer or Nd:YAG lasers. Direct write fabrication of microchannels has been demonstrated using a Nd:YAG [51], assembly of organic inks [52] and a femtosecond laser [45]. While direct write fabrication can be used to fabricate two- and three-dimensional microchannels directly in a range of different substrates, the drawback occurs when replication of the microfluidic system is required. Whereas the cost of a complex photomask is seen as prohibitive, the cost of a serial direct write fabrication method would preclude it from being used to replicate a commercial microfluidic device.

The advantage of developing a photomask becomes apparent when consideration is given to the method to be used for replicating the microchannels. Soft lithography with poly(dimethylsiloxane) (PDMS) is a desirable process as PDMS has been demonstrated to be a suitable material for biological [53, 54] microfluidic applications and for the ease with which microfluidic devices can be replicated in PDMS without destroying the master mould. Another advantage of using PDMS is that the method required to replicate the devices can be accomplished without expensive equipment compared with hot embossing.

As fabrication using femtosecond pulses is predominately a nonlinear process, the material ablation occurring within the focal region can be controlled by adjusting the pulse energy within the focal spot, known as 'thresholding'. This can result in the production of features with dimensions below the diffraction limit of the focusing optics used [32]. When layering a number of dissimilar materials with different optical breakdown thresholds, individual materials can be machined without affecting the surrounding materials. Femtosecond pulse lasers have been used to directly write a pattern into a gold and chromium coated quartz substrate in order to produce photomasks [55].

In this section it is demonstrated that femtosecond pulse lasers can be used to selectively etch the aluminium metal layer on a PMMA substrate in order to fabricate a binary photomask for the rapid prototyping of microfluidic devices using soft lithography. This method of producing photomasks can simplify the number of steps required to fabricate a microfluidic device and therefore reduce the time and cost associated with developing new devices.

The removal of the Al layer occurs in the focal spot of the objective lens where the energy density is sufficient to directly ionise the Al layer. From Fig. 9.12(a) it can be seen that the Rayleigh range (z_R) of a 0.25 NA objective lens extends beyond the metal

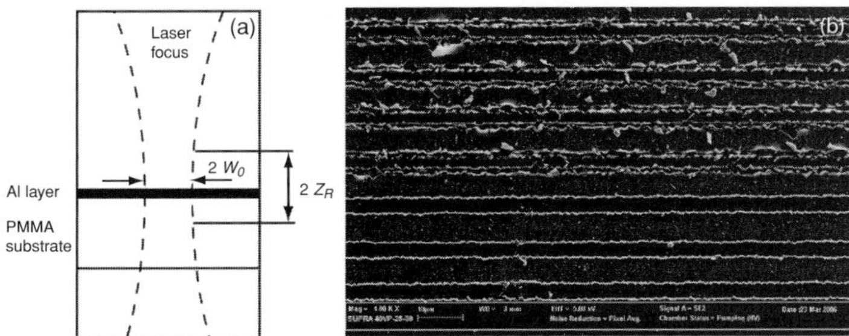

Fig. 9.12 (a) An illustration showing the focused femtosecond laser beam and its focus position relative to the metal layer. The beam waist in the focus is represented by ω_0 and the Rayleigh range is indicated by z_R. (b) SEM image of a series of lines fabricated at different energies. The top four lines are fabricated with a fluence (5 J/cm^2) above the threshold for ionising the PMMA substrate, while the bottom three lines are fabricated with a fluence (0.7 J/cm^2) where only the Al layer is ionised. Reprinted with permission from Ref. [56], D. Day and M. Gu, *Opt. Express* **14**, 10753 (2006). © 2006, Optical Society of America.

layer into the PMMA substrate. By controlling the energy in the focal spot and utilising the different optical and mechanical properties of the two materials, a threshold can be determined whereby the Al layer is removed without destroying the underlying PMMA layer. The Rayleigh range and focus diameter ($2\omega_0$) for the 0.25 NA objective lens are 11 μm and 2 μm, respectively.

A photomask is fabricated under atmospheric conditions, which results in the redeposition of some of the removed material. It has been shown that micromachining of different materials under a vacuum results in improved ionisation of the target material and little or no redeposition or debris [57]. However, fabricating under vacuum conditions increases the time and complexity of the fabrication of the photomask and is only required when fabricating nanometre size structures.

9.2.3.1 Effect of energy thresholding

The ability to select different materials for removal using femtosecond pulse laser fabrication is a unique characteristic of the physical processes associated with a femtosecond temporal pulse. The direct ionisation of a specific material in the focal region within single or multiple pulses results in the removal of that material with limited temperature build up or transfer to the surrounding material.

A scanning electron microscope (SEM) image (Fig. 9.12(b)) of a series of lines fabricated in the sample at different energies shows that, by controlling the energy per pulse, the metal layer can be removed without damaging the underlying PMMA substrate. The energy density required to ablate the PMMA when the objective was focused on the surface of the photmask was 5 J/cm^2. When the energy density is reduced to 0.7 J/cm^2 the Al layer is removed while leaving the PMMA substrate unaffected. At energies below the ionisation threshold modification of the optical properties of PMMA can still occur [58]. The width of the line generated by the removal of the Al layer is 3.8 μm, while the width of the channel ablated in the PMMA is 1.9 μm. The use of thresholding when fabricating photomasks using femtosecond pulse lasers

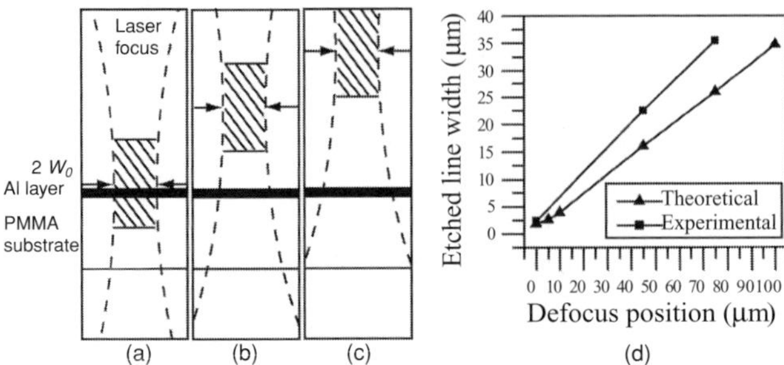

Fig. 9.13 Schematic diagram representing the defocus position of the objective lens and the limit of the Rayleigh range with respect to the Al layer. (a) The objective is focused on the metal layer, (b) the objective lens is defocused by 45 μm and (c) the objective lens is defocused by 75 μm. The Rayleigh range is indicated by the hashed region. (d) Illustrates the theoretical and experimental values for the width of the etched line as a function of defocus position. Reprinted with permission from Ref. [56], D. Day and M. Gu, *Opt. Express* **14**, 10753 (2006). © 2006, Optical Society of America.

is important as any damage to the PMMA substrate reduces the quality of the UV transmission which in turn affects the fabrication of the master. A feature that can be seen in Fig. 9.12(b) is the ripple along the edges of the fabricated regions. The ripple is due to the overlap between successive pulses as the sample is translated. In order to remove the ripples more overlap between the successive pulses would be required, which can be achieved by slowing the translation speed of the sample. Another characteristic that can be observed is the lack of debris surrounding the fabrication of the metal layer. For the fabrication conditions used to remove the Al layer, the energy density was sufficient to almost completely ionise the metal layer, limiting the debris or redeposition of Al.

9.2.3.2 Effect of defocusing

By controlling the focal position of the focused femtosecond pulses with respect to the surface of the metal layer, the amount of metal that is ionised per pulse can be changed. As it is the energy density that is most significant in terms of whether the material is ionised or not, any change in the focal position will also change the energy density incident on the metal layer. Using this method of defocusing, two effects can be achieved. The first is that a larger area of material can be removed per pulse and second; defocusing changes the tolerances associated with removing the metal while not affecting the substrate. It can be seen from Fig. 9.13 that by moving the focal position away from the sample surface the area that is irradiated by the beam is increased. The defocus distance (Δz) is given as the distance between the focused beam waist and the metal layer. Figure 9.13(a) illustrates focused conditions ($\Delta z = 0$) while Figs. 9.13(b) and (c) illustrate 45 μm and 75 μm defocus conditions. Figure 9.13(d) illustrates the theoretical and experimental width of an etched line based on the defocus position. Also, within the focal spot the energy density is a Gaussian distribution, with the highest energy located at the centre of the focal spot, whereas the energy density outside of the Rayleigh range is relatively uniform across the beam in comparison to that in the

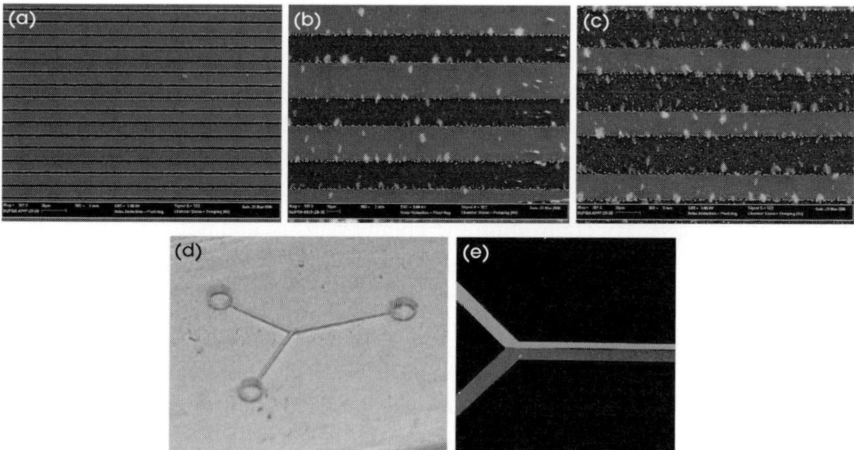

Fig. 9.14 Fabrication of a photomask using defocusing. SEM images of a series of lines fabricated at different objective lens defocus positions: (a) $\Delta z = 0$ μm, (b) $\Delta z = 45$ μm and (c) $\Delta z = 75$ μm. (d) A sealed y-junction microfluidic device. The length of the microfluidic channels from the inlets to the outlet is 13.5 mm and the channel width and depth are both 100 μm. (e) A fluorescence image showing the laminar flow produced in the y-junction microfluidic device. Reprinted with permission from Ref. [56], D. Day and M. Gu, *Opt. Express* **14**, 10753 (2006). © 2006, Optical Society of America.

focal spot. By defocusing the fabrication beam, more control over the energy density irradiating the sample can be achieved, which reduces the likelihood of ionising the substrate.

The effect of defocusing on the removal of an Al layer can be seen in Fig. 9.14. A series of lines were fabricated where the metal layer was removed under (a) focused, (b) 45 μm defocus and (c) 75 μm defocus conditions. The widths of the lines fabricated as a result of the focus, 45 μm defocus and 75 μm defocus conditions are 2.4 μm, 22.5 μm and 35.4 μm, respectively. Theoretical predictions of the beam width corresponding to the three defocus positions are 1.9 μm ($\Delta z = 0$), 16 μm ($\Delta z = 45$ μm) and 26 μm ($\Delta z = 75$ μm). The energy density used to remove the Al layer was 0.7 J/cm^2 and the sample was translated at a speed of 800 m/s. From Figs. 9.14(a), (b) and (c) it can be seen that the energy density is significant enough to ionise the metal layer without affecting the substrate, however, in Fig. 9.14(c) the energy density is reduced to 0.3 J/cm^2 such that not all the metal layer can be ionised. It is clear that the edges of the remaining metal were melted rather than ionised. This is also confirmed by the significant quantity of debris that can be seen in the image.

Fabrication of a photomask by selectively removing the metal layer from a PMMA substrate can be used to increase the speed of development of microfluidic devices. An example of a device that is fabricated using this method of fabricating a photomask is shown in Fig. 9.14(d) and (e). The completed and sealed y-junction microfluidic device can be seen in Fig. 9.14(d) with a fluorescence image of the laminar flow shown in Fig. 9.14(e). The final channel dimensions are width 100 μm, depth 100 μm and length 13.5 mm. By controlling the defocus during the fabrication process the time taken to produce the photomask can be reduced. For the case of the y-junction photomask the fabrication time was less than 5 minutes.

The production of photomasks for microfluidic devices is an expensive exercise both in time and cost, which does not allow any flexibility for iterative development cycles. The ability to selectively remove a metal layer from an optically transparent substrate using femtosecond pulse lasers provides a means of directly transferring the pattern to the photomask. By controlling the energy density and defocus position of the focused femtosecond pulses the photomask can be generated in minutes, which is suitable for producing a master for replication using PDMS. The ability to change the focus position and therefore change the amount of the metal layer removed ensures that both high and low resolution features can be achieved in the photomask. This method of producing a photomask also provides a method for modifying a photomask which could lead to shorter microfluidic device development times.

9.3   Femtosecond fabricated microenvironments

The development of tools that have enabled live cell imaging have led to the improved understanding of cellular processes such as migration [59] and cell–cell communications [60]. Many of the advances in knowledge of the molecular activities of cells are related to epithelial, fibroblasts and neuronal cells, all of which are relatively large with slow movement which is suitable for long term imaging. On the other hand, lymphocytes provide unique challenges for live cell imaging that have restricted some studies of lymphocyte biology. Lymphocytes, in particular mature T and B cells and thymoctyes, can migrate up to 25 µm/min [61] meaning that they are likely to leave the field of view of the microscope after a few minutes. Also T cells have a tendency to undergo homotypic interactions *in vitro*, which complicates single cell imaging and analysis by forming large clumps of cells [62].

The activation of T cells through the presentation of antigens by professional antigen presenting cells (APC) involves morphological rearrangements within the T cell and the development of a structure at the interface called the immunological synapse [63]. Imaging of T cell activation *in vivo* has revealed that T cells can remain contacted with APCs for hours [64], even during division which occurs more than 20 hours after activation [65]. Recent observations of asymmetric cell division, a process that can dictate cell fate [66], have led to the requirement to develop new approaches for *in vitro* imaging of T cell interactions over hours and days. There have been several approaches developed to overcome the problems of mobility and clumping associated with observing antigen presentation, such as the following.

- Functionalised surfaces that trap T cells. Coating the slide or coverslip surface with pol-L-lysine can bind lymphocyte surface proteins and therefore trap the cells [67], however, adhesion of the lymphocyte to the surface prohibits migration and the normal processes of interaction [68]. This is likely to impact upon the signalling process and provide misleading information.

- Cells embedded in a collagen matrix. The APC and T cells are embedded into a three-dimensional collagen matrix for imaging of molecular reorganisation [62], however, this approach is not necessarily physiologically relevant as collagen is not present in the environment in which naïve T cells are activated [69].
- Antigen presentation from a planar bilayer. The APC is replaced by a planar bilayer in which the molecules that are typically presented by an APC are arrayed in a lipid bilayer [70]. This method enables the dynamic interaction between T cells and the antigen presenting molecules, however, a systematic comparison using real APCs is required to assess the physiological relevance.

One possible solution is to isolate single T cells in individual chambers, however, culture of isolated T cells in large volumes of medium prevents cytokine signalling and culture in small volumes cannot be maintained due to the exhaustion of the medium. In this section we will show the development of a novel technique in which T cells are physically separated by a polymer microgrid but share a culture medium to enable long term culture and imaging. The microgrids enable *in vitro* imaging of lymphocyte migration, cell division and response to antigen presentation in a more efficient and relevant physiological context.

9.3.1   Fabrication of microgrids

Presented here is the fabrication of the cell microgrids for containing lymphocytes within a field of view and to prevent excessive cell clumping [71]. The chambers in the grids have walls but no lids so they effectively operate as individual wells, physically confining the T cells, while allowing for the exchange of cytokines and growth factors between cells, maintaining viability and functional integrity.

The microgrids are produced by casting PDMS from a femtosecond laser micromachined mould. A femtosecond laser is used to selectively etch material from a PMMA substrate in order to produce a negative three-dimensional mould of the structure. The amplified femtosecond pulse laser directly ionises the substrate material, allowing for direct laser writing of a three-dimensional mould with little damage to the surrounding material. To replicate the microgrids PDMS is poured over the mould, covered with a glass microscope slide and cured. Once cured the microgrids are removed from the PMMA mould, as illustrated in Fig 9.15(a). Using PDMS for the microgrids enables the user to easily transfer and insert the microgrids into traditional cell culture ware. In this research square microgrids with the length of one side ranging from 30 µm to 250 µm and depths ranging from 20 µm to 60 µm are fabricated as either small (Fig. 9.15(b)) or large (Fig. 9.15(c)) arrays. Two types of PDMS grid have been produced, the first is a pure PDMS microgrid and the second is a multi-layered microgrid structure that included a fibronectin layer covering only the bottom of the wells.

To test whether the microgrids could contain T cells within the field of view for an extended period, time lapse imaging using the T cell line MLA-144 [72] was conducted. Over a 20 hour time period control cells without microgrids (Fig. 9.16(a1)) formed large cell aggregates, which moved rapidly and were often lost from the field of view (see

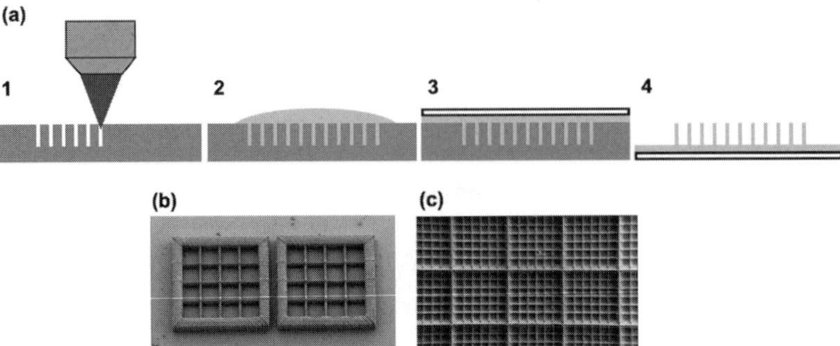

Fig. 9.15 Fabrication and replication method for microgrids. (a) A microstructured mould is created using a femtosecond pulse laser to etch microscale features into a PMMA substrate (1). PDMS is then poured into the mould (2) and a second substrate placed over the PDMS and then cured (3). After curing the microgrids are removed from the mould (4) and transferred to traditional cell culture ware. Arrays of cells with sizes ranging from 30 µm to 250 µm square and 20 µm to 60 µm deep are produced. (b) An SEM image of a small array of 125 µm × 125 µm by 60 µm deep grids containing 32 wells. (c) SEM image of a large array of 80 µm × 80 µm by 60 µm deep grids containing 900 wells [71].

arrow). In contrast, cells in microgrids aggregated less and remained within the field of view (Fig. 9.16(a2)). We found that a minimum 60 µm high wall was required for T cells. Walls of about 20 µm provided some containment, but the cells occasionally jumped to the adjacent cell microgrid. To determine whether the microgrids are compatible with fluorescent imaging, time lapse imaging of MLA-144 cells that had been engineered to express proteins fused to either green or red fluorescent proteins [73] was conducted. Although some polymers exhibited strong autofluorescence and were not compatible with fluorescence imaging, the PDMS used in these grids is not autofluorescent and enables clear visualisation of protein trafficking (Fig. 9.16(b)).

9.3.2   Effect of long term imaging in microgrids

To demonstrate that the microgrids do not impact upon cell viability or proliferation, activated primary human T cells were cultured in the microgrids in an eight well chamber slide for up to six days, and their viability and proliferation were compared with cells cultured in a 96 well plate (Fig. 9.17). Cells were incubated with propridium iodide (PI) to indicate cells undergoing cell death (uptake of the dye indicates a loss of membrane integrity associated with cell death). Cell proliferation is also monitored by measuring the fluorescence intensity of cells that had been previously labelled with the fluorescent dye carboxyfluorescein diacetate (CFSE). Each halving of the CFSE fluorescence corresponds to a cell division (Fig. 9.17, right column). Both the control and microgrid samples demonstrate that more than six divisions occurred over the culture period. It should also be noted that both cell death and cell proliferation were identical regardless of whether the cells had been cultured in microgrids or not. This indicates that even after six days' culture, the microgrids do not impact upon either T cell survival or proliferation.

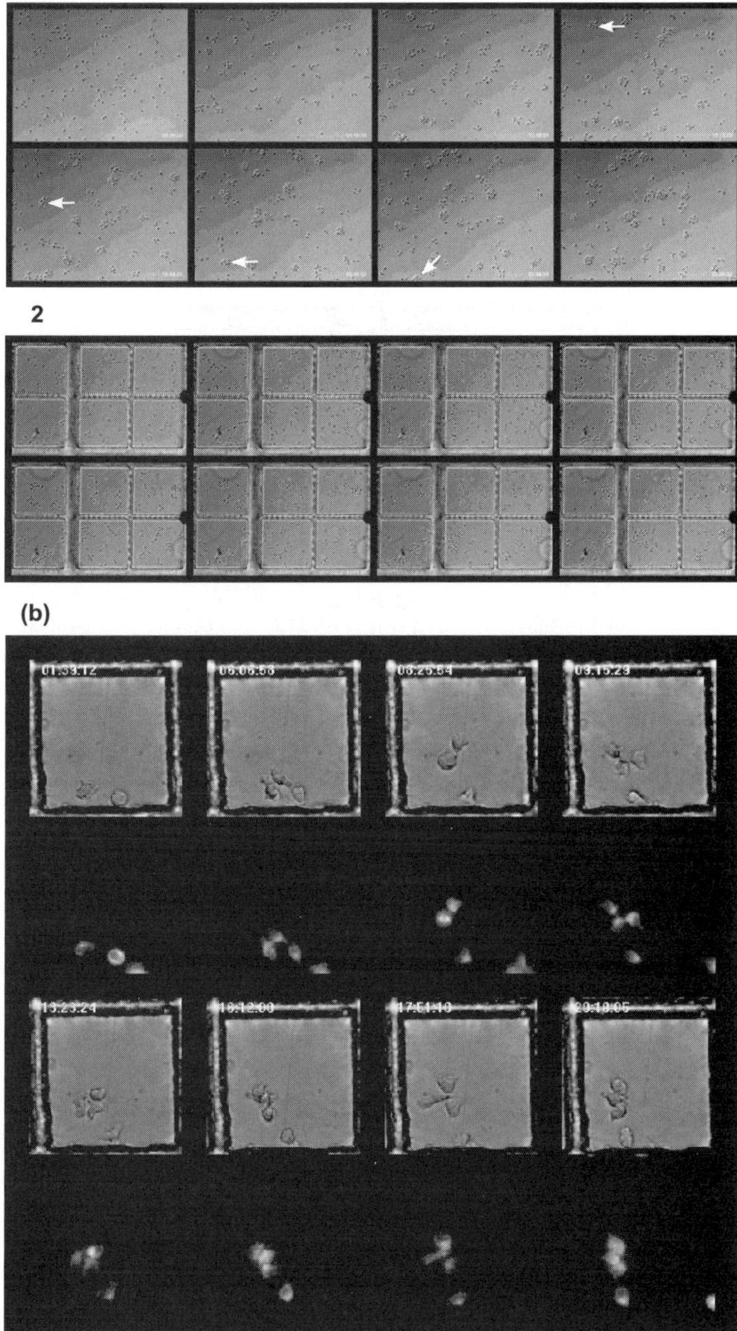

Fig. 9.16 Microgrids contain T cells within the field of view, and enable DIC and fluorescent time lapse imaging. (a) MLA cells were cultured in 8-well chamber slides without (1) or with (2) microgrids, and imaged using DIC. Each panel represents a frame from a time lapse video. Panels represent images taken at the following time points: (1) 0 hr 0 min, 4 hr 48 min, 12 hr 32 min, 18 hr 12 min, 18 hr 40 min, 20 hr 40 min, 21 hr 32 min and 22 hr 20 min, (2) 0 hr 0 min, 2 hr 40 min, 5 hr 20 min, 8 hr 0 min, 10 hr 40 min, 13 hr 20 min, 16 hr 0 min and 21 hr 20 min. (b) MLA cells expressing tubulin fused to Cherry fluorescent protein were imaged at the following time intervals: 1 hr 39 min, 6 hr 6 min, 8 hr 25 min, 9 hr 15 min, 13 hr 23 min, 16 hr 12 min, 17 hr 51 min and 20 hr 10 min [71].

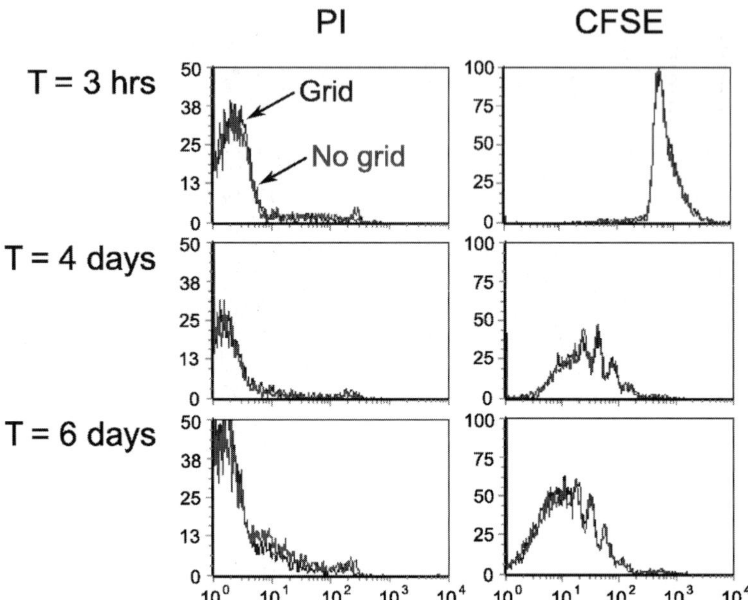

Fig. 9.17 Culture in microgrids has no effect on cell viability or proliferation. Primary human PBMC were activated with antibody to CD3 and interleukin-2 for seven days, and then cultured for the time points indicated in a 96-well plate either with or without 50 μm square microgrids. The two lines are almost perfectly overlapped. Cells were either stained after harvesting with propidium iodide for detection of death (left column, values over 10 represent dead cells), or were prestained with CFSE for detection of cell proliferation (each consecutive two-fold reduction in fluorescence indicates a cell division) [71].

Also demonstrated is whether the microgrids can sustain T cell interactions with dendritic cells, the cell type that most commonly presents antigen to naïve T cells. In the microgrids the dendritic cells adhere to the PDMS on the base of the microgrids, and interact with T cells over many hours in these conditions (Fig. 9.18). The microgrids can also sustain the growth of the stromal cell line required for *in vitro* thymocyte differentiation, OP9-DL1 [74]. The adherence of the stromal cells to the PDMS depended upon its prior coating with fibronectin. However, coating by incubation of the microgrids with soluble fibronectin is problematic as the stromal cells can migrate extensively between the individual microgrids by crawling over the fibronectin-coated walls. There we developed a technique for coating the floors, but not the walls, with fibronectin. Using this approach enabled the stromal cells to adhere to the floors but not the walls of the microgrids, and prevented their migration over walls. This approach allowed for visualisation of the interaction between the stromal cells and developing thymocytes (Fig. 9.19).

What is described here is an approach for containing non-adherent cells within the field of view during microscopy, without compromising cell survival or function. Our method for functionalising the floor of the microgrids without altering the walls enables the adaptation of this approach for many types of cell–cell interaction. The microgrids can be modified in size and height according to the different requirements for different

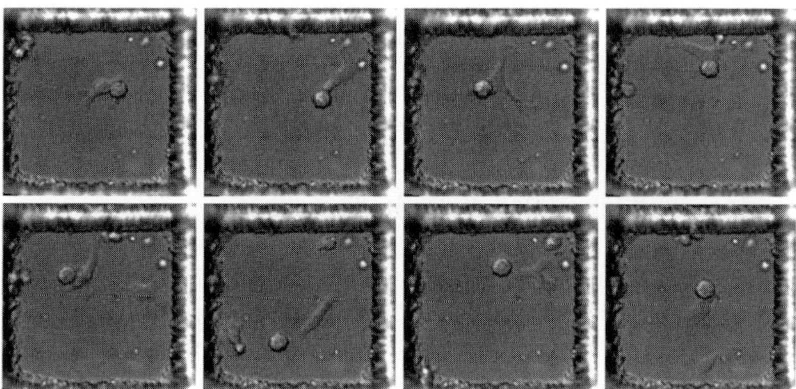

Fig. 9.18 Microgrids enable visualisation of the interaction of naïve T cells with antigen presenting cells. Dendritic cells were seeded into an eight well chamber slide containing 125 μm grids, and pulsed with SIINFEKL peptide. After 24 hours, naïve T cells purified from the spleen of an OT-1 transgenic mouse were added to the chamber slide, and imaged for 15 hours. Panels represent DIC images captured at the following time points: 11 hr 56 min, 15 hr 52 min, 18 hr 20 min, 20 hr 14 min, 20 hr 15 min, 21 hr 30 min, 21 hr 54 min and 22 hr 56 min [71].

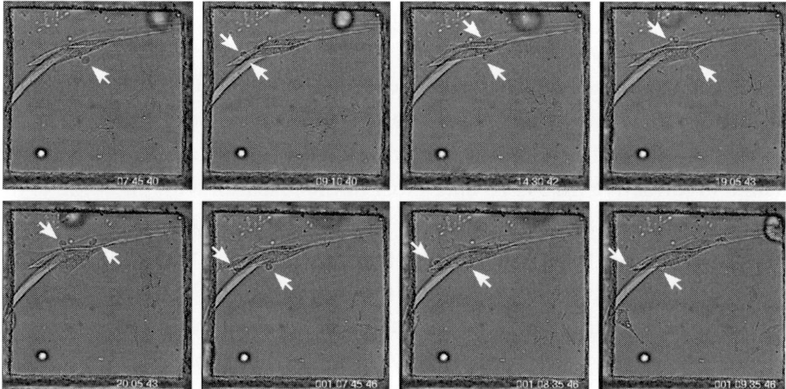

Fig. 9.19 Fibronectin coated microgrids for prolonged imaging of a thymocyte interacting with a stromal cell. OP9-DL1 stromal cells were seeded into an eight well chamber slide containing 250 μm grids, adhered overnight, and then overlayed with *in vitro*-generated DN3 thymocytes and imaged for up to 24 hours. Panels represent DIC images captured at the following time points: 7 hr 45 min, 9 hr 10 min, 14 hr 30 min, 19 hr 5 min, 20 hr 5 min, 31 hr 45 min, 32 hr 35 min and 33 hr 35 min [71].

cell types. In this research, naïve T cell interactions with dendritic cells were most effectively imaged in grids of 125 μm^2, and thymocyte-stromal interactions in 250 μm^2.

The fabrication of the microgrids now enables the long term imaging and analysis of the activities of mature lymphocytes and thymocytes. Indeed, cell microgrids enable imaging of naïve T cell activation, a process not previously studied intensively, as alternative techniques were not compatible with long term imaging of antigen presentation. It is predicted that the microgrids will also enable imaging of other non-adherent cells such as hematopoietic stem cells. An important attribute of the microgrids is that cells and their progeny can be contained for days, and through multiple generations. This enables the correlation of molecular activities with, for instance, proliferation, death and differentiation of the progeny.

9.4   Femtosecond fabricated microfluidic sensor

Microfluidics has received an increasing amount of attention since the 1990s, as it is rapidly transforming the way with which biological, chemical and physical research is being conducted. The continued development of functional micro-total-analysis systems (μTAS) or lab-on-a-chip devices for processes such as high-throughput screening, clinical analysis or biochemical synthesis will ultimately require the integration of sensing elements for detection and analysis. Integrated optical sensors are capable of achieving high sensitivity and high throughput with independent non-contact sensing techniques using either labelled or label-free protocols. While the range of optical sensing technologies compatible with microfluidic devices is based on methods such as absorbance, fluorescence or surface plasmon resonance, it is the ability to detect changes in refractive index associated with the binding of biomaterials that is generating significant interest due to the potential to monitor biochemical reactions without the need for fluorescently tagged molecules [75, 76].

Photonic crystal structures have recently been demonstrated as optical biosensors [77, 78]. Due to the highly localised confinement of the coupled light, photonic crystal sensors can be incorporated into microfluidic devices to facilitate localised measurements of the change in refractive index. To date most of the research with photonic crystal sensors in microfluidic devices has been limited to one-dimensional or two-dimensional photonic crystal sensors [79]. At present the only three-dimensional photonic crystals used for biosensing have been fabricated by self-assembly [80]. Three-dimensional photonic crystals based on the woodpile geometry or face-centred-tetragonal lattice (fct) have been an active research topic not only because of the ability of opening up a complete bandgap but also of their great flexibility in tuning the photonic bandgap effect by manipulating the geometrical parameters through the fabrication parameters (scanning speed, laser power, laser wavelength or pulse duration) [44].

In this section we present an integrated 3D photonic crystal microfluidic sensor fabricated with a femtosecond laser for optical sensing. A 3D photonic crystal microfluidic sensor based on femtosecond laser fabricated void channels in a polymer substrate is used [81]. When the 3D photonic crystal sensor is integrated with a microfluidic channel, a shift in the bandgap and bandgap defect peak positions are observed according to changes in the refractive index of the fluid.

The sensor and microfluidic channel are fabricated in two steps using femtosecond laser direct writing based on the techniques described in Section 9.2. First, a 3D photonic crystal based on void channels is fabricated in a precured Norland Optical Adhesive (NOA 63) polymer substrate. In the second process, a microfluidic channel is fabricated by etching a channel in the polymer substrate with an amplified femtosecond pulse laser

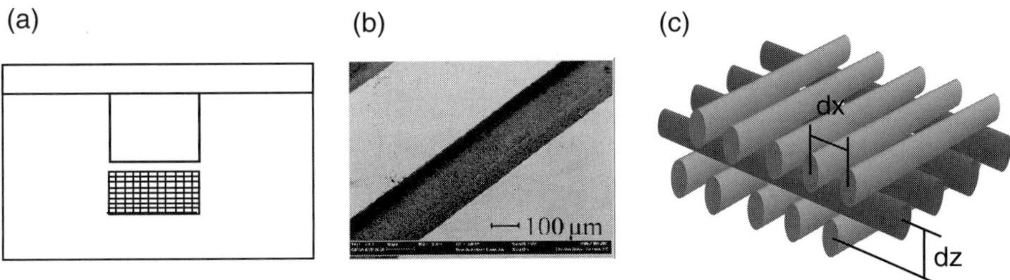

Fig. 9.20 (a) A schematic diagram of the integrated 3D photonic crystal microfluidic sensor. (b) An SEM image of a femtosecond laser etched microchannel above the photonic crystal and (c) an illustration of the unit cell for the woodpile photonic crystal structure, where δz and δx are the layer spacing and in-plane spacing, respectively. Reprinted with permission from Ref. [81], J. Wu, D. Day and M. Gu, *App. Phys. Lett.* **92**, 071108 (2008). © 2008, American Institute of Physics.

beam focused through a 0.25 NA objective lens. The fabrication energy is 15 μJ per pulse with a sample translation speed of 1,500 μm/s. By scanning the amplified beam across the polymer substrate a microchannel can be etched into the surface directly above the photonic crystal. After the fabrication of the 3D photonic crystals and microfluidic channels, the channels were sealed with a layer of poly(dimethylsiloxane) (PDMS) polymer. The PDMS was prepared by spin-coating at 1,200 RPM on a glass slide and cured at 75°C for 20 minutes, after which it was transferred to the microfluidic polymer substrate. Poly(dimethylsiloxane) was chosen because of its good elasticity, optical transparency and biocompatibility. An illustration of the sensor is shown in Fig. 9.20(a) and an SEM image of the etched channel above the sensor is shown in Fig. 9.20(b). The distance separating the photonic crystal and the microfluidic channel was less than 10 μm. Refractive index fluids were introduced into one end of the microfluidic channel, with capillary force drawing the fluid into the remainder of the channel. After each optical measurement a washing procedure was used to rinse the channel before introducing another refractive index fluid. The response of the photonic crystal microfluidic sensors were characterised with Fourier transform infrared (FTIR) spectroscopy.

In this experiment, 24 layer woodpile photonic crystals, without and with a planar defect, were fabricated [82]. The range of the physical parameters for the woodpile photonic crystals were $\delta x = 1.3 - 1.4$ μm for the in-plane spacing and $\delta z = 1.4$–1.5 μm for the layer spacing. An illustration of the unit cell is shown in Fig. 9.20(c). For the photonic crystal with a planar defect, the 13th through 24th layers are displaced by a distance $\Delta d = 0.3$ μm creating a microcavity defect between the 12th and 13th layers of the photonic crystal, resulting in a Fabry–Perot (FP) cavity. The properties of the planar defect can be tailored in order to produce a FP resonance peak in the photonic crystal bandgap [82].

Figure 9.21 demonstrates that photonic crystals can be integrated into a microfluidic device without negatively affecting the performance of the photonic crystal. A photonic band gap with a suppression of the infrared transmission of 85% in the stacking direction can still be achieved (see Fig. 9.21(a)) for a 3D woodpile photonic crystal. The photonic

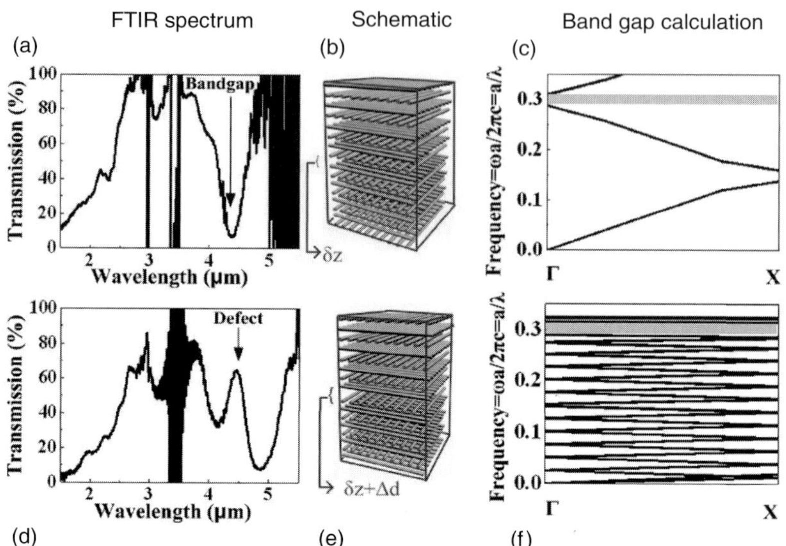

Fig. 9.21 (a) FTIR spectrum of the integrated 24 layer photonic crystal sensor with $\delta z = 1.5$ µm and $\delta x = 1.3$ µm. The photonic band gap in the stacking direction produces an 85% suppression of the infrared transmission at 4.693 µm. (b) An illustration of the 24 layer photonic crystal structure. (c) Photonic band structure calculation for the photonic crystal showing the photonic band gap at the normalised frequency of 0.2981. (d) FTIR spectrum of the integrated 24 layer photonic crystal sensor with $\delta z = 1.5$ µm, $\delta x = 1.3$ µm and $\Delta d = 0.3$ µm. The planar defect results in a FP resonance peak at 4.441 µm. (e) An illustration of the photonic crystal with a planar defect between the 12th and 13th layers. (f) The photonic band structure calculation showing the band gap and defect resonance located at the normalised frequency of 0.315. Reprinted with permission from Ref. [81], J. Wu, D. Day and M. Gu, *App. Phys. Lett.* **92**, 071108 (2008). © 2008, American Institute of Physics.

crystal is a 24 layer woodpile void channel structure with a layer spacing $\delta z = 1.5$ µm and an in-plane spacing $\delta x = 1.3$ µm, as shown in Fig. 9.21(b). The experimentally measured photonic bandgap of the microfluidic photonic crystal sensor corresponds quite well with the theoretically predicted band calculation along the stacking direction of $\Gamma - X'$, shown in Fig. 9.21(c). The centre of the stop gap illustrated by the grey region in Fig. 9.21(c) is located at a frequency of 0.2981 which correlates with a wavelength of 4.693 µm that is consistent with the experimentally measured bandgap. For the photonic crystal sensor with a planar defect, the suppression of the infrared transmission of 80% is achieved with a FP resonance peak positioned in the middle of the bandgap (see Fig. 9.21(d)). The photonic crystal with a planar defect is a 24 layer woodpile void channel structure with a layer spacing $\delta z = 1.5$ µm, an in-plane spacing $\delta x = 1.4$ µm and a defect spacing of $\Delta d = 0.3$ µm, as illustrated in Fig. 9.21(e). The FP resonance peak is located at a frequency of 0.315 which corresponds to a wavelength of 4.441 µm. The experimental measurements of the bandgap and defect are again confirmed by the theoretical band calculations along the $\Gamma - X'$ direction, as shown in Fig. 9.21(f). The agreement between the experimental measurements and theoretical predictions of the bandgap confirm that the photonic crystal and microfluidic channel

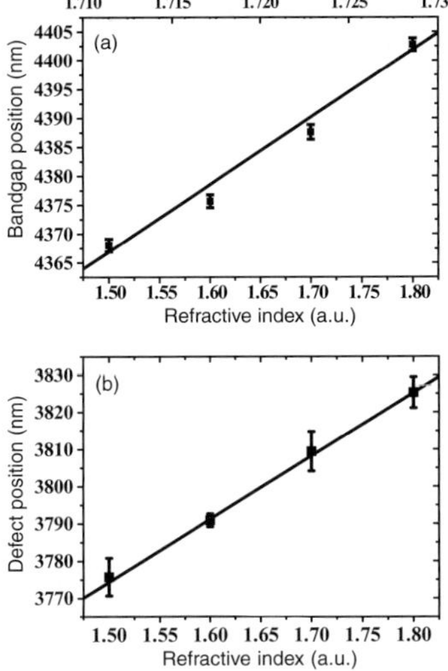

Fig. 9.22 The shift in the position of the photonic bandgap (a) and defect resonance (b) as the index of refraction of the liquid in the microchannel is increased. The averaged refractive index of the photonic crystal sensor according to Eq. 9.3 is displayed on the secondary x-axis. Reprinted with permission from Ref. [81], J. Wu, D. Day and M. Gu, *App. Phys. Lett.* **92**, 071108 (2008). © 2008, American Institute of Physics.

can be integrated into a single piece of polymer substrate. Figure 9.21 also demonstrates that the surface roughness produced by the laser machining of the channel does not affect the performance of the photonic crystal.

To demonstrate the ability of the 3D photonic crystal to sense changes in refractive index a series of index matching fluids were introduced into the microfluidic channel. Figure 9.22(a) shows the measured position of the photonic bandgap as a function of the index of refraction of the liquid in the microchannel. The photonic crystal is a 24 layer woodpile void channel structure with a layer spacing $\delta z = 1.5$ µm and an in-plane spacing $\delta x = 1.3$ µm. As the index of refraction of the liquid is increased a linear shift in the bandgap position to longer wavelengths is recorded. A shift in the photonic bandgap of 35 nm is measured for a change in refractive index of the liquid of $\Delta n = 0.3$, which corresponds to a sensitivity of 8×10^{-3} in refractive index. The sensitivity of the sensor is defined as the detectable change in refractive index of the fluid as a function of the photonic bandgap shift ($\Delta n / \Delta \lambda$) in the FTIR spectrum of the photonic crystal. The resolution of the minimum detectable wavelength shift of the bandgap of 1 nm has been determined, which is limited by the width of the photonic bandgap and the accuracy of the FTIR for the region where the bandgap appears. The measurements of the bandgap position as a function of the refractive index of the liquid are displayed in Fig. 9.22(a) as the average of three repeated experiments with the error bars representing the variation of the measured value. The sensitivity of the 3D photonic crystal is an improvement on the published two-dimensional photonic crystal sensor which demonstrated a sensitivity of 1×10^{-2} [79]. To determine the bandgap position as a function of the average refractive index of the sensor and photonic crystal geometrical parameters we can interpret the change in average refractive index (n_{avg}) of the photonic crystal sensor based on the change in position of the bandgap, given by [83]

$$m\lambda_{\mathrm{gap}} \approx 2\delta z n_{\mathrm{avg}} \approx 2\delta z[n - A/(\delta z \delta x)(n - 1)], \qquad (9.3)$$

where λ_{gap} is the bandgap position, A is the cross section of the void channel and n is the effective refractive index of the polymer material. The average refractive index of the photonic crystal sensor is displayed on the secondary x-axis (top of the frame) in Fig. 9.22(a). As the relationship between the average refractive index and the position of the bandgap is known, this sensor can be used to determine the absolute value of the refractive index rather than a relative value to within an accuracy of 8×10^{-3}.

As the photonic bandgap can be relatively broad with respect to the change in position of the bandgap, a photonic crystal with a planar defect was fabricated in order to utilise the resulting FP cavity resonance. The position and finesse of the FP cavity can be tuned by the geometrical properties of the photonic crystal and planar defect. Figure 9.22(b) shows the measured position of the FP resonance as the refractive index of the liquid in the microchannel is changed. The photonic crystal with a planar defect is a 24 layer woodpile void channel structure with a layer spacing $\delta z = 1.4$ µm, an in-plane spacing $\delta x = 1.3$ µm and a defect spacing of $\Delta d = 0.3$ µm. The measurements of the bandgap position as a function of the refractive index of the liquid are displayed in Fig. 9.22(b) as the average of three repeated experiments with the error bars representing the variation of the measured value. As the index of refraction of the liquid is increased a linear shift of the resonance position to longer wavelengths is detected. A shift of 50 nm is recorded for a change in the index of refraction of the liquid of $\Delta n = 0.3$. This results in a sensitivity of 6×10^{-3}, which is slightly higher than that for the photonic bandgap measurement.

In summary, we have demonstrated the concept of applying 3D photonic crystals to detecting changes of refractive index in a microfluidic device. The 3D photonic crystals were integrated into a microfluidic device using femtosecond direct laser fabrication of the photonic crystals and microfluidic channels. Shifts in the photonic bandgap and defect resonance positions as a result of a change in the refractive index of the liquid in the microfluidic channel are measured with Fourier transform infrared spectroscopy. A change of 6×10^{-3} in the refractive index can be detected using the photonic crystal with a planar defect. This method of sensing changes in refractive index is able to measure both a relative shift in the refractive index and the absolute value of the refractive index as a function of the bandgap position.

9.5   Summary

This chapter has introduced the versatility of the femtosecond pulse laser beyond its initial imaging applications. The ability to directly interact with cells has been demonstrated through the use of targeted cell membrane microporation, which has the potential to lead to new biological and medical advances through directed cell transfection. As well as that, a short exposure to a femtosecond pulse laser has been shown to generate a localised stress gradient seen through the expansion and contraction of neighbouring cells. This capacity for manipulating the cellular microenvironment could lead to advances in tissue repair, specifically bone mineralisation, by artificially applying localised mechanical forces on cells.

Microfluidics, or lab-on-a-chip devices, is a new area of technology development that will have a significant impact on biological and biomedical research in the future. It has been shown that the femtosecond pulse laser is capable of fabricating lab-on-a-chip devices with fewer steps and with a shorter fabrication time then current fabrication technology. The use of the femtosecond pulse laser for microfabrication has also been demonstrated in this chapter with an emphasis on the design and fabrication of devices for biological applications. The fabrication of microscopic landscapes or microgrids is a technological advancement which is enabling live cell studies of biological applications previously not possible.

In this chapter it has been shown that the femtosecond pulse laser has added many new dimensions to biophotonics research beyond the traditional imaging applications. Indeed, femtosecond biophotonics has now achieved the ability to be utilised as a dynamic tool for imaging, manipulation and fabrication to create the next generation of technology for biological and biomedical applications.

References

[1] J. White and E. H. Stelzer. Photobleaching GFP reveals protein dynamics inside live cells. *Nature*, **9**:61–65, 1999.

[2] K. König, P. T. C. So, W. W. Mantulin, B. J. Tromberg and E. Gratton. Cellular response to near-infrared femtosecond laser pulses in two-photon microscopes. *Opt. Lett.*, **22**:135–136, 1997.

[3] K. König, T. W. Becker, P. Fischer, I. Riemann and K.-J. Halbhuber. Pulse-length dependence of cellular response to intense near-infrared laser pulses in multiphoton microscopes. *Opt. Lett.*, **24**:113–115, 1999.

[4] M. W. Berns, J. Aist, J. Edwards *et al.* Laser microsurgery in cell and developmental biology. *Science*, **213**:505–513, 1982.

[5] N. Shen, M. Colvin, F. Genin *et al.* Using femtosecond laser subcellular surgery to study cell biology. *Biophys. J.*, **86**:520A–520A, 2004.

[6] K. König, I. Riemann, P. Fischer and K.-J. Halbhuber. Laser tweezers are sources of two-photon excitation. *Cell. Mol. Biol.*, **44**:721–733, 1998.

[7] B. Agate, C. T. A. Brown, W. Sibbett and K. Dholakia. Femtosecond optical tweezers for *in-situ* control of two-photon fluorescence. *Opt. Express*, **12**:3011–3017, 2004.

[8] K. König, I. Riemann, P. Fischer and K.-J. Halbhuber. Intracellular nanosurgery with near-infrared femtosecond laser pulses. *Cell. Mol. Biol.*, **45**:195–201, 1999.

[9] W. Watanabe, N. Arakawa, S. Matsunaga, T. Higashi, K. Fukui, K. Isobe and K. Itoh. Femtosecond laser disruption of subcellular organelles in a living cell. *Opt. Express*, **12**:4203–4213, 2004.

[10] H. Koester, D. Baur, R. Uhl and S. Hell. Ca^{2+} fluorescence imaging with pico- and femtosecond two-photon excitation: signal and photodamage. *Biophys. J.*, **77**:2226–2236, 1999.

[11] S. Mohanty, M. Sharma and P. Gupta. Laser-assisted microinjection into targeted animal cells. *Bioimages*, **25**:895–899, 2003.

[12] U. Tirlapur and K. König. Cell biology: targeted transfection by femtosecond laser. *Nature*, **418**:290–291, 2002.

[13] N. Smith, K. Fujita, T. Kaneko, K. Katoh, O. Nakamura, S. Kawata and T. Takamatsu. Generation of calcium waves in living cells by pulsed-laser-induced photodisruption. *App. Phys. Lett.*, **79**:1208–1210, 2001.

[14] A. Ichihara, T. Tanaami, K. Isozaki *et al.* High-speed confocal fluorescence microscopy using a nipkow disk scanner with microlenses for 3d-imaging of single fluorescent molecule in real time. *Bioimages*, **4**:57–63, 1996.

[15] D. Day, C. Cranfield and M. Gu. High speed fluorescence imaging and intensity profiling of femtosecond induced calcium transients. *Int. J. Biomed. Img.*, **2006**:93438, 2006.

[16] H. Tashjian, Y. Yasumura, L. Levine, G. H. Sato and M. L. Parker. Establishment of clonal strains of rat pituitary tumor cells that secrete growth hormone. *Endocrinology*, **82**:342–352, 1968.

[17] A. J. ElHaj, M. A. Wood, P. Thomas and Y. Yang. Controlling cell biomechanics in orthopaedic tissue engineering and repair. *Pathol. Biol.*, **53**:581–589, 2005.

[18] M. A. Wood, Y. Yang, P. Thomas and A. J. ElHaj. Using dihydropyridine-release strategies to enhance load effects in engineered human bone constructs. *Tissue Eng.*, **12**:2489–2497, 2006.

[19] C. Wang, N. Chahine, C. Hung and G. Ateshian. Optical determination of anisotropic material properties of bovine articular cartilage in compression. *J. Biomech.*, **36**:339–353, 2003.

[20] C. Cranfield, Z. Bomzon, D. Day, M. Gu and S. Cartmell. Mechanical strains induced in osteoblasts by use of point femtosecond laser targeting. *Int. J. Biomed. Img.*, **2006**:21304, 2006.

[21] S. H. Cartmell, B. D. Porter, A. J. Garcia and R. E. Guldberg. Effects of media perfusion rate on cell seeded 3d bone constructs *in vitro*. *Tissue Eng.*, **9**:1197–1203, 2003.

[22] T. McKinley and B. Bay. Trabecular bone strain changes associated with subchondral stiffening of the proximal tibia. *J. Biomech.*, **36**:155–163, 2003.

[23] M. Knight, Z. Bomzon, E. Kimmel, A. Sharma, D. Lee and D. Bader. Chondrocyte deformation induces mitochondrial distortion and heterogeneous intracellular strain fields. *Biomech. & Model Mechanobio.*, **5**:180–191, 2006.

[24] S. Hu, J. Chen, B. Fabry *et al.* Intracellular stress tomography reveals stress focusing and structural anisotropy in cytoskeleton of living cells. *Am. J. Physiol.*, **285**:C1082–C1090, 2003.

[25] Y. Wang and A. Cuitino. Full-field measurements of heterogeneous deformation patterns on polymeric foams using digital image correlation. *Int. J. Sol. & Struct.*, **39**:3777–3796, 2002.

[26] Z. Bomzon, M. M. Knight, D. L. Bader and E. Kimmel. Mitochondrial dynamics in chondrocytes and their connection to the mechanical properties of the cytoplasm. *J. Biomech. Eng.*, **128**:674–680, 2006.

[27] X. Chen, C. M. Macica, K. W. Ng and A. E. Broadus. Stretch-induced PTH-related protein gene expression in osteoblasts. *J. Biomech. Eng.*, **20**:1454–1461, 2005.

[28] K. Fong, S. Warren, E. Loboa *et al.* Mechanical strain affects dura mater biological processes: implications for immature calvarial healing. *Plast. Reconstr. Surg.*, **112**:1312–1327, 2003.

[29] T. Yaakobi, L. Maltz and U. Oron. Promotion of bone repair in cortical bone of the tibia in rats by low energy laser (HeNe) irradiation. *Calcif. Tissue Int.*, **59**:297–300, 1996.

[30] A. Stein, D. Benayahu, L. Maltz and U. Oron. Low-level laser irradiation promotes proliferation and differentiation of human osteoblasts *in vitro*. *Photomedicine and Laser Surgery*, **23**:161–166, 2005.

[31] D. E. Ingber. Tensegrity: the architectural basis of cellular mechanotransduction. *Ann. Rev. Physiol.*, **59**:575–599, 1997.

[32] C. B. Schaffer, A. O. Jamison and E. Mazur. Morphology of femtosecond laser-induced structural changes in bulk transparent materials. *App. Phys. Lett.*, **84**:1441–1443, 2004.

[33] D. Ashkenasi, G. Muller, A. Rosenfeld *et al.* Fundamentals and advantages of ultrafast micro-structuring of transparent materials. *App. Phys. A*, **77**:223–228, 2003.

[34] Z. Wu, H. Jiang, Z. Zhang, Q. Sun, H. Yang and Q. Gong. Morphological investigation at the front and rear surfaces of fused silica processed with femtosecond laser pulses in air. *Opt. Express*, **10**:1244–1249, 2002.

[35] E. N. Glezer, M. Milosavljevic, L. Huang, R. J. Finlay, T. H. Her, J. P. Callan and E. Mazur. Three-dimensional optical storage inside transparent materials. *Opt. Express*, **21**:2023–2025, 1996.

[36] K. M. Davis, K. Miura, N. Sugimoto and K. Hirao. Writing waveguides in glass with a femtosecond laser. *Opt. Lett.*, **21**:1729–1731, 1996.

[37] M. Masuda, K. Sugiola, Y. Cheng *et al.* 3-d microstructuring inside photosensitive glass by femtosecond laser excitation. *App. Phys. A*, **76**:857–860, 2003.

[38] Y. Cheng, K. Sugioka and K. Midorikawa. Microfluidic laser embedded in glass by three-dimensional femtosecond laser microprocessing. *Opt. Lett.*, **29**:2007–2009, 2004.

[39] Y. Li, K. Itoh, W. Watanabe, K. Yamada, D. Kuroda, J. Nishii and Y. Jiang. Three-dimensional hole drilling of silica glass from the rear surface with femtosecond laser pulses. *Opt. Lett.*, **26**:1912–1914, 2001.

[40] M. S. Giridhar, K. Seong, A. Schulzgen, P. Khulbe, N. Peyghambarian and M. Mansuripur. Femtosecond pulsed laser micromachining of glass substrates with application to microfluidic devices. *Appl. Opt.*, **43**:4584–4589, 2004.

[41] D. J. Hwang, T. Y. Choi and C. P. Grigoropoulos. Liquid-assisted femtosecond laser drilling of straight and three-dimensional microchannels in glasss. *App. Phys. A*, **79**:605–612, 2004.

[42] D. J. Day. *Three-Dimensional Bit Optical Data Storage in a Photorefractive Polymer*. Ph.D. thesis, School of Biophysical Sciences and Electrical Engineering, Swinburne University of Technology, Australia, 2001.

[43] D. Day and M. Gu. Formation of voids in a doped polymethylmethacrylate polymer. *Appl. Opt.*, **41**:1852–1856, 2002.

[44] M. Ventura, M. Straub and M. Gu. Void-channel microstructures in resin solids as an efficient way to photonic crystals. *App. Phys. Lett.*, **82**:1649–1651, 2003.

[45] D. Day and M. Gu. Microchannel fabrication in PMMA based on localized heating by nanojoule high repetition rate femtosecond pulses. *Opt. Express*, **13**:5939–5946, 2005.

[46] S. Eaton, F. Yoshino, L. Shah, H. Zhang, S. Ho, P. Herman and E. S. Rogers. Thermal heating effects in writing optical waveguides with 0.1-5 MHz repetition rate. *Proc. SPIE*, **5713**:35–42, 2005.

[47] C. H. Fan, J. Sun and J. P. Longtin. Plasma absorption of femtosecond laser pulses in dielectrics. *J. Heat Transf.*, **124**:275–283, 2002.

[48] M. J. Madou. *Fundamentals of Microfabrication: the Science of Miniaturisation*. Boca Raton, CRC Press, 2002.

[49] S. Rizvi. *Handbook of Photomask Manufacturing Technology*. Boca Raton, CRC Press, 2005.

[50] D. C. Duffy, J. C. McDonald, O. J. Schueller and G. Whitesides. Rapid prototyping of microfluidic systems in poly(dimethyl siloxane). *Anal. Chem.*, **70**:4974–4984, 1998.

[51] D. Lim, Y. Kamotani, B. Cho, J. Mazumder and S. Takayama. Fabrication of microfluidic mixers and artificial vasculatures using a high-brightness diode-pumped Nd:YAG laser direct write method. *Lab Chip*, **3**:318–323, 2003.

[52] D. Therriault, S. R. White and J. Lewis. Chaotic mixing in three-dimensional microvascular networks fabricated by direct-write assembly. *Nature Mat.*, **2**:265–271, 2003.

[53] J. C. McDonald, D. C. Duffy, J. R. Anderson, D. Chiu, H. Wu, O. J. Schueller and G. Whitesides. Fabrication of microfluidic systems in poly(dimethyl siloxane). *Electroph.*, **21**:27–40, 2000.

[54] S. K. Sia and G. Whitesides. Microfluidic devices fabricated in poly(dimethyl siloxane) for biological studies. *Electroph.*, **24**:3563–3576, 2003.

[55] K. Venkatakrishman, B. K. Ngoi, P. Stanley, L. E. Lim, B. Tan and N. R. Sivakumar. Laser writing techniques for photomask fabrication using a femtosecond laser. *App. Phys. A*, **74**:493–496, 2002.

[56] D. Day and M. Gu. Femtosecond fabricated photomasks for fabrication of microfluidic devices. *Opt. Express*, **14**:10753–10758, 2006.

[57] K. Venkatakrishman, P. Stanley and L. E. Lim. Femtosecond laser ablation of thin films for the fabrication of binary photomasks. *J. Micromech. Microeng.*, **12**:775–779, 2002.

[58] S. Sowa, W. Watanabe, T. Tamaki, J. Nishii and K. Itoh. Symmetric waveguides in poly(methyl methacrylate) fabricated by femtosecond laser pulses. *Opt. Express*, **14**:291–297, 2005.

[59] D. Dormann and C. J. Weijer. Imaging of cell migration. *Embo. J.*, **25**:3480–3493, 2006.

[60] R. Yasuda. Imaging spatiotemporal dynamics of neuronal signaling using fluorescence resonance energy transfer and fluorescence lifetime imaging microscopy. *Curr. Opin. Neurobiol.*, **16**:551–561, 2006.

[61] J. B. Beltman, A. F. Maree and R. J. de Boer. Spatial modelling of brief and long interactions between T cells and dendritic cells. *Immunol. Cell Biol.*, **85**:306–314, 2007.

[62] M. Gunzer, A. Schäfer, S. Borgmann *et al.* Antigen presentation in extracellular matrix: interactions of T cells with dendritic cells are dynamic, short lived, and sequential. *Immunity*, **13**:323–332, 2000.

[63] M. F. Krummel. Testing the organization of the immunological synapse. *Curr. Opin. Immunol.*, **19**:460–462, 2007.

[64] S. E. Henrickson and U. H. von Andrian. Single-cell dynamics of T-cell priming. *Curr. Opin. Immunol.*, **19**:249–258, 2007.

[65] S. Stoll, J. Delon, T. M. Brotz and R. N. Germain. Dynamic imaging of T cell-dendritic cell interactions in lymph nodes. *Science*, **296**:1873–1876, 2002.

[66] J. T. Chang, V. R. Palanivel, I. Kinjyo *et al.* Asymmetric T lymphocyte division in the initiation of adaptive immune responses. *Science*, **315**:1687–1691, 2007.

[67] T. A. Potter, K. Grebe, B. Freiberg and A. Kupfer. Formation of supramolecular activation clusters on fresh *ex vivo* CD8+ T cells after engagement of the T cell antigen receptor and CD8 by antigen-presenting cells. *Proc. Natl. Acad. Sci. USA*, **98**:12624–12629, 2001.

[68] X. Wei, B. J. Tromberg and M. D. Cahalan. Mapping the sensitivity of T cells with an optical trap: polarity and minimal number of receptors for Ca(2+) signaling. *Proc. Natl. Acad. Sci. USA*, **96**:8471–8476, 1999.

[69] M. L. Dustin, P. M. Allen and A. S. Shaw. Environmental control of immunological synapse formation and duration. *Trends Immunol.*, **22**:192–194, 2001.

[70] A. L. Demond and J. T. Groves. Interrogating the T cell synapse with patterned surfaces and photoactivated proteins. *Curr. Opin. Immunol.*, **19**:722–727, 2007.

[71] D. Day, K. Pham, M. Ludford-Menting, J. Oliaro, D. Izon, S. M. Russell and M. Gu. A method for producing microgrid arrays for prolonged imaging of highly motile T cells. *Immunol. Cell Biol.*, **87**:154–158, 2009.

[72] H. Rabin, R. F. Hopkins III and F. W. Ruscetti. Spontaneous release of a factor with properties of T cell growth factor from a continuous line of primate tumor T cells. *J. Immunol.*, **127**:1852–1856, 1981.

[73] M. Chalfie, Y. Tu, G. Euskirchen, W. W. Ward and D. C. Prasher. Green fluorescent protein as a marker for gene expression. *Science*, **263**:802–805, 1994.

[74] R. de Pooter and J. C. Zuniga-Pflucker. T-cell potential and development *in vitro*: the OP9-DL1 approach. *Curr. Opin. Immunol.*, **19**:163–168, 2007.

[75] M. A. Cooper. Optical biosensors in drug discovery. *Nat. Rev. Drug Discov.*, **1**:515–527, 2002.

[76] C. Monat, P. Domachuck and B. Eggleton. Integrated optofluidics: a new river of light. *Nat. Phot.*, **1**:106–114, 2007.

[77] C. J. Choi and B. T. Cunningham. Single-step fabrication and characterization of photonic crystal biosensors with polymer microfluidic channels. *Lab Chip*, **6**:1373–1380, 2006.

[78] B. Cunningham, P. Li, B. Lin and J. Pepper. Colorimetric resonant reflection as a direct biochemical assay technique. *Sens. Actuators B: Chem.*, **81**:316–328, 2002.

[79] D. Erickson, T. Rockwood, T. Emery, A. Scherer and D. Psaltis. Nanofluidic tuning of photonic crystal circuits. *Opt. Lett.*, **31**:59–61, 2006.

[80] H. J. Kima, S. Kima, H. Jeon *et al.* Fluorescence amplification using colloidal photonic crystal platform in sensing dye-labeled deoxyribonucleic acids. *Sens. Actuators B: Chem.*, **124**:147–152, 2007.

[81] J. Wu, D. Day and M. Gu. A microfluidic refractive index sensor based on an integrated three-dimensional photonic crystal. *App. Phys. Lett.*, **92**:071108, 2008.

[82] M. J. Ventura, M. Straub and M. Gu. Planar cavity modes in void channel polymer photonic crystals. *Opt. Express*, **13**:12767–12773, 2005.

[83] M. Straub, M. Ventura and M. Gu. Multiple higher-order stop gaps in infrared polymer photonic crystals. *Phys. Rev. Lett.*, **91**:0434901, 2003.

Index